Methods in Molecular

Series Editor
John M. Walker
School of Life and Medical Sciences
University of Hertfordshire
Hatfield, Hertfordshire, UK

For further volumes:
http://www.springer.com/series/7651

For over 35 years, biological scientists have come to rely on the research protocols and methodologies in the critically acclaimed *Methods in Molecular Biology* series. The series was the first to introduce the step-by-step protocols approach that has become the standard in all biomedical protocol publishing. Each protocol is provided in readily-reproducible step-by-step fashion, opening with an introductory overview, a list of the materials and reagents needed to complete the experiment, and followed by a detailed procedure that is supported with a helpful notes section offering tips and tricks of the trade as well as troubleshooting advice. These hallmark features were introduced by series editor Dr. John Walker and constitute the key ingredient in each and every volume of the *Methods in Molecular Biology* series. Tested and trusted, comprehensive and reliable, all protocols from the series are indexed in PubMed.

Plant Gravitropism

Methods and Protocols

Second Edition

Edited by

Elison B. Blancaflor

Utilization and Life Sciences Office, Kennedy Space Center, Merritt Island, FL, USA

Editor
Elison B. Blancaflor
Utilization and Life Sciences Office
Kennedy Space Center
Merritt Island, FL, USA

ISSN 1064-3745 ISSN 1940-6029 (electronic)
Methods in Molecular Biology
ISBN 978-1-0716-1679-6 ISBN 978-1-0716-1677-2 (eBook)
https://doi.org/10.1007/978-1-0716-1677-2

This Humana imprint is published by the registered company Springer Science+Business Media, LLC, part of Springer Nature.
The registered company address is: 1 New York Plaza, New York, NY 10004, U.S.A.

Preface

The first edition of *Plant Gravitropism: Methods and Protocols* was published in 2015. This second edition builds on the first edition with several updated chapters from the first edition and new chapters that describe methodologies to study the fascinating phenomenon of how plants readjust their growth in response to gravity. Gravitropism is used by plants to optimize their growth by ensuring their roots grow downward for proper anchorage in soils and to facilitate the acquisition of water and nutrients. The upward growth of shoots is also illustrative of how gravitropism enables the plant to maximize light absorption by photosynthetic tissues. Studies of gravitropism have deepened our understanding of the cellular, molecular, and biochemical networks that plants use to translate environmental stimuli into a growth response. Therefore, the protocols described herein continue to have broad applications for gaining insight into other plant physiological processes. A major difference between the first and second editions is that the second edition has an expanded list of chapters that deal with plants in space or the use of microgravity analogs to study plant biological phenomenon. After all, a major motivation for studying gravitropism is to work toward delivering plant-biology-based solutions to human missions to explore space—not just agricultural applications on Earth. I hope this volume will serve as a useful companion volume to the first edition.

I thank all of the authors for their contributions, especially at a time of the global COVID-19 pandemic, which presented a major challenge. I again thank Dr. John Walker, series editor, for his guidance with this project.

Merritt Island, FL, USA — *Elison B. Blancaflor*

Contents

Contributors

UTKU AVCI • *Faculty of Agriculture, Department of Agricultural Biotechnology, Eskisehir Osmangazi University, Eskisehir, Turkey*
RICHARD BARKER • *Department of Botany, University of Wisconsin, Madison, WI, USA*
PROMA BASU • *Department of Environmental and Plant Biology, Ohio University, Athens, OH, USA; Interdisciplinary Program in Molecular and Cellular Biology, Ohio University, Athens, OH, USA; Department of Molecular, Cell and Systems Biology, University of California, Riverside, CA, USA*
ADITI BHAT • *Department of Biology, Pennsylvania State University, University Park, PA, USA*
ELISON B. BLANCAFLOR • *NASA John F. Kennedy Space Center, Merritt Island, FL, USA*
WOLFGANG BUSCH • *Plant Molecular and Cellular Biology Laboratory, Salk Institute for Biological Studies, La Jolla, CA, USA*
ASHLEY E. CANNON • *Unites States Department of Agriculture, Agricultural Research Service, Wheat Health, Genetics, and Quality Research Unit, Pullman, WA, USA*
JINKE CHANG • *Ministry of Education Key Laboratory of Cell Activities and Stress Adaptations, School of Life Sciences, Lanzhou University, Lanzhou, China*
HUGO CHAUVET-THIRY • *Aix-Marseille Université, CNRS, IUSTI, Marseille, France; DISCO Beamline, Synchrotron Soleil, L'Orme des merisiers, Gif-sur-Yvette, France*
SABRINA CHIN • *Department of Botany, University of Wisconsin, Madison, WI, USA*
MALGORZATA CISKA • *Centro de Investigaciones Biológicas Margarita Salas—CSIC, Madrid, Spain*
MÉLANIE DECOURTEIX • *Université Clermont Auvergne, INRAE, PIAF, Clermont-Ferrand, France*
MARTA DEL BIANCO • *Centre for Plant Sciences, University of Leeds, Leeds, UK; Italian Space Agency, Rome, Italy*
CODY L. DEPEW • *Department of Biology, Pennsylvania State University, University Park, PA, USA*
COLLEEN J. DOHERTY • *Department of Molecular and Structural Biochemistry, North Carolina State University, Raleigh, NC, USA*
CALEB FITZGERALD • *Department of Biological Systems Engineering, University of Wisconsin, Madison, WI, USA; Black Gallina Consulting, Seattle, WA, USA*
JÉRÔME FRANCHEL • *Université Clermont Auvergne, INRAE, PIAF, Clermont-Ferrand, France*
JIŘÍ FRIML • *Institute of Science and Technology (IST) Austria, Klosterneuburg, Austria*
SIMON GILROY • *Department of Botany, University of Wisconsin, Madison, WI, USA*
CHRISTIAN GOESCHL • *GREENPASS GmbH, Vienna, Austria*
FÉLIX P. HARTMANN • *Université Clermont Auvergne, INRAE, PIAF, Clermont-Ferrand, France*
KARL H. HASENSTEIN • *Department of Biology, University of Louisiana, Lafayette, LA, USA*
JESSICA L. HELLEIN • *NASA John F. Kennedy Space Center, Merritt Island, FL, USA*
RAÚL HERRANZ • *Centro de Investigaciones Biológicas Margarita Salas—CSIC, Madrid, Spain*

Takayuki Hoson • *Department of Biology, Graduate School of Science, Osaka City University, Osaka, Japan*

Christina M. Johnson • *NASA John F. Kennedy Space Center, Merritt Island, FL, USA; Universities Space Research Association, Washington, DC, USA*

Motoshi Kamada • *Advanced Engineering Services Co., Ltd., Tsukuba, Ibaraki, Japan*

Khaled Y. Kamal • *Centro de Investigaciones Biológicas Margarita Salas – CSIC, Madrid, Spain*

Stefan Kepinski • *Centre for Plant Sciences, University of Leeds, Leeds, UK*

John Z. Kiss • *Department of Biology, University of North Carolina Greensboro, Greensboro, NC, USA*

Jürgen Kleine-Vehn • *Faculty of Biology, Center for Integrative Biological Signalling Studies (CIBSS), Department of Molecular Plant Physiology (MoPP), University of Freiburg, Freiburg, Germany*

Colin P. S. Kruse • *Department of Environmental and Plant Biology, Ohio University, Athens, OH, USA; Interdisciplinary Program in Molecular and Cellular Biology, Ohio University, Athens, OH, USA; Bioscience Division, Los Alamos National Laboratory, Los Alamos, NM, USA*

Nathalie Leblanc-Fournier • *Université Clermont Auvergne, INRAE, PIAF, Clermont-Ferrand, France*

Howard G. Levine • *NASA John F. Kennedy Space Center, Merritt Island, FL, USA*

Jia Li • *Ministry of Education Key Laboratory of Cell Activities and Stress Adaptations, School of Life Sciences, Lanzhou University, Lanzhou, China*

Lanxin Li • *Institute of Science and Technology (IST) Austria, Klosterneuburg, Austria*

Johnathan Lombardino • *Department of Botany, University of Wisconsin, Madison, WI, USA; Microbiology Doctoral Training Program, University of Wisconsin, Madison, WI, USA*

Darron R. Luesse • *Department of Biological Sciences, Southern Illinois University, Edwardsville, IL, USA*

Aránzazu Manzano • *Centro de Investigaciones Biológicas Margarita Salas—CSIC, Madrid, Spain*

Shouhei Matsumoto • *Department of Biology, Graduate School of Science, Osaka City University, Osaka, Japan*

F. Javier Medina • *Centro de Investigaciones Biológicas Margarita Salas—CSIC, Madrid, Spain*

Alexander Meyers • *Department of Environmental and Plant Biology, Ohio University, Athens, OH, USA; Program in Molecular and Cellular Biology, Ohio University, Athens, OH, USA*

Jerry Miao • *Department of Biochemistry, University of Wisconsin, Madison, WI, USA; Collaborative Science Environment, Madison, WI, USA*

Nathan Miller • *Department of Botany, University of Wisconsin, Madison, WI, USA*

Gabriele B. Monshausen • *Department of Biology, Pennsylvania State University, University Park, PA, USA*

Bruno Moulia • *Université Clermont Auvergne, INRAE, PIAF, Clermont-Ferrand, France*

Jin Nakashima • *Noble Research Institute, LLC, Ardmore, OK, USA*

Srujana Neelam • *NASA John F. Kennedy Space Center, Merritt Island, FL, USA; Universities Space Research Association, Washington, DC, USA*

TAKEHIKO OGURA • *Plant Molecular and Cellular Biology Laboratory, Salk Institute for Biological Studies, La Jolla, CA, USA*
IMARA Y. PERERA • *Department of Plant and Microbial Biology, North Carolina State University, Raleigh, NC, USA*
STÉPHANE PLOQUIN • *Université Clermont Auvergne, INRAE, PIAF, Clermont-Ferrand, France*
JEFFREY T. RICHARDS • *AECOM Management Services Inc., LASSO, Microgravity Simulation Support Facility, Kennedy Space Center, Merritt Island, FL, USA*
STANLEY J. ROUX • *Department of Molecular Biosciences, The University of Texas, Austin, TX, USA*
SURUCHI ROYCHOUDHRY • *Centre for Plant Sciences, University of Leeds, Leeds, UK*
ANNA MARIA J. RUBY • *NASA John F. Kennedy Space Center, Merritt Island, FL, USA*
TANYA SABHARWAL • *Department of Molecular Biosciences, The University of Texas at Austin, Austin, TX, USA*
NATHAN SCINTO-MADONICH • *Graduate Field of Plant Biology, Cornell University, Ithaca, NY, USA*
TATSIANA SHYMANOVICH • *Department of Biology, University of North Carolina Greensboro, Greensboro, NC, USA*
KOUICHI SOGA • *Department of Biology, Graduate School of Science, Osaka City University, Osaka, Japan*
TAIT SORENSON • *NASA John F. Kennedy Space Center, Merritt Island, FL, USA*
SARAH J. SWANSON • *Department of Botany, University of Wisconsin, Madison, WI, USA*
JOSEPH S. TOLSMA • *Genetics Program, North Carolina State University, Raleigh, NC, USA*
JACOB J. TORRES • *AECOM Management Services Inc., LASSO, Microgravity Simulation Support Facility, Kennedy Space Center, Merritt Island, FL, USA*
MIGUEL A. VALBUENA • *Centro de Investigaciones Biológicas Margarita Salas—CSIC, Madrid, Spain*
JACK J. W. A. VAN LOON • *DESC (Dutch Experiment Support Center), Amsterdam UMC location VUmc and Academic Centre for Dentistry Amsterdam (ACTA), Vrije Universiteit Amsterdam, Department of Oral and Maxillofacial Surgery/Pathology, Amsterdam Movement Sciences, Amsterdam, The Netherlands; European Space Agency (ESA), Technology Center (ESTEC), TEC-MMG, Noordwijk, The Netherlands*
CULLEN S. VENS • *Department of Botany, University of Wisconsin, Madison, WI, USA; Department of Computer Science, University of Wisconsin, Madison, WI, USA*
ALICIA VILLACAMPA • *Centro de Investigaciones Biológicas Margarita Salas—CSIC, Madrid, Spain*
SASCHA WAIDMANN • *Faculty of Biology, Center for Integrative Biological Signalling Studies (CIBSS), Department of Molecular Plant Physiology (MoPP), University of Freiburg, Freiburg, Germany*
MATTHEW WESTPHALL • *Department of Forest and Wildlife Ecology, University of Wisconsin, Madison, WI, USA; i3 Product Development, Sun Prairie, WI, USA*
CHRIS WOLVERTON • *Department of Botany and Microbiology, Ohio Wesleyan University, Delaware, OH, USA*
JULIA WOODALL • *NASA John F. Kennedy Space Center, Merritt Island, FL, USA*
SARAH E. WYATT • *Department of Environmental and Plant Biology, Ohio University, Athens, OH, USA; Interdisciplinary Program in Molecular and Cellular Biology, Ohio University, Athens, OH, USA*

SACHIKO YANO • *Space Environment Utilization Center, Japan Aerospace Exploration Agency, Sengen, Tsukuba, Japan*
YE ZHANG • *NASA John F. Kennedy Space Center, Merritt Island, FL, USA*
YUZHOU ZHANG • *Institute of Science and Technology (IST) Austria, Klosterneuburg, Austria*

Chapter 1

Plant Gravitropism: From Mechanistic Insights into Plant Function on Earth to Plants Colonizing Other Worlds

Sabrina Chin and Elison B. Blancaflor

Abstract

Gravitropism, the growth of roots and shoots toward or away from the direction of gravity, has been studied for centuries. Such studies have not only led to a better understanding of the gravitropic process itself, but also paved new paths leading to deeper mechanistic insights into a wide range of research areas. These include hormone biology, cell signal transduction, regulation of gene expression, plant evolution, and plant interactions with a variety of environmental stimuli. In addition to contributions to basic knowledge about how plants function, there is accumulating evidence that gravitropism confers adaptive advantages to crops, particularly under marginal agricultural soils. Therefore, gravitropism is emerging as a breeding target for enhancing agricultural productivity. Moreover, research on gravitropism has spawned several studies on plant growth in microgravity that have enabled researchers to uncouple the effects of gravity from other tropisms. Although rapid progress on understanding gravitropism witnessed during the past decade continues to be driven by traditional molecular, physiological, and cell biological tools, these tools have been enriched by technological innovations in next-generation omics platforms and microgravity analog facilities. In this chapter, we review the field of gravitropism by highlighting recent landmark studies that have provided unique insights into this classic research topic while also discussing potential contributions to agriculture on Earth and beyond.

Key words Agriculture, Genomics, Microgravity, Spaceflight, Root development, Tropisms

1 Introduction

In Robert Fulghum's book, *All I Really Need to Know I Learned in Kindergarten*, he wrote: "Be aware of wonder. Remember the little seed in the Styrofoam cup? The roots go down and the plant goes up, and nobody really knows why or how" [1]. Although this specific passage was part of a book meant to convey Fulghum's reflections on love, mortality, and other aspects of life in general, it brings into focus a classical topic in botany that has been studied by scientists for more than two centuries, including Charles Darwin's work on plant movements [2]. Roots growing down and shoots growing up is a phenomenon called gravitropism (geotropism in the earlier literature). In the past decade, gravitropism has been the

Elison B. Blancaflor (ed.), *Plant Gravitropism: Methods and Protocols*, Methods in Molecular Biology, vol. 2368, https://doi.org/10.1007/978-1-0716-1677-2_1,

subject of several review articles, but most have focused on how this process is regulated at the cellular and molecular levels (e.g., [3–10]). While mechanistic insights into how plants modulate their development is a significant outcome of research on gravitropism, an increasing number of studies are now investigating how knowledge about this plant environmental response can lead to practical agricultural applications [11, 12]. Furthermore, as efforts to advance human space exploration push forward, various space agencies worldwide have resumed support of research that seeks to better understand how plants grow in reduced gravity (i.e., microgravity). Gravitropism remains at the forefront of plant biological research. The rapid advances in the field have been realized, in large part, because of the rich collection of methods developed by laboratories around the world that study gravitropism both on Earth and in space. Some of these methods are described in the following chapters of this volume. As an introduction to these chapters, we review here a selection of literature that has further advanced the field. We begin by highlighting studies that have led to new insights into fundamental plant gravity response mechanisms while also mentioning established players. We then discuss how such knowledge has been applied to agriculture and space exploration.

2 Brief Overview of Gravitropism and Insights from Omics Studies

Gravitropism has traditionally been described as a process that consists of three sequential events: gravity perception, gravity signal transduction, and gravity response. Aerial organs such as inflorescence stems [13], shoots [14–16], hypocotyls [17], cereal pulvini [18, 19], and coleoptiles [20] respond to gravity negatively, whereas below-ground organs such as roots [21, 22] and peanut gynophores [23, 24] respond positively toward gravity (Fig. 1a). Although gravitropism in aerial and below-ground organs lead to growth in opposite directions, it is widely accepted that they a share common gravity-sensing mechanisms. In higher plants, gravity sensing is explained primarily by the starch-statolith hypothesis, which involves the sedimentation of amyloplasts (starch-filled organelles) that function as statoliths either in the endodermis of stems or the columella of roots (Fig. 1b). The physical movement of statoliths is then translated into biochemical signals that lead to differential organ growth [25–27] (Fig. 1b). Statolith sedimentation is thought to disrupt the actin cytoskeleton network and distort the endoplasmic reticulum and activate mechanosensitive ion channels at the plasma membrane [28–31]. This is followed by the rapid increase in secondary messengers as inositol-1,4,5-triphosphate (InP_3), cytoplasmic calcium (Ca^{2+}), reactive oxygen species, and alkalization of cytoplasmic pH in root columella cells

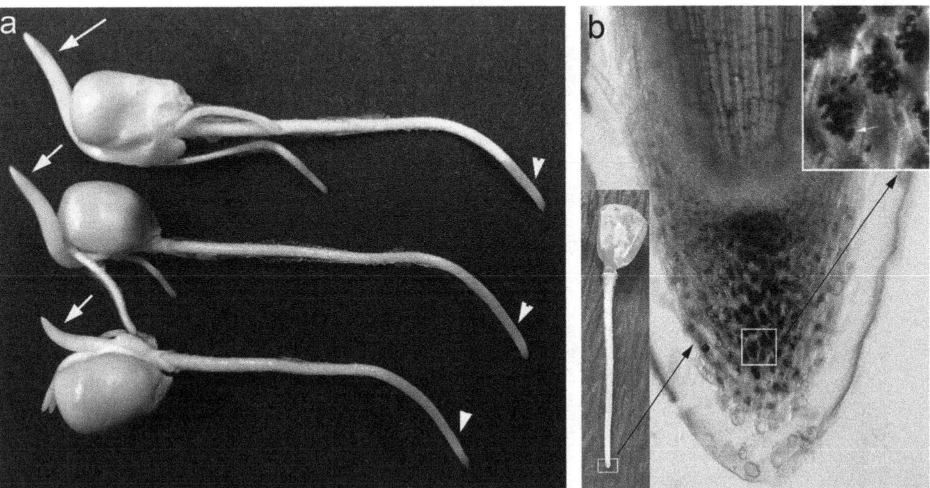

Fig. 1 Gravitropism in seedlings of maize (*Zea mays*). (**a**) Coleoptiles and roots of 4-day-old *Z. mays* seedlings reoriented horizontally grow upward (arrows) and downward (arrowheads), respectively. (**b**) Median longitudinal section of the root cap of *Z. mays* seedling reveal starch-filled amyloplasts (arrows, top right inset) in the columella cells (white square), which function as statoliths. Statolith sedimentation is widely believed to be part of the gravity-sensing machinery of roots of higher plants

and acidification of the root cap apoplast [32–35]. The resulting reorientation of organ growth following gravity sensing and signal transduction is explained by the Cholodny-Went theory [36], which involves the redistribution of auxin in the root tip or stem via auxin efflux (e.g., PIN2, PIN3) and influx (e.g., AUX1) carrier relocalization so that auxin accumulates in the lower side of the organ [37–39]. The auxin gradient between the top and bottom flanks results in asymmetrical growth with the root bending toward gravity or stem growth against gravity and eventually straightening when the statoliths recalibrate [17, 40].

Gravitropism is an extremely complex process, and there are still gaps in understanding how different pathways intersect to coordinate the steps from gravity perception to the growth response. For instance, is there a hierarchy to how secondary messengers are transduced? How do different secondary messengers directly and/or indirectly alter gene, protein, and metabolite expression and so forth? Work spanning two decades banked on the idea that omics approaches (i.e., transcriptomics, proteomics, and metabolomics) could help construct a bigger picture on how gravitropism works. These omic studies are summarized in Table 1 and are discussed in this section.

Statoliths in columella cells begin to sediment soon after the root is reoriented with respect to the gravity vector and gravity stimulation for about one minute is enough to initiate a growth response in *Arabidopsis* roots [25]. Other secondary messengers such as InP_3 was demonstrated to elevate within 10 s after

Table 1
Summary of transcriptomics, proteomics, and metabolomics studies that explored differentially expressed genes and proteins in ground-based gravitropism research

Method	Plant	Treatment	Tissue type	Time	Summary	References
Transcriptomics						
Arabidopsis 8.3 k microarray	*Arabidopsis* ecotype Columbia, 3 weeks old	Dark grown, rotated 90° or 360°	Whole seedlings	0, 15, 30 min	• 141 gravity-specific gene changes within 30 min of gravistimulation • In descending order of enrichment, unclassified proteins, oxidative stress/ plant defense, metabolism, transcription, cell wall/ plasma membrane, signal transduction, heat shock proteins, ethylene-responsive element binding factors, calcium binding proteins, energy and cytoskeleton genes showed expression changes	[41]
Arabidopsis whole genome microarray	*Arabidopsis* ecotype Columbia, 7 days old	Dark grown, rotated 135°	Root apices	2, 5, 15, 30, 60 min	• ↑ in transcription factors (*MYB, RingH2, KNAT1, HFR, NAM*), cell wall modifying enzymes (xyloglucan endotransglycosylases, pectinesterases), transporters (ions, sugar, purines), specific histone isoforms, RNA polymerase subunits, environmental stress, and ACC synthase genes. Some specific genes such as *AtPEN1* involved in brassinosteroid biosynthesis and *SAMT* homolog which may be involved in methyl salicylate and methyl jasmonate pathway for systemic and local defenses • ↓ in chromatin organization and modification genes (DNA helicase, chromomethylase, chromosome	[21]

					condensation protein, histones H2A and H3, DNA gyrase, SNF2 domain helicases, SET-domain regulators, and replication factors), *PIN2*, ACC oxidase genes • Rapid ↑/↓ in auxin-responsive genes (*SAUR*, *IAAs*)	
Oryza sativa whole genome microarray	*Oryza sativa* sp. japonica, 3 weeks old	Light grown, rotated 90°	Upper and lower flanks of shoot base	0.5 and 6 h	• 167 differentially expressed genes at 0.5 h. These were involved in ethylene, cytokinin, and secondary metabolism genes • 1202 differentially expressed genes at 6 h. Transcripts were involved in hormone metabolism and transportation (e.g., auxin, GA, BR) and cell wall expansion • 48 genes (e.g., auxin induced, TCA cycle, glycolysis, cell wall degrading enzymes, and secondary metabolism genes) were differentially expressed at both time points • Upper and lower flanks at 0.5 and 6 h differed in transcripts involved in photosynthesis ↑, DNA categories ↑, glycolysis ↓, fermentation ↓, lipid metabolism ↓ and cell organization ↓ • Genes common with *Arabidopsis* studies ([21, 41]) were cytochrome P450, expansin, xyloglucan endotransglycosylases, pectinesterases, auxin-responsive genes, and *PIN2* • Genes common in upper and lower flank study with *Brassica* ([17]) were expansins, *SAUR*, *IAAs*, and glycosyl-related genes	[15]

(continued)

Table 1
(continued)

Method	Plant	Treatment	Tissue type	Time	Summary	References
RNASeq	*Zea mays* HN17 and *la1-ref* mutant, 4 weeks old	Light grown	Third above ground nodes	N/A	• Global transcriptome profile comparison between WT and *la1-ref* nodes revealed that 646 genes were upregulated and 285 genes were downregulated • Differentially expressed genes were annotated to RNA regulation, transport and hormone metabolism with functional categories involved in 10 genes in Ca^{2+} signaling, 21 genes in lipid metabolism, seven genes in pH/H^{+} regulation, 12 light signaling genes, 13 auxin response genes, and 15 auxin transport-related genes • RNASeq results complemented other physiological evidences to inform about the role of *ZmLA1* in shoot gravitropism by mediating auxin transport and auxin signaling genes	[42]
Arabidopsis 2 Oligo Microarray	*Arabidopsis thaliana* ecotype Columbia, 10 days old	Light grown, rotated 90°	Upper and lower flanks of inflorescence stems	10 and 30 min	• Lower flank showed increased expressions in 30 genes which included *SAURs,SAUR*-like, *ACC, IAAs,* glucanase,, kinase regulator, and gibberellin oxidase genes • Detailed analysis on *IAA* genes showed that *IAA5* was highly upregulated in lower flank at 30 min post-gravistimulation, whereas *IAA2* reduced in upper flanks from 15 to 60 min post-gravistimulation	[13]

Agilent 4 × 44 k arrays	*Arabidopsis thaliana* ecotype Columbia, *pin2* and *pin3*, 5 days old	Fixation chamber of Carbocryonix, constant 1 *g* in ground or parabolic flight with series of 1 *g* and 1.8 *g* or parabolic flight with series of 1 *g*, 1.8 *g* and *ug*	Whole root	16 or 31 parabolas	• Genes upregulated in microgravity in all genotypes were involved in signal transduction and defense response, protein metabolism, lipid transport, carbohydrate metabolism, stress response, cytoskeleton organization, cell wall processes, etc. • Genes downregulated in microgravity in all genotypes were involved in heat response, MAPK cascade, protein folding, response to abiotic stress, ethylene signaling, abscisic acid response, ozone response, ER-nucleus signaling pathway, circadian rhythm, protein folding, etc. • Authors suggested that regulation of auxin-responsive genes in WT during microgravity depends on PIN3 and PIN2-mediated response were downstream of PIN3. *pin2* mutants upregulated genes involved in polar auxin transport in microgravity such as AGC protein kinase family WAG1 and WAG2 and PIN3	[43]
RNASeq	*Oryza sativa* Zhonghua 11, 11 days old	Light grown, rotated 90°	1.5 cm shoot bases	0, 0.25, 0.5, 1, 1.5, 2, 3, 4, 5, 6, 7, and 8 h	• Transcriptomic changes between control and gravistimulated shoots were distinct starting at 3 h time point onwards • Genes rapidly activated after gravistimulation (<3 h) were significantly enriched for protein processing in endoplasmic reticulum and unfolded protein binding • Genes that were differentially expressed at both early stages (<3 h) and late stages (>3 h) were enriched in plant hormone signal transduction and transcription regulator activity • Gene clusters at early and late time points displayed a hierarchical	[44]

(continued)

Table 1
(continued)

Method	Plant	Treatment	Tissue type	Time	Summary	References
					relationship, whereby the early genes upregulated and downregulated different sets of genes in later time stages and some genes that are differentially expressed in all time points • Gravistimulation-induced gene clusters were significantly enriched with auxin-activated genes (e.g., *YUCCA6, SAURs* except *SAUR8*, 5 *IAA* genes, *ARF7, ALDH2a*) and vice versa for gravistimulation repressed genes (e.g., *YUCCA3, YUCCA7, IAA25, ARF2, ARF9, ALDH3B1*). Auxin biosynthesis and signaling are likely important in shoot graviresponse under light • Transcription factor genes upregulated at 0.25, 0.5, and 1 h included *HSFA2D, MADS57*, and *EPR1* • Detailed studies on *hsfa2d* mutants with increased tiller angle revealed that *HSFA2D* regulates auxin redistribution in shoot via *LA1* pathway, which included *WOX6* and *WOX11* transcription factors	
Proteomics						
1 DE-GE with LC-MS/MS	*Arachis hypogaea* cv. Luhua 14, 6 months old	Light grown (stage 1) and dark grown (stages 2 and 3)	Tip of stage 1 (aerial gynophores), stage 2 (gynophores submerged in soil for 3 days) and stage 3 (mature dark-	N/A	• 2766 proteins detected in stage 1. 574 proteins are specific to this stage • 2518 proteins detected in stage 2. 307 proteins are specific to this stage • 2280 proteins detected in stage 3. 185 proteins specific to this stage	[24]

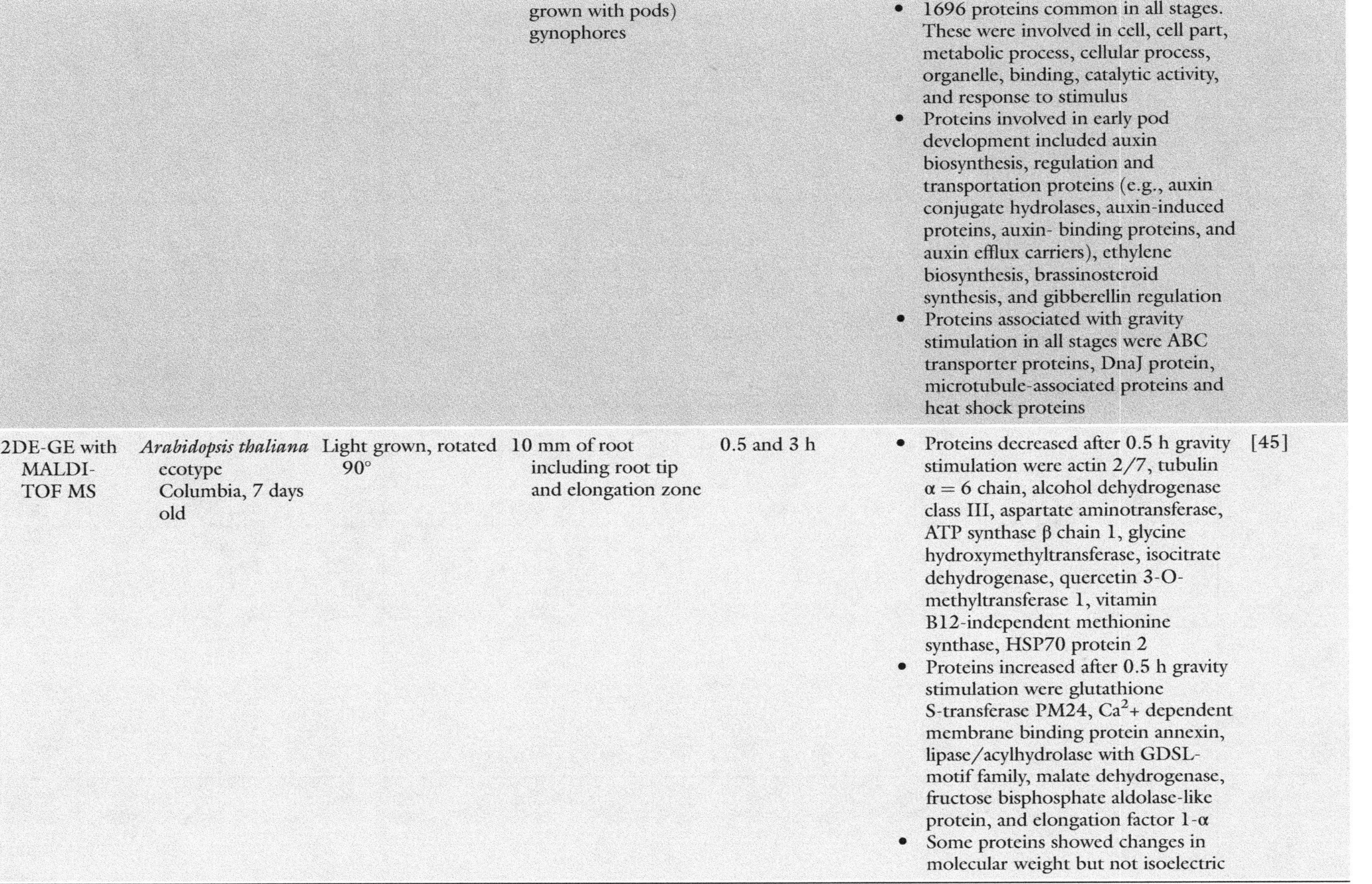

Method	Plant	Treatment	Tissue	Time	Findings	Ref.
			grown with pods) gynophores		• 1696 proteins common in all stages. These were involved in cell, cell part, metabolic process, cellular process, organelle, binding, catalytic activity, and response to stimulus • Proteins involved in early pod development included auxin biosynthesis, regulation and transportation proteins (e.g., auxin conjugate hydrolases, auxin-induced proteins, auxin- binding proteins, and auxin efflux carriers), ethylene biosynthesis, brassinosteroid synthesis, and gibberellin regulation • Proteins associated with gravity stimulation in all stages were ABC transporter proteins, DnaJ protein, microtubule-associated proteins and heat shock proteins	
2DE-GE with MALDI-TOF MS	*Arabidopsis thaliana* ecotype Columbia, 7 days old	Light grown, rotated 90°	10 mm of root including root tip and elongation zone	0.5 and 3 h	• Proteins decreased after 0.5 h gravity stimulation were actin 2/7, tubulin $\alpha = 6$ chain, alcohol dehydrogenase class III, aspartate aminotransferase, ATP synthase β chain 1, glycine hydroxymethyltransferase, isocitrate dehydrogenase, quercetin 3-O-methyltransferase 1, vitamin B12-independent methionine synthase, HSP70 protein 2 • Proteins increased after 0.5 h gravity stimulation were glutathione S-transferase PM24, Ca^2+ dependent membrane binding protein annexin, lipase/acylhydrolase with GDSL-motif family, malate dehydrogenase, fructose bisphosphate aldolase-like protein, and elongation factor 1-α • Some proteins showed changes in molecular weight but not isoelectric	[45]

(continued)

Table 1 (continued)

Method	Plant	Treatment	Tissue type	Time	Summary	References
					point in 0.5 and 3 h time point. Authors postulated that protein modifications may occur to the protein molecule due to gravity, and/or post-translational modification occurred	
iTRAQ with LC-MS/MS	*Arabidopsis thaliana* ecotype Columbia, grown until inflorescence were 8–10 cm long	Light grown, rotated 90° at 4 °C (control plants was mechanically stimulated at 4 °C)	Top 4 cm of inflorescence stems	2 and 4 min	• Proteins that were differentially expressed (<35%) at 2 and 4 min after gravistimulation were predicted to be chloroplast, plastid and cytosol proteins. Some proteins have predicted functions in catalytic activity or metal binding • Specific diferentially expressed proteins included (1) heat shock protein 81-2 (HSP81-2), (2) glutathione S-transferase PH19 (GSTF9), (3) vacuolar ATP synthase subunit A, (4) ADP, ATP, (5) ATP synthase CF1 beta subunit and (6) glutathione S-transferase TAU20 • Further analysis on *hsp81-2* and *gstf9* mutants showed defects in gravity response in roots (curvature, length, skewing) and inflorescence (curvature) • HSP81-2 may modulate Ca^{2+} flux for signal transduction and may interact with J-domain containing proteins • GSTF9 may be involved in hormone synthesis as other glutathione transferase interacted with an auxin-induced protein	[16]

2DE-GE with LC-MS	*Arabidopsis thaliana* ecotype Columbia and *pin2* mutant, 6 days old	Dark grown for 24 h prior to horizontal clinorotation or hypergravity or stationary control	5–10 mm root tips	12 h for horizontal clinorotation or 30 min for hypergravity samples	• Differentially expressed proteins in WT and *pin2* under clinorotation and hypergravity included TCP-1 chaperonin-like protein, ATP synthase beta chain 2, mevalonate diphosphate decarboxylase, isocitrate dehydrogenase, NAD^+-like protein, and glutathione-dependent formaldehyde dehydrogenase class III ADH. These proteins likely represent general stress responses and PIN2-independent gravity response • A protein differentially expressed in only WT but not *pin2* under clinorotation and hypergravity was putative malate oxidoreductase. This protein may be involved in auxin mediated step • Proteins uniquely changed in *pin2* but not WT under clinorotation and hypergravity were probable ubiquitin-like protein, alpha-galactosidase-like protein, and hydroxyacylglutathione hydrolase cytoplasmic. Proteins may be involved in auxin-mediated steps	[22]
2DE-GE with LC-MS/MS	*Arabidopsis thaliana* ecotype Estland, 4 days old	Light grown, rotated 90°	3–5 mm root tips	12 min	• Observed several protein spots that increased after gravistimulation • Identified one of the spots as ADK protein • *ADK1* and *ADK2* transcripts were unaltered in gravistimulation experiment ([21]), indicating that ADK proteins undergoes post-transcriptional modification • Further studies found with *adk1-1* and	[46]

(continued)

Table 1 (continued)

Method	Plant	Treatment	Tissue type	Time	Summary	References
					adk2-1 identified ADK1 as the active isoform in root gravity response. ADK1 is likely involved in gravity signal transduction via production of polyamine and ethylene precursors in SAM pathway	
Metabolomics						
GC-MS	*Arabidopsis thaliana* ecotype Columbia, 7 days old	Light grown, rotated 90° with application of ethylene inhibitors, 10μM AVG or 10 μM salicylic acid, or buffer control	Whole seedlings or 2 mm root tips	60 min	• Gravistimulated seedlings had increased mannose, stachyose, thanolamine, alanine but decreased fatty acids and most amino acids. In root tips, several amino acids increased, whereas hexoses, malate, and oxalate decreased • Salicylic acid treatment in gravistimulated seedlings increased stearic and palmitic acids and decreased valine, threonine, serine, aspartic and glutamic acids but did not significantly change metabolites between vertically grown and gravistimulated seedlings • AVG treatment in gravistimulated seedlings slightly increased glycine, proline, and serine and decreased some amino acids but did not significantly change metabolites between vertically grown and gravistimulated seedlings. In gravistimulated and AVG-treated root tips, treatment with AVG resulted in metabolic changes more similar to vertically treated roots • Metabolic changes due to gravistimulation were more pronounced in root tips than whole seedlings	[47]

					• Sugar accumulation in root tips during gravistimulation is likely used to build cell wall polymers, whereas changes in fatty acids (palmitic and stearic acid) may be due to their roles in signaling • Salicylic acid may affect the ethylene pathway or exert direct effects • Inhibition of the ethylene biosynthesis via AVG was likely linked to the root growth rate and gravitropic bending during gravitropism response. However, authors did not measure root bending response	
Transcriptomics and metabolomics						
RNA-Seq, LC-MS/MS	*Camellia sinesis* cv. Qiqu (QQ) and Lianyuanqiqu (LYQQ) (zigzag stems), and Meizhan (MZ) (straight stem)	Light grown	Stems between the first and third leaves from the apical bud	N/A	• Genes associated in zigzag formation were associated with (1) cell wall synthesis and cell expansion, (2) transcription factor genes, (3) auxin, jasmonic acid, salicylic acid metabolism and transport, (4) protein processing and transport on ER and vesicles, (5) vacuolar sorting and membrane proteins, and (6) cell division control proteins. Specific genes included downregulation of *PIN3, LAZY, LAZY 1-like, VILLIN2,* and actin-depolymerizing factor 2 in QQ and LLQQ • Metabolites increased in zigzag genotypes vs straight MZ were quercetin O-acetylhexoside, methyl gallate, D-pantothenic acid and L-glutamic, while fustin, 10-formyl-THF, skimming, LysoPC, 2-methylsuccinic acid, 2-isopropylmalate and caffeine were decreased	[14]

(continued)

Table 1 (continued)

Method	Plant	Treatment	Tissue type	Time	Summary	References
					• Flavonols may be regulated for auxin transport control. t*ZIG* and *SGR* genes were not significantly expressed (unlike *Arabidopsis zig* mutants) but changes in vesicular and vacuolar proteins were reported	
GC-TOF, *Arabidopsis* whole genome microarray	*Arabidopsis thaliana* ecotype Landsberg erecta, 7 days old	Dark grown, rotated 90°	Whole seedling	24 h	• Gravistimulated plants showed significant increase in aspartate, GABA, asparagine, phenylalanine, ribose, glucose, and ornithine • 147 genes were upregulated and 192 genes were downregulated after gravistimulation. These are mostly associated with functional categories of metabolism, subcellular localization, protein with binding function and unclassified proteins • Specific genes that were upregulated included xyloglucan endotransglycosylase 7, GHMP kinase-related, lactoylglutathione lyase family protein/ glyoxalase I family protein, threonine aldolase 1, dark inducible 6, GCN5-related N-acetyltransferase family protein and protein kinase family protein • Specific genes that were downregulated included cellulose synthase-like A15, lipoxygenase 2, glycosyl hydrolase family 1 protein, and chalcone synthase	[48]

- Most differentially expressed genes and metabolites are not unique to gravity stimulus, also overlapped with blue and red light stimulation. Authors suggested that gravitropism and phototropism generally induced stress responses

ACC 1-amino-cyclopropane-1-carboxylate, *ADK* adenosine 5′-phosphotransferase, *AVG* L-α-{2aminoethoxyvinyl] glycine, *ALDH* aldehyde dehydrogenase, *DE-GE* dimensional gel electrophoresis, *EPR* early phytochrome responsive, *ER* endoplasmic reticulum, *GABA* gamma aminobutyric acid, *GC* gas chromatography, *iTRAQ* isobaric tags for relative and absolute quantification, *LC* liquid chromatography, *MALDI* matrix-assisted laser desorption, *MS* mass spectrometry, *N/A* not applicable, *SAM* *S*-adenosyl-L-methionine, *SAMT* *S*-adenosyl-L-Met:carboxymethyltransferase, *TOF* time of flight

gravistimulation in cereal pulvini [18, 19] while alkalinization of cytoplasmic pH in *Arabidopsis* columella cells began in less than 1 min [32] and cytoplasmic Ca^{2+} spikes around 25 s while peaking at 90 s [35]. Thereafter, auxin redistribution toward the lower side that precedes asymmetric growth typically occurs around one to one and a half hours onward during *Arabidopsis* root and tobacco stem bending [40, 49]. Therefore, the considerations of omics studies should account for time in relation to gravistimulation, including the: (1) presentation time, which is the time to elicit a gravity signal, (2) perception time, which is the duration for gravity signal to be perceived and transduced prior to bending response, and (3) response time, whereby the organ shows asymmetrical growth.

Genes that are likely involved in gravity signal perception and/or transduction include those that encode a pentacyclic triterpene synthase (At*PEN1*), an unclassified protein (At2g16005), a Cys protease, a *S*-adenosyl-L-Methionine (SAM) carboxymethyltransferase homolog, and a major latex-related protein [21]. These were reported to be significantly upregulated in *Arabidopsis* root apices within two minutes [21]. In addition, the authors also identified increases in the expression of genes encoding other transcription factors after 5–15 min of reorientation, such as *HFR1, AtHB-12,* and *KNAT1,* which possibly are responsible for the early regulation of growth responses. By around 30 min, other gravity-specific genes change their expressions, including genes involved in oxidative stress/plant defense, metabolism, cell wall/plasma membrane, signal transduction, heat shock proteins, ethylene-responsive element-binding factors, calcium-binding proteins, energy, cytoskeletal genes, chromatin organization/modification, endoplasmic reticulum, NA^+/H^+ exchange, and protein binding [21, 41, 44]. A key finding in different transcriptomic works that is consistent with the Cholodny-Went theory is the enrichment of auxin-inducible and auxin-responsive genes at later time points such as *SAURs, SAUR-like, YUCCAs, IAAs, IAA OXIDASEs, PIN2,* and *Auxin Response Factors* (*ARFs*) (Table 1) [13–15, 21, 42, 44]. Other hormones, including brassinosteroids, ethylene, cytokinin, and gibberellic acid, also have been implicated in gravity response through transcriptomics studies [13, 15, 21]. It is also important to highlight the advantage of including systematic time points, as done by Zhang et al. [44], who included 12 time points. Through these studies, the authors were able to construct hierarchical gene clustering to demonstrate how gene clusters in the earlier time points influenced later gene clusters.

The results from whole genome transcriptomic analysis has also been useful in mining gravity-specific genes for detailed analysis. For example, Zhang et al. [44] identified the rapid induction of *HSFA2D* and used reverse genetics to determine that this gene controls rice tiller angle in conjunction with *LAZY1* (*LAZY1* also regulates asymmetric auxin pathway to determine rice tiller angle) to control auxin redistribution. The advent and accessibility of newer transcriptomic pipelines, particularly RNA-Seq, has enabled researchers to include RNA-Seq to complement physiological data by the genome-wide effect of a mutated gene [42]. Nonetheless, the boon and bane of transcriptomics lies in its ability to only detect *de novo* synthesis. Changes in proteins (post-transcriptionally and post-translationally) can occur without changes in mRNA abundance, such as phosphorylation [50], and ubiquitin tagging (e.g., *PIN2* gene) [51], can be missed in transcriptomics and should be accompanied with proteomic studies. Moreover, the dynamic pH fluctuations that occur in columella and root cap cells during early gravity signal transduction likely affected protein structure (e.g., folding and cleavage) as Kamada et al. [45] observed changes in isoelectric point but not molecular weight of gravity-related proteins.

Proteomics approaches have uncovered over 30% of early gravity-related proteins (two and four minutes) that were predicted to be located in the chloroplast, vacuole or plastid with functions in catalytic activity or metal binding, which is likely reflected in gravity signal transduction processes [16]. Young et al. [46] found adenosine kinase (ADK) protein in *Arabidopsis* roots 12 min after gravistimulation and discovered that ADK1 functions in the SAM pathway after cytoplasmic alkalization of the columella cells, but before auxin redistribution. Furthermore, the role of SAM pathway in gravitropism has previously been reported by the induction of an *SAM* transferase homolog in *Arabidopsis* roots around two minutes after gravistimulation [21]. Another common denominator in proteomics and transcriptomics studies described in Table 1 is the differential expression of proteins/genes such as heat shock proteins (HSPs) and glutathione S-transferases (GSTs) [16, 24, 41, 45]. Although HSPs are typically induced in response to abiotic and biotic stresses [52, 53], the levels of some gravity-related HSPs, such as HSP70-2 and HSP81-2, were reduced [16, 45]. Moreover, *hsp81-2 Arabidopsis* mutants exhibited developmental defects such as agravitropic growth and shorter and skewed roots [16], while *Arabidopsis* transgenics that over-expressed *HSP81-2* showed decreased salt and drought stress, but increased Ca^{2+} tolerance [54]. These studies suggest that abundance of HSP81-2 is tightly regulated for normal development and may modulate Ca^{2+} flux for gravitropic signaling [16]. The other proteins of interest, GSTs, were found to be increased in *Arabidopsis* roots [45] or differentially expressed in *Arabidopsis* inflorescence [16] during gravity response. GSTs have been mostly described in xenobiotic detoxification and sulfur-containing phytochemical production such as glucosinolates [55, 56]. *Brassica rapa* that are grown in hypergravity produced significantly less glucosinolates in stems, roots and leaves [57]. These studies may uncover new roles for GSTs in gravitropism, perhaps in auxin regulation for bending response, as some GSTs can bind to auxin and participate in auxin transport [56] or act as sulfur-containing gravity signal transducers.

Currently, the use of metabolomics to study gravitropism is still under-utilized (Table 1). Pozhvanov et al.[47] found that sugars, amino acids and fatty acid compositions distinctively changed, especially in *Arabidopsis* root tips during gravistimulation. Such changes could be linked to cell wall building for root bending or function as signaling molecules. Moreover, the authors discovered the importance of the ethylene pathway in the gravity response, as the inhibition of ethylene biosynthesis with 10μM L-α-[2aminoethoxyvinyl] glycine abolished these metabolic changes. Millar and Kiss [48] also found changes in amino acids, sugars and phenylalanine in gravistimulated *Arabidopsis* seedlings. In addition, the authors [48] complemented their metabolomics data with microarray results. For example, phenylalanine, the precursor to flavonoid biosynthesis via chalcone synthase, was found to increase while chalcone synthase transcripts were downregulated. There is a huge untapped potential of studying the role of metabolites in gravitropism such as small peptides (GOLVENs, CLEs), glucosinolates, flavonoids, alkaloids, and some of the less-studied plant hormones, such as strigolactones and brassinosteroids, as mass spectrometry technologies have enabled the detection of multiple types of plant compounds. Further, metabolites that are significantly increased during gravistimulation can be enriched by fractionation and identified using nuclear-magnetic resonance [58] for physiological studies. In summary, the omics-based studies outlined in this section have enabled the generation of new hypotheses that could be tested in future studies.

3 Evolution of Fast Gravitropism

Perhaps one of the most exciting advances on gravitropism research in the past few years has been the new insights gained on how this process evolved from slow responses to gravity in lower land plants to fast responses observed in higher plants. Extant land plants are derived from streptophyte algae, which includes a variety of extant marine and terrestrial green algae [59]. A keystone adaptation to colonizing the terrestrial terrain was the evolution of roots for anchorage in the soil, and water and nutrient uptake [60]. Primitive land plants constituted petite herbaceous vegetation with small leafless plants and simpler rhizoid-based roots [61]. Some extant plants, such as the green algae, liverworts, mosses, and hornworts, still maintained rhizoids as their main rooting system, whereas the ferns developed rhizoids in the gametophyte stage before switching to true roots in the sporophyte stage, while seed plants only have true roots. Nonetheless, all rooting systems use statoliths, which sediment in relation to the gravity vector.

Roots have evolved several innovations that increased gravitropism speed. The first was the use of starch and plastid-based statoliths called amyloplasts, which we noted earlier (see Fig. 1). Rhizoids are devoid of amyloplasts [62] and instead utilize barium sulfate crystals with an organic mix of protein and carbohydrate moieties that function as statoliths [63–66]. These "simpler" statoliths mostly relied on the higher mass of the crystals to sediment as barium sulfate crystals themselves are inert and direct contact between sedimenting statoliths and mechanosensitive plasma membrane is sufficient to trigger the signal cascade for the gravity response [64, 65]. The evolution of amyloplasts likely originated in a lineage of ancestral vascular plants as the presence of amyloplasts in roots was observed in lycophytes and ferns [62]. Despite the evolution of amyloplasts in lycophytes and ferns, these amyloplasts had yet adapted to function as gravity-sensing statoliths, as Zhang et al. [62] found that the amyloplasts in the lycophyte, *Selaginella moellendorffii*, and the fern, *Ceratopteris richardii* were randomly distributed in the root cells and did not sediment after 180° reorientation. Moreover, *S. moellendorffii* and *C. richardii* showed slow gravitropic root bending. By contrast, seed plants that have evolved to use amyloplasts as statoliths had noticeably faster gravitropic root bending [62]. Perhaps the other advantage of switching to amyloplast-based statoliths is the expression of surface proteins for interactions with the actin cytoskeleton to regulate amyloplast sedimentation. Amyloplasts in *Arabidopsis* also express proteins on their surface, such as SGR9, a RING-type E3 ligase that controls the detachment of amyloplasts from actin filaments to equilibrate amyloplast between saltatory movements and sedimentation [67]. Despite the accumulating evidence for statolith-based gravity sensing, Edelmann [68] proposed that maize bypassed the need for statolith sedimentation to perceive gravity as the removal of tissues containing columella and endodermal cells did not affect gravity-bending responses in coleoptiles and roots. Monocots like maize typically have various root systems that consist of primary roots, seminal roots, crown roots, and brace roots that have different architecture and grow at different depths [69]. It would be interesting to determine if these root types employ alternative gravity-sensing mechanisms or whether differences in signaling cascades following the sedimentation of amyloplasts can explain their variable gravitropic setpoint angles.

The second modification that contributed to more efficient gravitropism is concentrating gravity-sensing columella cells to a more confined region at the root tip [62]. Basal vascular plants *S. moellendorffii* had amyloplasts distributed along the roots, but not in the apex, whereas *C. richardii* had amyloplasts distributed above and within the root apex [62]. By contrast, the seed plants, i.e., gymnosperm *Pinus taeda,* dicots *Gossypium arboreum* and *Arabidopsis,* and monocot *O. sativa,* had amyloplasts specifically localized only within columella cells in the root apex [62]. Columella cells have evolved to be essential to gravitropism in seed plants, so much so that the surgical removal of root apex (therein columella cells, too) [70] and the laser ablation of columella cells resulted in agravitropic roots. Additionally, columella cells in *A. thaliana* have evolved to be highly specialized, so much so that each story of columella cells contributed differentially to gravitropism [25].

Thirdly, changes in auxin, which is the primary signal to coordinate organ growth in response to gravity, played a crucial role in the evolution of fast gravitropism. Zhang et al. [62] found that multiple innovations in the transmembrane domains and hydrophilic loops of PIN2 auxin efflux transporters in seed plants enabled PIN2 to be localized in a polar manner toward the shoot ward side of epidermal cells. It is also interesting to note that in the same paper, the authors demonstrated that amyloplasts in *S. moellendorffii* and *C. richardii* were localized in similar locations in the root epidermis and cortex as PIN2 proteins in seed plants. Perhaps this denotes that *S. moellendorffii* and *C. richardii* used a more site-specific signal transduction of auxin in lieu of auxin transport from columella to the elongation zone. In addition, regulations in auxin transport, auxin synthesis, conjugation, and degradation in seed plants are far more complex than those of the liverwort, *Marchantia* spp.; moss, *P. patens;* and lycophyte, *S. moellendorffii* (reviewed in [71]). This may be attributable to more precise auxin regulation in the root for gravity response although this has not been well studied. Moreover, the formation of the quiescent center in seed plants like maize may contribute additional auxin regulation in the root tip by controlling auxin oxidation [72].

Lastly, there is evidence that changes in the signal transduction pathway are the key to fast gravitropism in flowering plants. Limbach et al. [64] postulated that *Chara* rhizoids have shorter and simpler signal transduction and transmission pathways, as the authors can easily discriminate between gravity susception and perception in the rhizoids. Furthermore, the lack of evidence supporting the ligand-gravireceptor model in flowering plants [73] suggests that flowering plants require a more complex signaling pathway for rapid transduction of the gravity signal. This may be linked to the functional adaptation of amyloplasts as statoliths. Amyloplasts in *Arabidopsis* have a specialized plastid-associated gravity transducer, Translocon of Outer Membrane of Chloroplasts (TOC) complex [74, 75]. TOC complex interacts with altered response to ALTERED RESPONSE TO GRAVITY1 (ARG1) protein and its close homology ARL2 present on the plasma membrane and endoplasmic reticulum to co-mediate gravity signal transduction. Single mutants of *pgm, toc132, toc120, toc34,* and *arg1* are not as severely agravitropic as *arg1 pgm, arg1 toc132, arg1 toc120, and arg1 toc34* double mutants [74, 75]. This suggests that the synergistic interaction between TOC complex and ARG1 amplifies the gravity signal during signal transduction post-amyloplast sedimentation. This also raises the question of whether the strength and/or persistence of other second messengers such as Ca^{2+} and $InSP_3$ can explain faster gravitropism in flowering plants.

In summary, roots have undergone numerous iterations to achieve the efficiency in gravitropism seen in land plants today. More research is required to understand the evolution of gravitropism, as it is integral in informing how roots evolve and develop. This is especially important for future agricultural applications in designing roots that are better at growing deeper, whether on Earth or in space.

4 New Insights on the Role of Auxin and Other Hormones in Gravitropism

Several recent publications have revealed an even more complex role for auxin beyond its function in asymmetric organ growth. For example, Zhang et al. [76] found that auxin plays dual roles in gravity perception and gravity response in *A. thaliana* roots. They demonstrated that auxin concentration altered the expression of starch biosynthesis genes *STARCH SYNTHASE 4 (SS4), PHOS PHOGLUCOMUTASE (PGM),* and *ADENOSINE D IPHO SPHATE GLUCOSE PYROPHOSPHORYLASE (ADG1)* via TIR/AFB auxin receptor pathway. Furthermore, increased local a uxin concentration in the roots either by exogenous application of synthetic auxins 1-napthaleneacetic acid and N-1-naphthylphthala mic acid or the use of *pin2* and *pin3/4/7* mutants resulted in the upregulation of *PGM, ADG1,* and *SS4* transcripts resulting in larger amyloplasts and faster bending responses. This mechanism may re flect a positive feedback loop to enhance gravity perception when the root is responding to gravity to further speed up gravitropism although the comparison of spatial and temporal regulation of *IAA CARBOXYL METHYLTRANSFERASE1* (*IAMT1*) (see below) during graviperception and graviresponse is currently lacking. How ever, it is unclear if auxin's effect on starch biosynthesis also occurs in amyloplasts in the endodermis.

The importance of auxin homeostasis in modulating gravitropism is becoming clearer as a study by Abbas et al. [77] showed that perturbations in an auxin homeostasis gene affected gravitropism. They found that auxin accumulation during the gravity-bending response in *A. thaliana* hypocotyl is regulated by *IAMT1* that converts IAA to methyl-IAA (me-IAA). Me-IAA is required for the establishment of auxin gradients after gravistimulation by suppressing *PIN3* expression and promoting PIN3 lateralization to the inner side of endodermal cells in the upper flank of hypocotyls to reduce polar auxin transport [77]. Furthermore, Mellor et al. [78] demonstrated that low to normal auxin concentrations in *A. thaliana* are modulated by auxin oxidation via *DIOXYGENASE FOR AUXIN OXIDATION 1* (*DAO1*), whereas high auxin levels were controlled by increased *GRETCHEN HAGEN 3* expression and auxin conjugation. While there has not been a study that investigates the role of *DAO* genes in gravitropism, the localization of DAO proteins suggests that they may be involved in gravitropism. DAO1 protein is asymmetrically localized in the lower flank of *A. thaliana* apical hook, whereas DAO2 is localized weakly in the root cap [79]. This is an interesting facet to auxin regulation that may be related to the evolution of fast gravitropism in land plants as mentioned in the preceding section. The multiple layers of auxin regulation, including local synthesis and polar auxin transport, highlights the importance of precise auxin accumulation for gravitropism.

The crosstalk of auxins with other hormones is also important in the coordination of plant growth during gravitropism. Both ethylene and auxin have synergistic effects in the inhibition of root elongation (reviewed in Muday et al. [80]). Ethylene regulates auxin biosynthesis and local auxin transport in roots to activate local auxin response to inhibit root elongation during root growth [80–82], whereas auxin upregulates ethylene production via *1-AMINOCYCLOPROPANE-1-CARBOXYLATE SYNTHASE (ACS)* genes in etiolated seedlings [83, 84]. Nziengui et al. [85] discovered that the folate precursor, *para*- aminobenzoic acid (PABA) influenced auxin via ethylene signaling, but independent of the folate pathway. PABA enhances root bending during gravistimulation by promoting asymmetric auxin accumulation in the epidermis, which requires the activation of auxin transcription factors ARF7 and ARF19 and ethylene biosynthesis and signaling, as well as increasing local auxin response [85]. Interestingly, there is evidence showing that compounds with an amino benzene ring and a carboxylic acid side chains including PABA, endosidin 8, and anthranilic acid can alter polar auxin transport via *WEAKLY ETHYLENE INSENSITIVE (WEI)* pathway by modifying basal polarization of PIN proteins [85–87].

In addition to ethylene, cytokinin also contributes to root gravitropism although this occurs differentially in the primary root and lateral roots (LRs). Cytokinin accumulates on the bottom flank of gravistimulated primary roots, similar to auxin accumulation, to contribute to root bending [88]. Moreover, the reduction of cytokinin levels by the irreversible degradation of *CYTOKININ OXIDASE (CKX) 2* and *3* caused primary roots to become more agravitropic [89]. Cytokinin in primary roots also acts on auxin transport by relocalizing PIN3 and PIN7 in columella cells [89]. By contrast, new research by Waidmann et al. [90] demonstrated that cytokinin performed a different function in lateral roots by acting as an anti-gravity signal to set a larger gravitropic set angle on lateral roots. They found that *CKX2* was upregulated at the root tip of LRs during stage II and stage III of LR formation, during which the LR root sets its gravitropic angle and continued to mature, respectively. Furthermore, cytokinin signaling was more asymmetrical in stage II LRs, with higher cytokinin signaling in the upper flank than the lower flank occurring during gravitropic bending [90]. This asymmetry declined when LRs reached the maturation stage III. Unlike the primary root, cytokinin in LRs did not directly affect PIN3 localization although pharmacological interference of auxin transport with N-1-naphtylphthalamic acid (NPA) disrupted cytokinin asymmetry in stage II LRs [90].

The work by Waidmann et al. [90] in implicating CKX2 as an anti-gravity signal was the result of examining 215 natural *A. thaliana* accessions for variation in lateral root gravitropic set point angle (GSA) and conducting Genome-Wide Association Studies (GWAS). Through this approach, it was found that a single nucleotide polymorphism in *CKX2* caused a single amino acid substitution that reduced CKX2 catalytic activity, leading to more horizontal lateral root growth. Interestingly, this inactive *CKX2* allele is preferentially expressed in accessions north of Sweden, where frequent snowfall leads to hypoxic conditions. This led the authors to propose that the inactive *CKX2* allele may have been selected to mitigate the harmful effects of hypoxia by promoting more horizontal lateral root growth.

The power of GWAS to uncover novel players was further demonstrated by evaluating the sensitivity of *A. thaliana* accessions to the NPA. Through this work, the *EXOCYST70A3 (EXO70A3)* was found to be involved in root gravitropism by regulating PIN4 protein distribution in columella cells [91]. Allelic variation in *EXO70A3* was associated with the differential responses of natural accessions to NPA. Similar to *CKX2*, allelic variation of *EXO70A3* was proposed to give rise to root systems that can be adapted to a particular environment. In the case of *EXO70A3*, differences in root system architecture conferred by *EXO70A3* allelic variation could be indicative of an adaptation for more efficient capture of soil surface water [91].

A group of carotenoid-derived phytohormones called strigolactones (SLs) was recently discovered to reduce gravitropic response in rice tillers by inhibiting the auxin biosynthesis gene *OsIAA20* in the bottom flank of gravistimulated shoots [92]. This, in turn, reduced auxin asymmetry during gravitropic bending response in the shoots, resulting in a spread-out shoot architecture [92]. Even though SLs can deplete PIN proteins from the plasma membrane to regulate auxin transport during the shoot branching inhibition process [93], Sang et al. [92] did not find evidence of SLs directly affecting auxin transport during gravistimulation because defects in SL biosynthetic and signaling pathways did not alter lateral auxin transport. This mechanism is proposed to function to aid shoots in heat avoidance to reduce thermal damage [92]. The regulation of graviresponse by SLs may be specific only to shoots, as the effective dose (>10μM) for the SL analog GR24 required to cause agravitropic root growth is too high to show physiological relevance [93].

The above-ground shoot organs exhibit various tropistic responses such as phototropism, gravitropism, thermotropism, and thigmotropism. In 1996, Fukaki et al. made two seminal discoveries regarding shoot gravitropism: 1) gravitropic bending of *A. thaliana* inflorescence was sensitive to temperature, even when gravity perception was intact and 2) *SHOOT GRAVITROPISM* genes *(SGRs)* are involved in shoot gravitropic responses. Recent work by Kim et al. [94] found that the *SGR5* gene contributes to the gravitropism and thermotropism signal integration in *A. thaliana* inflorescence. They demonstrated that *SGR5* undergoes alternative splicing to produce two isoforms, SGR5α and SGR5β, that are temperature dependent, with the SGR5β variant being more prominent at temperatures above 37 °C. SGR5α is the full-sized functional variant, which is involved in early steps of gravity perception, whereas SGR5β is the truncated non-active form. The increase in SGR5β protein relative to SGR5α at high temperatures resulted in reduced gravitropic response in *A. thaliana* inflorescence. When the SGR5β variant is present at high levels, they form heterodimers with SGR5α to disrupt its DNA binding in the nucleus. The authors proposed that this mechanism likely enables inflorescence stems to reduce gravitropic curvature to protect the organs from heat damage.

The studies reviewed here represent a small fraction of new breakthroughs in gravitropism. Auxin is undoubtedly central to gravitropism with roles expanding into gravity perception, gravity signaling, and gravity response, so much so that other hormones and pathways co-regulate and crosstalk with auxin, likely to act as a "fail-safe" mechanism in the event of failures in auxin regulation. As we begin to better understand the different pieces of the puzzle of gravitropism, we can begin to construct a clearer picture on how to modulate shoot and root architecture and function to design the ideal plant ideotype for cultivation in Earth and beyond.

5 Insights on Gravitropism and Other Plant Movements Enabled by Research in Microgravity

The bulk of our knowledge about gravitropism has come from experiments conducted in Earth-based laboratories. However, because of gravity's continuous presence on Earth, a more in-depth understanding of its effects on biological systems requires a near-weightless environment. For several decades, the space shuttle and facilities on the International Space Station (ISS) have addressed this requirement. The organizational and technical difficulties associated with spaceflight experiments also have led researchers to rely on a variety of microgravity simulators such as clinostats for their work [95, 96]. This section highlights a selection of research in the microgravity environment of space and microgravity simulators on Earth that have led to a better understanding of directional plant organ growth and gravitropism.

The study of Ca^{2+} signaling is one example of how the use of short periods of microgravity provided by parabolic flights can advance knowledge about gravitropism. As noted in Subheading 2, ground-based studies have demonstrated changes in cytosolic Ca^{2+} in plants that have been reoriented at 1 *g*. These Ca^{2+} changes consisted of an initial transient increase followed by a second sustained increase. Elevations in cytosolic Ca^{2+} have been proposed to be part of the second messenger cascade that leads the plant to redirect its growth after it is reoriented [35, 97]. An argument can be made, however, that the act of rotating the plant presents a mechanical stimulus that triggers a Ca^{2+} response independent of gravity. Through parabolic flights and a ground-based device that rotates *Arabidopsis* seedlings without a resulting change in their orientation, it was shown that the second, sustained increase in cytosolic $Ca^{2+}$is the true signal associated with gravity sensing and signaling [98]. In a recent study of *Brassica napus* seedling roots on the ISS, cytosolic Ca^{2+} changes were shown to occur in gravity-sensing columella cells even with minimal displacement of statoliths [99]. However, this ISS study relied on fixing roots and imaging Ca^{2+} precipitates that formed to reach such conclusions. Because cytosolic Ca^{2+}changes occur within seconds after the plant is exposed to a stimulus, it is unclear whether Ca^{2+} imaged through chemical fixation is indicative of genuine cellular signaling events. For the future, it will be exciting to deploy plant lines expressing live cell Ca^{2+} reporters on the ISS. This will not only require imaging hardware, but also protocols for handling plants expressing these reporters in space. Various fluorescence imaging systems such as the light microscopy module have been used to monitor auxin-based fluorescent sensors on the ISS [100]. Thus, imaging genetically encoded Ca^{2+} fluorescent sensors on the ISS is a possibility.

As noted in preceding sections, changes in the direction of plant organ growth can be triggered by a range of environmental factors such as light, moisture, nutrient, and salinity gradients [73, 101, 102]. However, because of the ubiquitous nature of gravity on Earth, it can mask some of the effects of these other stimuli on plant growth. Microgravity analogs and research on the ISS have enabled researchers to minimize the influence of gravity to better understand other tropisms and plant movements. One particular type of plant movement that has benefited from research in microgravity is the phenomenon of circumnutation. Circumnutation has been described as endogenous oscillatory growth patterns in plants, and hypothesized to be under the tight control of gravity [103]. According to this hypothesis, plant organs exhibit circumnutation because they continuously attempt to realign with the gravity vector as they grow [104]. The dependence of circumnutation on gravity is supported by work with dicots on Earth. For example, mutants that are defective in gravity sensing are compromised in their ability to circumnutate [105–107]. Experiments in space have reinforced conclusions from these ground-based studies. For instance, while low-amplitude circumnutations were observed in *Arabidopsis* stems on the ISS, application of gravitational acceleration using a centrifuge onboard the ISS increased the amplitude of circumnutations [108]. Moreover, a recent study on the ISS showed that gravity also played a role in monocot circumnutation through studies with rice. On the ISS, coleoptiles of wild-type rice seedlings showed reduced circumnutation compared to seedlings on Earth. The amplitude of circumnutation of wild-type rice seedlings on the ISS was restored to almost Earth-like levels when provided with gravitational force on a centrifuge. On the other hand, the agravitropic *lazy-1* mutant, which had defects in circumnutation on Earth, also had diminished circumnutation in microgravity. These defects were not restored to wild-type levels despite centrifugation [109]. The attenuation of circumnutation in *Arabidopsis* inflorescence stems observed on the ISS was validated by studies on the Chinese recoverable satellite SJ-10 and spacelab TG-2. In this study, inhibition of circumnutation in space was shown to be influenced by photoperiod [110].

Root waving and skewing particularly in *Arabidopsis* are other growth phenomena that have been studied in microgravity. Waving and skewing are most apparent in roots that are in contact with hard agar surfaces. Waving refers to a series of periodic undulations in the growth of the root tips, while skewing describes the deviation of the root axis from vertical growth making the roots appear slanted [111]. Several molecular components of root waving and skewing have been discovered, including roles for the microtubule cytoskeleton, endomembranes, and G proteins [112–114]. Root waving and skewing have been proposed to be driven by a combination of root tip contact with the agar surface, circumnutation, and gravity. Unlike circumnutation, however, research on the Space Shuttle and ISS now indicate that root skewing is independent of gravity because roots still skewed during spaceflight [115–117]. The occurrence of root skewing in microgravity suggests that mechanical signaling may play a more predominant role in the skewing response compared to gravity. Mutants to the *SPIRAL1 (SPR1)* gene (*spr1*), which encodes a protein that binds to microtubules, exhibited stronger root skewing phenotypes on Earth [112, 118]. Interestingly, the root skewing phenotypes of *spr1* mutants were enhanced on the ISS [119]. Enhanced root skewing and coiling responses were also observed in mutants to the root-expressed vegetative actin isoform, *ACTIN2* [116]. Taken together, these spaceflight results point to the cytoskeleton as an important component of plant mechanical signaling in microgravity and on Earth.

With regard to the cytoskeleton, several studies have shown that reorienting plants treated with actin antagonists or plants with mutations in genes encoding components of the acto-myosin system exhibit enhanced shoot and root gravitropism [28, 120, 121]. Recent research using clinostats now indicate that these actin-compromised plants may be exhibiting inhibited autotropic straightening rather than a promotion of gravitropism. Negating the effects of gravity by rotating plants with a clinostat enabled the organ autotropic straightening response to be manifested more clearly compared to reorienting the plant at 1 *g* [122, 123]. In this regard, it is likely that the enhanced root coiling of *Arabidopsis* actin mutants observed in microgravity can also be explained by inhibited autotropic straightening [116]. Although the exact mechanisms by which actin regulates autotropic root straightening are unknown, there is recent evidence that the hormone brassinolide may be involved in this process [122].

Hydrotropism (i.e., growth of roots toward higher moisture [124, 125]) and phototropism (the growth of plant organs toward or away from a light source [126]) are two tropisms that are masked by the gravitational force on Earth. As such, additional insights into these two tropistic plant responses have been uncovered by research in space. A couple of recent experiments on the ISS looked at hydrotropism in particular. In one study, the hydrotropic response of cucumber roots in space was found to be stronger compared to roots of Earth-grown plants. Like gravitropism, this ISS experiment showed that hydrotropism is modulated by changes in auxin redistribution facilitated by relocalization of PIN auxin efflux carriers [127]. It should be noted, however, that the involvement of auxin in hydrotropism is still under debate [128]. In another recent ISS experiment, carrot roots exhibited strong growth toward a disodium phosphate source (called chemotropism), which was found to overcome the influence of hydrotropism [129]. While these recent ISS studies have shed some light on interactions among different tropisms, they have mostly confirmed major findings about hydrotropism made on Earth. Like work on gravitropism, hydrotropism research could advance further by including hydrotropism mutants discovered on Earth in future spaceflight experiments [130–133].

Perhaps the plant tropism that has advanced the most from the unique microgravity environment of space is phototropism. In this regard, hypocotyls of *Arabidopsis* on the ISS displayed a novel red light phototropic response that was masked on Earth [134]. In addition to confirming the red light phototropic response, a follow-up study on the ISS revealed that roots exhibited a positive blue light phototropic response [135]. Blue light-induced phototropic responses observed in space were accompanied by changes in the expression of genes associated with photosynthesis, light perception, starch metabolism, carotenoid biosynthesis, and cell wall/ membrane remodeling [136, 137]. Taken together, the microgravity environment in space and the use of microgravity simulators have proven to be valuable complement to many Earth-based studies of tropisms.

6 Gravitropism for Optimizing Resource Acquisition Efficiency in Crop Plants

As discussed in the preceding sections, research on gravitropism has increased our fundamental understanding about how plants modulate their development through sensory systems located in specific cell types, and how environmental stimuli perceived in these cell types are translated into growth responses. Because gravity has such a profound influence on above- and below-ground plant architecture, gravitropism has received increased attention in crop improvement strategies. In this regard, the focus has been on enhancing water and nutrient foraging and optimizing light capture for more efficient photosynthesis [11, 138, 139]. Our review on gravitropism up to this point has for the most part highlighted experiments conducted with *A. thaliana* in artificial gel substrates and controlled environmental conditions. In this section, we shift our focus to gravitropism in agriculturally relevant plant species, and briefly review how knowledge about this fundamental plant environmental response has been applied for crop improvement.

Plant architecture is specified by several parameters, such as branching frequency, growth rate, organ shape, surface area, and growth angle. From our review thus far, it is obvious that growth angle is the reason gravitropism has generated increased interest for crop improvement. Crop root systems have received most of this attention since root growth angle is a major driver that defines the spatial distribution of roots in soils [140]. While root growth angle is the trait most closely tied to gravitropism, downstream traits resulting from root growth angle are the ones with more direct relevance for crops. One such trait is rooting depth, as this trait not only can increase the availability of water and nutrient resources to crops, but also help improve soil health through better carbon storage [141]. With regard to crop resource availability, access to water in drought-prone areas is one immediate benefit of having deep-rooted plants. One noteworthy study that demonstrated the benefit of root growth angle for drought-induced stress involved the positional cloning of the quantitative trait locus (QTL) in rice (*Oryza sativa*) called *deeper rooting 1* (*DRO1*). The rice cultivar IR64, which had shallow root growth angles, was found to be more susceptible to drought. IR64 had a single base pair deletion in the *DRO1* gene resulting in a premature stop codon. On the other hand, a near-isogenic line (NIL) with full-length *DRO1* had steeper root angles and a higher frequency of deep roots. As a result, this rice NIL was able to avoid drought by gaining access to water located in deeper soil zones [12]. Consistent with observations in rice, overexpressing the plum (*Prunus domesta*) *DRO1* was found to increase rooting depth in container-grown plum seedlings. However, it is not known if steep root growth angles are directly responsible for the increased rooting depth in plum or whether plum *DRO1* overexpressing lines are drought tolerant [142]. *DRO1* belongs to the same family as the *A. thaliana LAZY* genes involved in gravity signal transduction [8]. Taken together, studies with rice and plum provide examples of how a component of the plant gravitropic response pathway can be applied to growing crops in water-limited conditions.

The ability to better access nitrogen is another benefit provided by deep-rooted crops since nitrogen is a highly mobile nutrient and leached into deeper soil zones [143]. In studies of 108 inbreed maize (*Zea mays*) lines, brace and crown root angles become steeper by about 18° under nitrogen-limited conditions [144]. Simulation modeling of *Z. mays* root growth showed that axial root growth angle enhances nitrogen capture from the soil. Optimal nitrogen capture was dependent on soil type and precipitation. For instance, under low rainfall, nitrogen is located mostly in the upper soil layers. Under such conditions, cultivars with shallow roots performed better, but not as well under a high rainfall scenario where nitrogen is leached to deeper soil layers. Cultivars with both shallow and steep roots performed well under both conditions, supporting the hypothesis that root growth angle is a major determinant of nitrogen capture [145]. The underlying molecular mechanisms that specify root growth angle under different nitrogen conditions are not fully understood. However, work with ammonium stress in *A. thaliana* implicates components of the gravity signaling pathway, including ARG1, AUX1, and PIN [146].

Simulation modeling with *Z. mays* roots mentioned above indicates that the contribution of root growth angle to resource capture is complex and highlights tradeoffs between shallow versus steep root growth angles. While steep root angles are beneficial for water during drought and leached nitrogen in areas of high rainfall, they might not confer any advantage when foraging for immobile resources that remain on the top soil. This point is best illustrated in studies of phosphate nutrition in plants. Unlike nitrogen, phosphate is immobile and accumulates in the top soil. Like other major nutrients, poor access to phosphate is a major constraint for crop productivity [147]. In this regard, studies have shown that plants with shallow root angles that enable foraging of nutrients at or close to the soil surface are more beneficial when phosphate pools are limited. For example, in the common bean (*Phaseolus vulgaris*), availability of phosphate influenced the angle of basal roots. The gravitropic response under different phosphate conditions was genotype dependent. Genotypes that were phosphate inefficient had deeper roots while those that were phosphate efficient produced shallow roots under phosphate stress [148]. Interestingly, comparative studies between lateral roots of *A. thaliana* and *P. vulgaris*, and *P. vulgaris* basal roots showed that GSA in response to phosphate deprivation was root type dependent. Phosphate deficiency caused lateral roots to assume a more vertical GSA, while basal root GSA became more horizontal, the latter of which was consistent with previous studies. The differences in response to phosphate levels between root types is unknown. One proposed mechanism is that root types differ in their sensitivity to auxin [149]. This hypothesis is supported by studies showing that root architectural responses to limited phosphate involve components of the auxin signaling machinery [150, 151]. More relevant to the gravitropism, GSA of *pin3* and starchless *A. thaliana* mutants differed from that of wild-type under phosphate stress [152].

In *O. sativa*, the *rice morphology determinant* (*RMD*) gene was shown to be a component of crown root angle regulation in response to phosphate. Mutants to *RMD* have faster root gravitropism and steeper root growth angles regardless of phosphate levels. The RMD protein localizes to statoliths in the columella cells and is an actin-binding protein that belongs to the formin family of actin regulatory proteins [150]. There is evidence that *RMD* modulates auxin transport and redistribution by specifying recycling of PIN proteins at the plasma membrane [153]. Recently, RMD was also shown to function in light-dependent stem gravitropism in rice [154]. These studies indicate that RMD is a key molecular link between gravitropism and actin-auxin signaling that could be exploited for breeding crops with desired growth angles. It is worth noting, however, that the utility of gravitropism as a breeding target for optimizing plant resource acquisition should be evaluated for synergism with other traits and across contrasting environments due to tradeoffs noted earlier [145, 155]. In *P. vulgaris*, for example, long and dense root hairs combined with shallow roots increased above-ground biomass by almost 300% under phosphate stress [156]. Studies for trait synergism and validation under different environments will be key to developing cultivars that are more efficient at capturing soil resources.

7 Conclusions

As a prequel to the following chapters of this volume, we have provided an overview of gravitropism. Although this review is by no means comprehensive, we hope that we have presented a flavor of how different approaches are enabling us to gain new insights into this classical field of plant biology. Here we provided examples of how omics technologies have been applied to the problem of gravitropism. However, the ceiling for implementing more omics approaches has become higher with the introduction of more elaborate instrumentation. For example, as sequencing and mass spectrometry capacities advance, there are more opportunities to integrate different omics within the same experiment, paired with sophisticated mapping to elucidate changes in one pathway comprehensively from RNA, protein, and metabolites or find changes in multiple pathways. There is a lot of potential to include more epigenomic work using whole genome bisulfite sequencing, as there are strong indications of methylation changes during gravitropism based on the downregulation of chromatin-related genes [21] and differentially methylated cytosines in spaceflight *Arabidopsis* DNA versus ground DNA although effects of cosmic irradiation and other stresses during spaceflight cannot be excluded [157]. The emergence of new single-cell omics techniques such as single-cell RNA-Seq [158–161], single-cell bisulfite sequencing [162–164], and single cell and live single-cell mass spectrometry for proteomics and metabolomics [165, 166] can be applied to describe cell-specific changes, particularly in columella cells (albeit with limitations of protoplasting columella cells). The technique of gravity perception and response in a single cell can be combined with conventional omics to inform on single-cell changes. The development of omics data repositories for *in silico* analysis, such as the Test of Arabidopsis Space Transcriptome (TOAST) database [167], will also be instrumental in finding highly conserved genes, proteins, and metabolites involved in gravitropism.

Other studies highlighted here are those that have addressed the evolution of gravitropism. It is interesting that the starch-statolith hypothesis and the Cholodny-Went theory, which have been the hallmarks for explaining gravity sensing and response, respectively, have been central to the transition of land plants from slow to fast gravitropism [62]. In this regard, it has become more apparent through work with natural accessions and GWAS that plant adaptation to specific environments is driven by allelic variation in auxin- and cytokinin-related genes that regulate gravitropism [90, 91]. Although such insights continue to be gained from model plants such as *A. thaliana*, studies in agriculturally relevant crops indicate that gravitropism could potentially be a breeding target that could provide adaptive advantages, particularly when considered with other phenes and under low input agricultural systems [168]. Finally, space and facilities that simulate microgravity will continue to be a valuable resource as researchers attempt to translate basic knowledge about gravitropism to guide the development of crop cultivars and design of enclosed plant habitats for advanced human life support systems during future space exploration missions.

Acknowledgments

Research on gravitropism in the authors' laboratory is supported by the National Aeronautics and Space Administration (NASA grants 80NSSC19K0129 and 80NSSC18K1462).

References

1. Fulghum R (1989) All I really need to know I learned in Kindergarten. The Random House Publishing Group, New York
2. Darwin C, Darwin F (1880) The power of movement in plants. John Murry, London
3. Baldwin KL, Strohm AK, Masson PH (2013) Gravity sensing and signal transduction in vascular plant primary roots. Am J Bot 100 (1):126–142. https://doi.org/10.3732/ajb.1200318
4. Hashiguchi Y, Tasaka M, Morita MT (2013) Mechanism of higher plant gravity sensing. Am J Bot 100(1):91–100. https://doi.org/10.3732/ajb.1200315
5. Morita MT (2010) Directional gravity sensing in gravitropism. Annu Rev Plant Biol 61:705–720. https://doi.org/10.1146/annurev.arplant.043008.092042
6. Sato EM, Hijazi H, Bennett MJ, Vissenberg K, Swarup R (2015) New insights into root gravitropic signalling. J Exp Bot 66 (8):2155–2165. https://doi.org/10.1093/jxb/eru515
7. Nakamura M, Nishimura T, Morita MT (2019) Gravity sensing and signal conversion in plant gravitropism. J Exp Bot 70 (14):3495–3506. https://doi.org/10.1093/jxb/erz158
8. Nakamura M, Nishimura T, Morita MT (2019) Bridging the gap between amyloplasts and directional auxin transport in plant gravitropism. Curr Opin Plant Biol 52:54–60. https://doi.org/10.1016/j.pbi.2019.07.005
9. Toyota M, Gilroy S (2013) Gravitropism and mechanical signaling in plants. Am J Bot 100 (1):111–125. https://doi.org/10.3732/ajb.1200408
10. Vandenbrink JP, Kiss JZ (2019) Plant responses to gravity. Semin Cell Dev Biol 92:122–125. https://doi.org/10.1016/j.semcdb.2019.03.011
11. Rich SM, Watt M (2013) Soil conditions and cereal root system architecture: review and considerations for linking Darwin and Weaver. J Exp Bot 64(5):1193–1208. https://doi.org/10.1093/jxb/ert043
12. Uga Y, Sugimoto K, Ogawa S, Rane J, Ishitani M, Hara N, Kitomi Y, Inukai Y, Ono K, Kanno N, Inoue H, Takehisa H, Motoyama R, Nagamura Y, Wu J, Matsumoto T, Takai T, Okuno K, Yano M (2013) Control of root system architecture by deeper rooting 1 increases rice yield under drought conditions. Nat Genet 45 (9):1097–1102. https://doi.org/10.1038/ng.2725
13. Taniguchi M, Nakamura M, Tasaka M, Morita MT (2014) Identification of gravitropic response indicator genes in *Arabidopsis* inflorescence stems. Plant Signal Behav 9(9): e29570. https://doi.org/10.4161/psb.29570
14. Cao H, Wang F, Lin H, Ye Y, Zheng Y, Li J, Hao Z, Ye N, Yue C (2020) Transcriptome and metabolite analyses provide insights into zigzag-shaped stem formation in tea plants (*Camellia sinensis*). BMC Plant Biol 20 (1):98. https://doi.org/10.1186/s12870-020-2311-z
15. Hu L, Mei Z, Zang A, Chen H, Dou X, Jin J, Cai W (2013) Microarray analyses and comparisons of upper or lower flanks of rice shoot

base preceding gravitropic bending. PLoS One 8(9):e74646. https://doi.org/10.1371/journal.pone.0074646

16. Schenck CA, Nadella V, Clay SL, Lindner J, Abrams Z, Wyatt SE (2013) A proteomics approach identifies novel proteins involved in gravitropic signal transduction. Am J Bot 100 (1):194–202. https://doi.org/10.3732/ajb.1200339
17. Esmon CA, Tinsley AG, Ljung K, Sandberg G, Hearne LB, Liscum E (2006) A gradient of auxin and auxin-dependent transcription precedes tropic growth responses. Proc Natl Acad Sci U S A 103(1):236–241. https://doi.org/10.1073/pnas.0507127103
18. Perera IY, Heilmann I, Boss WF (1999) Transient and sustained increases in inositol 1,4,5-trisphosphate precede the differential growth response in gravistimulated maize pulvini. Proc Natl Acad Sci U S A 96 (10):5838–5843. https://doi.org/10.1073/pnas.96.10.5838
19. Perera IY, Heilmann I, Chang SC, Boss WF, Kaufman PB (2001) A role for inositol 1,4,5-trisphosphate in gravitropic signaling and the retention of cold-perceived gravistimulation of oat shoot pulvini. Plant Physiol 125 (3):1499–1507. https://doi.org/10.1104/pp.125.3.1499
20. Parker KE, Briggs WR (1990) Transport of indole-3-acetic acid during gravitropism in intact maize coleoptiles. Plant Physiol 94 (4):1763–1769. https://doi.org/10.1104/pp.94.4.1763
21. Kimbrough JM, Salinas-Mondragon R, Boss WF, Brown CS, Sederoff HW (2004) The fast and transient transcriptional network of gravity and mechanical stimulation in the Arabidopsis root apex. Plant Physiol 136 (1):2790–2805. https://doi.org/10.1104/pp.104.044594
22. Tan C, Wang H, Zhang Y, Qi B, Xu G, Zheng H (2011) A proteomic approach to analyzing responses of *Arabidopsis thaliana* root cells to different gravitational conditions using an agravitropic mutant, *pin2* and its wild type. Proteome Sci 9(72):1–16. https://doi.org/10.1186/1477-5956-9-72
23. Xia H, Zhao C, Hou L, Li A, Zhao S, Bi Y, An J, Zhao Y, Wan S, Wang X (2013) Transcriptome profiling of peanut gynophores revealed global reprogramming of gene expression during early pod development in darkness. BMC Genomics 14:517. https://doi.org/10.1186/1471-2164-14-517
24. Zhao C, Zhao S, Hou L, Xia H, Wang J, Li C, Li A, Li T, Zhang X, Wang X (2015) Proteomics analysis reveals differentially activated pathways that operate in peanut gynophores at different developmental stages. BMC Plant Biol 15(1):188. https://doi.org/10.1186/s12870-015-0582-6
25. Blancaflor EB, Fasano JM, Gilroy S (1998) Mapping the functional roles of cap cells in the response of Arabidopsis primary roots to gravity. Plant Physiol 116(1):213–222. https://doi.org/10.1104/pp.116.1.213
26. Maccleery SA, Kiss JZ (1999) Plastid sedimentation kinetics in roots of wild-type and starch-deficient mutants of Arabidopsis. Plant Physiol 120(1):183–192. https://doi.org/10.1104/pp.120.1.183
27. Toyota M, Ikeda N, Sawai-Toyota S, Kato T, Gilroy S, Tasaka M, Morita MT (2013) Amyloplast displacement is necessary for gravisensing in Arabidopsis shoots as revealed by a centrifuge microscope. Plant J 76 (4):648–660. https://doi.org/10.1111/tpj.12324
28. Hou G, Mohamalawari DR, Blancaflor EB (2003) Enhanced gravitropism of roots with a disrupted cap actin cytoskeleton. Plant Physiol 131(3):1360–1373. https://doi.org/10.1104/pp.014423
29. Leitz G, Kang B-H, Schoenwaelder MEA, Staehelin LA (2009) Statolith sedimentation kinetics and force transduction to the cortical endoplasmic reticulum in gravity-sensing *Arabidopsis* Columella cells. Plant Cell 21 (3):843–860. https://doi.org/10.1105/tpc.108.065052
30. Zheng HQ, Staehelin LA (2001) Nodal endoplasmic reticulum, a specialized form of endoplasmic reticulum found in gravity-sensing root tip Columella cells. Plant Physiol 125(1):252–265. https://doi.org/10.1104/pp.125.1.252
31. Yoder TL, Zheng H-Q, Todd P, Staehelin LA (2001) Amyloplast sedimentation dynamics in maize Columella cells support a new model for the gravity-sensing apparatus of roots. Plant Physiol 125(2):1045–1060. https://doi.org/10.1104/pp.125.2.1045
32. Fasano JM, Swanson SJ, Blancaflor EB, Dowd PE, T-h K, Gilroy S (2001) Changes in root cap pH are required for the gravity response of the Arabidopsis root. Plant Cell 13 (4):907–922. https://doi.org/10.1105/tpc.13.4.907
33. Joo JH, Bae YS, Lee JS (2001) Role of auxin-induced reactive oxygen species in root gravitropism. Plant Physiol 126(3):1055–1060. https://doi.org/10.1104/pp.126.3.1055

34. Perera IY, Hung CY, Brady S, Muday GK, Boss WF (2006) A universal role for inositol 1,4,5-trisphosphate-mediated signaling in plant gravitropism. Plant Physiol 140 (2):746–760. https://doi.org/10.1104/pp.105.075119

35. Plieth C, Trewavas AJ (2002) Reorientation of seedlings in the earth's gravitational field induces cytosolic calcium transients. Plant Physiol 129(2):786–796. https://doi.org/10.1104/pp.011007

36. Went FW, Thimann KV (1937) Phytohormones. The Macmillan Company, New York

37. Band LR, Wells DM, Larrieu A, Sun J, Middleton AM, French AP, Brunoud G, Sato EM, Wilson MH, Peret B, Oliva M, Swarup R, Sairanen I, Parry G, Ljung K, Beeckman T, Garibaldi JM, Estelle M, Owen MR, Vissenberg K, Hodgman TC, Pridmore TP, King JR, Vernoux T, Bennett MJ (2012) Root gravitropism is regulated by a transient lateral auxin gradient controlled by a tipping-point mechanism. Proc Natl Acad Sci U S A 109 (12):4668–4673. https://doi.org/10.1073/pnas.1201498109

38. Marchant A, Kargul J, May ST, Muller P, Delbarre A, Perrot-Rechenmann C, Bennett MJ (1999) AUX1 regulates root gravitropism in Arabidopsis by facilitating auxin uptake within root apical tissues. EMBO J 18 (8):2066–2073. https://doi.org/10.1093/emboj/18.8.2066

39. Rakusová H, Abbas M, Han H, Song S, Hélène FJ (2016) Termination of shoot gravitropic responses by auxin feedback on PIN3 polarity. Curr Biol 26(22):3026–3032. https://doi.org/10.1016/j.cub.2016.08.067

40. Ottenschlager I, Wolff P, Wolverton C, Bhalerao RP, Sandberg G, Ishikawa H, Evans M, Palme K (2003) Gravity-regulated differential auxin transport from Columella to lateral root cap cells. Proc Natl Acad Sci U S A 100 (5):2987–2991. https://doi.org/10.1073/pnas.0437936100

41. Moseyko N, Zhu T, Chang HR, Wang X, Feldman LJ (2002) Transcription profiling of the early gravitropic response in Arabidopsis using high-density oligonucleotide probe microarrays. Plant Physiol 130(2):720–728. https://doi.org/10.1104/pp.009688

42. Dong Z, Jiang C, Chen X, Zhang T, Ding L, Song W, Luo H, Lai J, Chen H, Liu R, Zhang X, Jin W (2013) Maize LAZY1 mediates shoot gravitropism and inflorescence development through regulating auxin transport, auxin signaling, and light response. Plant Physiol 163(3):1306–1322. https://doi.org/10.1104/pp.113.227314

43. Aubry-Hivet D, Nziengui H, Rapp K, Oliveira O, Paponov IA, Li Y, Hauslage J, Vagt N, Braun M, Ditengou FA, Dovzhenko A, Palme K (2014) Analysis of gene expression during parabolic flights reveals distinct early gravity responses in *Arabidopsis* roots. Plant Biol 16:129–141. https://doi.org/10.1111/plb.12130

44. Zhang N, Yu H, Yu H, Cai Y, Huang L, Xu C, Xiong G, Meng X, Wang J, Chen H, Liu G, Jing Y, Yuan Y, Liang Y, Li S, Smith SM, Li J, Wang Y (2018) A core regulatory pathway controlling rice tiller angle mediated by the *LAZY1*-dependent asymmetric distribution of auxin. Plant Cell 30(7):1461–1475. https://doi.org/10.1105/tpc.18.00063

45. Kamada M, Higashitani A, Ishioka N (2005) Proteomic analysis of *Arabidopsis* root gravitropism. Biol Sci Space 19(3):148–154

46. Young LS, Harrison BR, Narayana Murthy UM, Moffatt BA, Gilroy S, Masson PH (2006) Adenosine kinase modulates root gravitropism and cap morphogenesis in Arabidopsis. Plant Physiol 142(2):564–573. https://doi.org/10.1104/pp.106.084798

47. Pozhvanov GA, Klimenko NS, Boilova TE, Shavarda AL, Medvedev SS (2017) Ethylene-dependent adjustment of metabolite profiles in *Arabidopsis thaliana* seedlings during gravitropic response. Russian J Plant Physiol 64 (6):906–918. https://doi.org/10.1134/s1021443717050090

48. Millar KDL, Kiss JZ (2013) Analyses of tropistic responses using metabolomics. Am J Bot 100(1):79–90. https://doi.org/10.3732/ajb.1200316

49. Li Y, Hagen G, Guilfoyle TJ (1991) An auxin-responsive promoter is differentially induced by auxin gradients during tropisms. Plant Cell 3(11):1167–1175. https://doi.org/10.1105/tpc.3.11.1167

50. Chang SC, Cho MH, Kim SK, Lee JS, Kirakosyan A, Kaufman PB (2003) Changes in phosphorylation of 50 and 53 kDa soluble proteins in graviresponding oat (*Avena sativa*) shoots. J Exp Biol 54 (384):1013–1022. https://doi.org/10.1093/jxb/erg104

51. Sieberer T, Seifert GJ, Hauser M-T, Grisafi P, Fink GR, Luschnig C (2000) Post-transcriptional control of the Arabidopsis auxin efflux carrier EIR1 requires AXR1. Curr Biol 10(24):1595–1598. https://doi.org/10.1016/s0960-9822(00)00861-7

52. Park C-J, Seo Y-S (2015) Heat shock proteins: a review of the molecular chaperones for plant immunity. Plant Pathol J 31 (4):323–333. https://doi.org/10.5423/PPJ.RW.08.2015.0150
53. Wang W, Vinocur B, Shoseyov O, Altman A (2004) Role of plant heat-shock proteins and molecular chaperones in the abiotic stress response. Trends Plant Sci 9(5):244–252. https://doi.org/10.1016/j.tplants.2004.03.006
54. Song H, Zhao R, Fan P, Wang X, Chen X, Li Y (2009) Overexpression of *AtHsp90.2, AtHsp90.5* and *AtHsp90.7* in Arabidopsis thaliana enhances plant sensitivity to salt and drought stresses. Planta 229(4):955–964. https://doi.org/10.1007/s00425-008-0886-y
55. Czerniawski P, Bednarek P (2018) Glutathione *S*-transferases in the biosynthesis of sulfur-containing secondary metabolites in Brassicaceae plants. Front Plant Sci 9:1639. https://doi.org/10.3389/fpls.2018.01639
56. Gullner G, Komives T, Király L, Schröder P (2018) Glutathione S-transferase enzymes in plant-pathogen interactions. Front Plant Sci 9:1836. https://doi.org/10.3389/fpls.2018.01836
57. Allen J, Bisbee PA, Darnell RL, Kuang A, Levine LH, Musgrave ME, Van Loon JJWA (2009) Gravity control of growth form in Brassica rapa and *Arabidopsis thaliana* (Brassicaceae): consequences for secondary metabolism. Am J Bot 96(3):652–660. https://doi.org/10.3732/ajb.0800261
58. Bingol K, Brüschweiler R (2017) Knowns and unknowns in metabolomics identified by multidimensional NMR and hybrid MS/NMR methods. Curr Opin Biotechnol 43:17–24. https://doi.org/10.1016/j.copbio.2016.07.006
59. Hetherington AJ, Dolan L (2018) Stepwise and independent origins of roots among land plants. Nature 561(7722):235–238. https://doi.org/10.1038/s41586-018-0445-z
60. Jones VAS, Dolan L (2012) The evolution of root hairs and rhizoids. Ann Bot 110 (2):205–212. https://doi.org/10.1093/aob/mcs136
61. Kenrick P, Strullu-Derrien C (2014) The origin and early evolution of roots. Plant Physiol 166(2):570–580. https://doi.org/10.1104/pp.114.244517
62. Zhang Y, Xiao G, Wang X, Zhang X, Friml J (2019) Evolution of fast root gravitropism in seed plants. Nat Commun 10(1):3480. https://doi.org/10.1038/s41467-019-11471-8
63. Kiss JZ (1994) The response to gravity is correlated with the number of statoliths in *Chara* rhizoids. Plant Physiol 105:937–940. https://doi.org/10.1104/pp.105.3.937
64. Limbach C, Hauslage J, Schäfer C, Braun M (2005) How to activate a plant gravireceptor. Early mechanisms of gravity sensing studied in Characean rhizoids during parabolic flights. Plant Physiol 139(2):1030–1040. https://doi.org/10.1104/pp.105.068106
65. Schröter K, Läuchli A, Sievers A (1975) Microanalytical identification of barium sulphate crystals in statoliths of *Chara* rhizoids (*Ch. fragilis*, desv.). Planta 122:213–225. https://doi.org/10.1007/BF00385269
66. Wang-Cahill F, Kiss JZ (1995) The statolith compartment in *Chara* rhizoids contains carbohydrate and protein. Am J Bot 82 (2):220–229. https://doi.org/10.1002/j.1537-2197.1995.tb11490.x
67. Nakamura M, Toyota M, Tasaka M, Morita MT (2011) An *Arabidopsis* E3 ligase, shoot gravitropism9, modulates the interaction between statoliths and F-actin in gravity sensing. Plant Cell 23(5):1830–1848. https://doi.org/10.1105/tpc.110.079442
68. Edelmann HG (2018) Graviperception in maize plants: is amyloplast sedimentation a red herring? Protoplasma 255:1877–1881. https://doi.org/10.1007/s00709-018-1272-7
69. Hochholdinger F, Yu P, Marcon C (2018) Genetic control of root system development in maize. Trends Plant Sci 23(1):79–88. https://doi.org/10.1016/j.tplants.2017.10.004
70. Monshaushen GB, Zieschang HE, Sievers A (1996) Differential proton secretion in the apical elongation zone caused by gravistimulation is induced by a signal from the root cap. Plant Cell Environ 19(12):1408–1414. https://doi.org/10.1111/j.1365-3040.1996.tb00019.x
71. Casanova-Sáez R, Voß U (2019) Auxin metabolism controls developmental decisions in land plants. Trends Plant Sci 24 (8):741–754. https://doi.org/10.1016/j.tplants.2019.05.006
72. Jiang K (2003) Quiescent center formation in maize roots is associated with an auxin-regulated oxidizing environment. Development 130(7):1429–1438. https://doi.org/10.1242/dev.00359
73. Muthert LWF, Izzo LG, van Zanten M, Aronne G (2020) Root tropisms:

investigations on earth and in space to unravel plant growth direction. Front Plant Sci 10:1807. https://doi.org/10.3389/fpls.2019.01807

74. Stanga JP, Boonsirichai K, Sedbrook JC, Otegui MS, Masson PH (2009) A role for the TOC complex in Arabidopsis root gravitropism. Plant Physiol 149(4):1896–1905. https://doi.org/10.1104/pp.109.135301

75. Strohm AK, Barrett-Wilt GA, Masson PH (2014) A functional TOC complex contributes to gravity signal transduction in Arabidopsis. Front Plant Sci 5:148. https://doi.org/10.3389/fpls.2014.00148

76. Zhang Y, He P, Ma X, Yang Z, Pang C, Yu J, Wang G, Friml J, Xiao G (2019) Auxin-mediated statolith production for root gravitropism. New Phytol 224(2):761–774. https://doi.org/10.1111/nph.15932

77. Abbas M, Hernández-García J, Pollmann S, Samodelov SL, Kolb M, Friml J, Hammes UZ, Zurbriggen MD, Blázquez MA, Alabadí D (2018) Auxin methylation is required for differential growth in *Arabidopsis*. Proc Natl Acad Sci U S A 115(26):6864–6869. https://doi.org/10.1073/pnas.1806565115

78. Mellor N, Band LR, Pěnčík A, Novák O, Rashed A, Holman T, Wilson MH, Voß U, Bishopp A, King JR, Ljung K, Bennett MJ, Owen MR (2016) Dynamic regulation of auxin oxidase and conjugating enzymes *AtDAO1* and *GH3* modulates auxin homeostasis. Proc Natl Acad Sci U S A 113 (39):11022–11027. https://doi.org/10.1073/pnas.1604458113

79. Zhang J, Lin JE, Harris C, Campos Mastrotti Pereira F, Wu F, Blakeslee JJ, Peer WA (2016) DAO1 catalyzes temporal and tissue-specific oxidative inactivation of auxin in *Arabidopsis thaliana*. Proc Natl Acad Sci U S A 113 (39):11010–11015. https://doi.org/10.1073/pnas.1604769113

80. Muday GK, Rahman A, Binder BM (2012) Auxin and ethylene: collaborators or competitors? Trends Plant Sci 17(4):181–195. https://doi.org/10.1016/j.tplants.2012.02.001

81. Stepanova AN, Yun J, Likhacheva AV, Alonso JM (2007) Multilevel interactions between ethylene and auxin in Arabidopsis roots. Plant Cell 19(7):2169–2185. https://doi.org/10.1105/tpc.107.052068

82. Ruzicka K, Ljung K, Vanneste S, Podhorska R, Beeckman T, Friml J, Benkova E (2007) Ethylene regulates root growth through effects on auxin biosynthesis and transport-dependent auxin distribution. Plant Cell 19(7):2197–2212. https://doi.org/10.1105/tpc.107.052126

83. Abel S, Nguyen MD, Chow W, Theologis A (1995) ASC4, a primary indoleacetic acid-responsive gene encoding 1-aminocyclopropane-1-carboxylate synthase in *Arabidopsis thaliana*. J Biol Chem 270 (32):19093–19099. https://doi.org/10.1074/jbc.270.32.19093

84. Woeste KE, Ye C, Kieber JJ (1999) Two Arabidopsis mutants that overproduce ethylene are affected in the posttranscriptional regulation of 1-aminocyclopropane-1-carboxylic acid synthase. Plant Physiol 119 (2):521–530. https://doi.org/10.1104/pp.119.2.521

85. Nziengui H, Lasok H, Kochersperger P, Ruperti B, Rébeillé F, Palme K, Ditengou FA (2018) Root gravitropism is regulated by a crosstalk between *para*-aminobenzoic acid, ethylene, and auxin. Plant Physiol 178 (3):1370–1389. https://doi.org/10.1104/pp.18.00126

86. Doyle SM, Haeger A, Vain T, Rigal A, Viotti C, Łangowska M, Ma Q, Friml J, Raikhel NV, Hicks GR, Robert S (2015) An early secretory pathway mediated by GNOM-LIKE 1 and GNOM is essential for basal polarity establishment in Arabidopsis thaliana. Proc Natl Acad Sci U S A 112(7):E806–E815. https://doi.org/10.1073/pnas.1424856112

87. Doyle SM, Rigal A, Grones P, Karady M, Barange DK, Majda M, Pařízková B, Karampelias M, Zwiewka M, Pěnčík A, Almqvist F, Ljung K, Novák O, Robert S (2019) A role for the auxin precursor anthranilic acid in root gravitropism via regulation of PIN—FORMED protein polarity and relocalisation in *Arabidopsis*. New Phytol 233:1420–1432. https://doi.org/10.1111/nph.15877

88. Aloni R, Langhans M, Aloni E, Ullrich CI (2004) Role of cytokinin in the regulation of root gravitropism. Planta 220(1):177–182. https://doi.org/10.1007/s00425-004-1381-8

89. Pernisova M, Prat T, Grones P, Harustiakova D, Matonohova M, Spichal L, Nodzynski T, Friml J, Hejatko J (2016) Cytokinins influence root gravitropism via differential regulation of auxin transporter expression and localization inArabidopsis. New Phytol 212(2):497–509. https://doi.org/10.1111/nph.14049

90. Waidmann S, Ruiz Rosquete M, Schöller M, Sarkel E, Lindner H, Larue T, Petřík I, Dünser K, Martopawiro S, Sasidharan R,

Novak O, Wabnik K, Dinneny JR, Kleine-Vehn J (2019) Cytokinin functions as an asymmetric and anti-gravitropic signal in lateral roots. Nat Commun 10(1):1–14. https://doi.org/10.1038/s41467-019-11483-4

91. Ogura T, Goeschl C, Filiault D, Mirea M, Slovak R, Wolhrab B, Satbhai SB, Busch W (2019) Root system depth in Arabidopsis is shaped by EXOCYST70A3 via the dynamic modulation of auxin transport. Cell 178 (2):400–412.e416. https://doi.org/10.1016/j.cell.2019.06.021

92. Sang D, Chen D, Liu G, Liang Y, Huang L, Meng X, Chu J, Sun X, Dong G, Yuan Y, Qian Q, Li J, Wang Y (2014) Strigolactones regulate rice tiller angle by attenuating shoot gravitropism through inhibiting auxin biosynthesis. Proc Natl Acad Sci U S A 111 (30):11199–11204. https://doi.org/10.1073/pnas.1411859111

93. Shinohara N, Taylor C, Leyser O (2013) Strigolactone can promote or inhibit shoot branching by triggering rapid depletion of the auxin efflux protein PIN1 from the plasma membrane. PLoS Biol 11(1):e1001474. https://doi.org/10.1371/journal.pbio.1001474

94. Kim J-Y, Ryu JY, Baek K, Park C-M (2016) High temperature attenuates the gravitropism of inflorescence stems by inducing *SHOOT GRAVITROPISM 5* alternative splicing in *Arabidopsis*. New Phytol 209(1):265–279. https://doi.org/10.1111/nph.13602

95. Herranz R, Valbuena MA, Manzano A, Kamal KY, Medina FJ (2015) Use of microgravity simulators for plant biological studies. Methods Mol Biol 1309:239–254. https://doi.org/10.1007/978-1-4939-2697-8_18

96. Kiss JZ, Wolverton C, Wyatt SE, Hasenstein KH, van Loon J (2019) Comparison of microgravity analogs to spaceflight in studies of plant growth and development. Front Plant Sci 10:1577. https://doi.org/10.3389/fpls.2019.01577

97. Toyota M, Furuichi T, Tatsumi H, Sokabe M (2008) Cytoplasmic calcium increases in response to changes in the gravity vector in hypocotyls and petioles of Arabidopsis seedlings. Plant Physiol 146(2):505–514. https://doi.org/10.1104/pp.107.106450

98. Toyota M, Furuichi T, Sokabe M, Tatsumi H (2013) Analyses of a gravistimulation-specific Ca2+ signature in Arabidopsis using parabolic flights. Plant Physiol 163(2):543–554. https://doi.org/10.1104/pp.113.223313

99. Bizet F, Pereda-Loth V, Chauvet H, Gerard J, Eche B, Girousse C, Courtade M, Perbal G, Legue V (2018) Both gravistimulation onset and removal trigger an increase of cytoplasmic free calcium in statocytes of roots grown in microgravity. Sci Rep 8(1):11442. https://doi.org/10.1038/s41598-018-29788-7

100. Ferl RJ, Paul A-L (2016) The effect of spaceflight on the gravity-sensing auxin gradient of roots: GFP reporter gene microscopy on orbit. NPJ Microgravity 2(1):15023. https://doi.org/10.1038/npjmgrav.2015.23

101. Böhmer M, Schleiff E (2019) Microgravity research in plants: a range of platforms and options allow research on plants in zero or low gravity that can yield important insights into plant physiology. EMBO Rep 20(7): e48541. https://doi.org/10.15252/embr.201948541

102. Yamazaki K, Ohmori Y, Fujiwara T (2020) A positive tropism of rice roots toward a nutrient source. Plant Cell Physiol 61 (3):546–553. https://doi.org/10.1093/pcp/pcz218

103. Migliaccio F, Tassone P, Fortunati A (2013) Circumnutation as an autonomous root movement in plants. Am J Bot 100(1):4–13. https://doi.org/10.3732/ajb.1200314

104. Bastien R, Bohr T, Moulia B, Douady S (2013) Unifying model of shoot gravitropism reveals proprioception as a central feature of posture control in plants. Proc Natl Acad Sci U S A 110(2):755–760. https://doi.org/10.1073/pnas.1214301109

105. Kim HJ, Kobayashi A, Fujii N, Miyazawa Y, Takahashi H (2016) Gravitropic response and circumnutation in pea (*Pisum sativum*) seedling roots. Physiol Plant 157(1):108–118. https://doi.org/10.1111/ppl.12406

106. Kitazawa D, Hatakeda Y, Kamada M, Fujii N, Miyazawa Y, Hoshino A, Iida S, Fukaki H, Morita MT, Tasaka M, Suge H, Takahashi H (2005) Shoot circumnutation and winding movements require gravisensing cells. Proc Natl Acad Sci U S A 102(51):18742–18747. https://doi.org/10.1073/pnas.0504617102

107. Tanimoto M, Tremblay R, Colasanti J (2008) Altered gravitropic response, amyloplast sedimentation and circumnutation in the Arabidopsis shoot gravitropism 5 mutant are associated with reduced starch levels. Plant Mol Biol 67(1–2):57–69. https://doi.org/10.1007/s11103-008-9301-0

108. Johnsson A, Solheim BG, Iversen TH (2009) Gravity amplifies and microgravity decreases circumnutations in Arabidopsis thaliana stems: results from a space experiment. New

Phytol 182(3):621–629. https://doi.org/10.1111/j.1469-8137.2009.02777.x

109. Kobayashi A, Kim HJ, Tomita Y, Miyazawa Y, Fujii N, Yano S, Yamazaki C, Kamada M, Kasahara H, Miyabayashi S, Shimazu T, Fusejima Y, Takahashi H (2019) Circumnutational movement in rice coleoptiles involves the gravitropic response: analysis of an agravitropic mutant and space-grown seedlings. Physiol Plant 165(3):464–475. https://doi.org/10.1111/ppl.12824

110. Wu Y, Xie J, Wang L, Zheng H (2020) Circumnutation and growth of inflorescence stems of Arabidopsis thaliana in response to microgravity under different photoperiod conditions. Life (Basel) 10(3):26. https://doi.org/10.3390/life10030026

111. Oliva M, Dunand C (2007) Waving and skewing: how gravity and the surface of growth media affect root development in Arabidopsis. New Phytol 176(1):37–43. https://doi.org/10.1111/j.1469-8137.2007.02184.x

112. Galva C, Kirik V, Lindeboom JJ, Kaloriti D, Rancour DM, Hussey PJ, Bednarek SY, Ehrhardt DW, Sedbrook JC (2014) The microtubule plus-end tracking proteins SPR1 and EB1b interact to maintain polar cell elongation and directional organ growth in Arabidopsis. Plant Cell 26(11):4409–4425. https://doi.org/10.1105/tpc.114.131482

113. Pandey S, Monshausen GB, Ding L, Assmann SM (2008) Regulation of root-wave response by extra large and conventional G proteins in Arabidopsis thaliana. Plant J 55(2):311–322. https://doi.org/10.1111/j.1365-313X.2008.03506.x

114. Roy R, Bassham DC (2017) TNO1, a TGN-localized SNARE-interacting protein, modulates root skewing in Arabidopsis thaliana. BMC Plant Biol 17(1):73. https://doi.org/10.1186/s12870-017-1024-4

115. Millar KD, Johnson CM, Edelmann RE, Kiss JZ (2011) An endogenous growth pattern of roots is revealed in seedlings grown in microgravity. Astrobiology 11(8):787–797. https://doi.org/10.1089/ast.2011.0699

116. Nakashima J, Liao F, Sparks JA, Tang Y, Blancaflor EB (2014) The actin cytoskeleton is a suppressor of the endogenous skewing behavior of Arabidopsis primary roots in microgravity. Plant Biol (Stuttg) 16:142–150. https://doi.org/10.1111/plb.12062

117. Paul A-L, Zupanska AK, Schultz ER, Ferl RJ (2013) Organ-specific remodeling of the Arabidopsis transcriptome in response to spaceflight. BMC Plant Biol 13(1):112. https://doi.org/10.1186/1471-2229-13-112

118. Sedbrook JC, Ehrhardt DW, Fisher SE, Scheible WR, Somerville CR (2004) The Arabidopsis sku6/spiral1 gene encodes a plus end-localized microtubule-interacting protein involved in directional cell expansion. Plant Cell 16(6):1506–1520. https://doi.org/10.1105/tpc.020644

119. Califar B, Sng NJ, Zupanska A, Paul AL, Ferl RJ (2020) Root skewing-associated genes impact the spaceflight response of Arabidopsis thaliana. Front Plant Sci 11:239. https://doi.org/10.3389/fpls.2020.00239

120. Yamamoto K, Kiss JZ (2002) Disruption of the actin cytoskeleton results in the promotion of gravitropism in inflorescence stems and hypocotyls of Arabidopsis. Plant Physiol 128(2):669–681. https://doi.org/10.1104/pp.010804

121. Kato T, Morita MT, Tasaka M (2010) Defects in dynamics and functions of actin filament in Arabidopsis caused by the dominant-negative actin fiz1-induced fragmentation of actin filament. Plant Cell Physiol 51(2):333–338. https://doi.org/10.1093/pcp/pcp189

122. De Bang L, Paez-Garcia A, Cannon AE, Chin S, Kolape J, Liao F, Sparks JA, Jiang Q, Blancaflor EB (2020) Brassinosteroids inhibit autotropic root straightening by modifying filamentous-actin organization and dynamics. Front Plant Sci 11(5):1–15. https://doi.org/10.3389/fpls.2020.00005

123. Okamoto K, Ueda H, Shimada T, Tamura K, Kato T, Tasaka M, Morita MT, Hara-Nishimura I (2015) Regulation of organ straightening and plant posture by an actin--myosin XI cytoskeleton. Nat Plants 1 (4):15031. https://doi.org/10.1038/nplants.2015.31

124. Dietrich D (2018) Hydrotropism: how roots search for water. J Exp Bot 69 (11):2759–2771. https://doi.org/10.1093/jxb/ery034

125. Moriwaki T, Miyazawa Y, Kobayashi A, Takahashi H (2013) Molecular mechanisms of hydrotropism in seedling roots of Arabidopsis thaliana (Brassicaceae). Am J Bot 100 (1):25–34. https://doi.org/10.3732/ajb.1200419

126. Liscum E, Nittler P, Koskie K (2020) The continuing arc toward phototropic enlightenment. J Exp Bot 71(5):1652–1658. https://doi.org/10.1093/jxb/eraa005

127. Morohashi K, Okamoto M, Yamazaki C, Fujii N, Miyazawa Y, Kamada M, Kasahara H, Osada I, Shimazu T, Fusejima Y, Higashibata A, Yamazaki T, Ishioka N, Kobayashi A, Takahashi H (2017) Gravitropism interferes with hydrotropism via

counteracting auxin dynamics in cucumber roots: clinorotation and spaceflight experiments. New Phytol 215(4):1476–1489. https://doi.org/10.1111/nph.14689

128. Shkolnik D, Krieger G, Nuriel R, Fromm H (2016) Hydrotropism: root bending does not require auxin redistribution. Mol Plant 9 (5):757–759. https://doi.org/10.1016/j.molp.2016.02.001

129. Izzo LG, Romano LE, De Pascale S, Mele G, Gargiulo L, Aronne G (2019) Chemotropic vs hydrotropic stimuli for root growth orientation in microgravity. Front Plant Sci 10:1547. https://doi.org/10.3389/fpls.2019.01547

130. Chang J, Li X, Fu W, Wang J, Yong Y, Shi H, Ding Z, Kui H, Gou X, He K, Li J (2019) Asymmetric distribution of cytokinins determines root hydrotropism in Arabidopsis thaliana. Cell Res 29(12):984–993. https://doi.org/10.1038/s41422-019-0239-3

131. Miyazawa Y, Takahashi A, Kobayashi A, Kaneyasu T, Fujii N, Takahashi H (2009) GNOM-mediated vesicular trafficking plays an essential role in hydrotropism of Arabidopsis roots. Plant Physiol 149(2):835–840. https://doi.org/10.1104/pp.108.131003

132. Saucedo M, Ponce G, Campos ME, Eapen D, Garcia E, Lujan R, Sanchez Y, Cassab GI (2012) An altered hydrotropic response (ahr1) mutant of Arabidopsis recovers root hydrotropism with cytokinin. J Exp Bot 63 (10):3587–3601. https://doi.org/10.1093/jxb/ers025

133. Shkolnik D, Nuriel R, Bonza MC, Costa A, Fromm H (2018) MIZ1 regulates ECA1 to generate a slow, long-distance phloem-transmitted Ca(2+) signal essential for root water tracking in Arabidopsis. Proc Natl Acad Sci U S A 115(31):8031–8036. https://doi.org/10.1073/pnas.1804130115

134. Millar KD, Kumar P, Correll MJ, Mullen JL, Hangarter RP, Edelmann RE, Kiss JZ (2010) A novel phototropic response to red light is revealed in microgravity. New Phytol 186 (3):648–656. https://doi.org/10.1111/j.1469-8137.2010.03211.x

135. Vandenbrink JP, Herranz R, Medina FJ, Edelmann RE, Kiss JZ (2016) A novel blue-light phototropic response is revealed in roots of Arabidopsis thaliana in microgravity. Planta 244(6):1201–1215. https://doi.org/10.1007/s00425-016-2581-8

136. Herranz R, Vandenbrink JP, Villacampa A, Manzano A, Poehlman WL, Feltus FA, Kiss JZ, Medina FJ (2019) RNAseq analysis of the response of *Arabidopsis thaliana* to fractional gravity under blue-light stimulation during spaceflight. Front Plant Sci 10:1529. https://doi.org/10.3389/fpls.2019.01529

137. Vandenbrink JP, Herranz R, Poehlman WL, Alex Feltus F, Villacampa A, Ciska M, Javier Medina F, Kiss JZ (2019) RNA-seq analyses of *Arabidopsis thaliana* seedlings after exposure to blue-light phototropic stimuli in microgravity. Am J Bot 106(11):1466–1476. https://doi.org/10.1002/ajb2.1384

138. Morris EC, Griffiths M, Golebiowska A, Mairhofer S, Burr-Hersey J, Goh T, von Wangenheim D, Atkinson B, Sturrock CJ, Lynch JP, Vissenberg K, Ritz K, Wells DM, Mooney SJ, Bennett MJ (2017) Shaping 3D root system architecture. Curr Biol 27(17): R919–R930. https://doi.org/10.1016/j.cub.2017.06.043

139. Paez-Garcia A, Motes CM, Scheible WR, Chen R, Blancaflor EB, Monteros MJ (2015) Root traits and phenotyping strategies for plant improvement. Plants (Basel) 4 (2):334–355. https://doi.org/10.3390/plants4020334

140. Uga Y, Kitomi Y, Ishikawa S, Yano M (2015) Genetic improvement for root growth angle to enhance crop production. Breed Sci 65 (2):111–119. https://doi.org/10.1270/jsbbs.65.111

141. Thorup-Kristensen K, Halberg N, Nicolaisen M, Olesen JE, Crews TE, Hinsinger P, Kirkegaard J, Pierret A, Dresboll DB (2020) Digging deeper for agricultural resources, the value of deep rooting. Trends Plant Sci 25(4):406–417. https://doi.org/10.1016/j.tplants.2019.12.007

142. Guseman JM, Webb K, Srinivasan C, Dardick C (2017) DRO1 influences root system architecture in Arabidopsis and Prunus species. Plant J 89(6):1093–1105. https://doi.org/10.1111/tpj.13470

143. Lynch JP (2019) Root phenotypes for improved nutrient capture: an underexploited opportunity for global agriculture. New Phytol 223(2):548–564. https://doi.org/10.1111/nph.15738

144. Trachsel S, Kaeppler SM, Brown KM, Lynch JP (2013) Maize root growth angles become steeper under low N conditions. Field Crop Res 140:18–31. https://doi.org/10.1016/j.fcr.2012.09.010

145. Dathe A, Postma JA, Postma-Blaauw MB, Lynch JP (2016) Impact of axial root growth angles on nitrogen acquisition in maize depends on environmental conditions. Ann Bot 118(3):401–414. https://doi.org/10.1093/aob/mcw112

146. Zou N, Li B, Chen H, Su Y, Kronzucker HJ, Xiong L, Baluska F, Shi W (2013) GSA-1/ ARG1 protects root gravitropism in Arabidopsis under ammonium stress. New Phytol 200(1):97–111. https://doi.org/10.1111/nph.12365
147. Vitousek PM, Naylor R, Crews T, David MB, Drinkwater LE, Holland E, Johnes PJ, Katzenberger J, Martinelli LA, Matson PA, Nziguheba G, Ojima D, Palm CA, Robertson GP, Sanchez PA, Townsend AR, Zhang FS (2009) Agriculture. Nutrient imbalances in agricultural development. Science 324 (5934):1519–1520. https://doi.org/10.1126/science.1170261
148. Liao H, Rubio G, Yan X, Cao A, Brown KM, Lynch JP (2001) Effect of phosphorus availability on basal root shallowness in common bean. Plant Soil 232(1–2):69–79
149. Roychoudhry S, Kieffer M, Del Bianco M, Liao CY, Weijers D, Kepinski S (2017) The developmental and environmental regulation of gravitropic setpoint angle in Arabidopsis and bean. Sci Rep 7:42664. https://doi.org/10.1038/srep42664
150. Huang G, Liang W, Sturrock CJ, Pandey BK, Giri J, Mairhofer S, Wang D, Muller L, Tan H, York LM, Yang J, Song Y, Kim Y-J, Qiao Y, Xu J, Kepinski S, Bennett MJ, Zhang D (2018) Rice actin binding protein RMD controls crown root angle in response to external phosphate. Nat Commun 9 (1):2346. https://doi.org/10.1038/s41467-018-04710-x
151. Wang X, Feng J, White PJ, Shen J, Cheng L (2020) Heterogeneous phosphate supply influences maize lateral root proliferation by regulating auxin redistribution. Ann Bot 125 (1):119–130. https://doi.org/10.1093/aob/mcz154
152. Bai H, Murali B, Barber K, Wolverton C (2013) Low phosphate alters lateral root setpoint angle and gravitropism. Am J Bot 100 (1):175–182. https://doi.org/10.3732/ajb.1200285
153. Li G, Liang W, Zhang X, Ren H, Hu J, Bennett MJ, Zhang D (2014) Rice actin-binding protein RMD is a key link in the auxin-actin regulatory loop that controls cell growth. Proc Natl Acad Sci U S A 111 (28):10377–10382. https://doi.org/10.1073/pnas.1401680111
154. Song Y, Li G, Nowak J, Zhang X, Xu D, Yang X, Huang G, Liang W, Yang L, Wang C, Bulone V, Nikoloski Z, Hu J, Persson S, Zhang D (2019) The rice actin-binding protein RMD regulates light-dependent shoot gravitropism. Plant Physiol 181(2):630–644. https://doi.org/10.1104/pp.19.00497
155. Rangarajan H, Postma JA, Lynch JP (2018) Co-optimization of axial root phenotypes for nitrogen and phosphorus acquisition in common bean. Ann Bot 122(3):485–499. https://doi.org/10.1093/aob/mcy092
156. Miguel MA, Postma JA, Lynch JP (2015) Phene synergism between root hair length and basal root growth angle for phosphorus acquisition. Plant Physiol 167 (4):1430–1439. https://doi.org/10.1104/pp.15.00145
157. Zhou M, Sng NJ, Lefrois CE, Paul A-L, Ferl RJ (2019) Epigenomics in an extraterrestrial environment: organ-specific alteration of DNA methylation and gene expression elicited by spaceflight in *Arabidopsis thaliana*. BMC Genomics 20(1):205. https://doi.org/10.1186/s12864-019-5554-z
158. Denyer T, Ma X, Klesen S, Scacchi E, Nieselt K, Timmermans MCP (2019) Spatiotemporal developmental trajectories in the *Arabidopsis* root revealed using high-throughput single-cell RNA sequencing. Dev Cell 48(6):840–852.e845. https://doi.org/10.1016/j.devcel.2019.02.022
159. Jean-Baptiste K, McFaline-Figueroa JL, Alexandre CM, Dorrity MW, Saunders L, Bubb KL, Trapnell C, Fields S, Queitsch C, Cuperus JT (2019) Dynamics of gene expression in single root cells of *Arabidopsis thaliana*. Plant Cell 31(5):993–1011. https://doi.org/10.1105/tpc.18.00785
160. Ryu KH, Huang L, Kang HM, Schiefelbein J (2019) Single-cell RNA sequencing resolves molecular relationships among individual plant cells. Plant Physiol 179(4):1444–1456. https://doi.org/10.1104/pp.18.01482
161. Shulse CN, Cole BJ, Ciobanu D, Lin J, Yoshinaga Y, Gouran M, Turco GM, Zhu Y, O'Malley RC, Brady SM, Dickel DE (2019) High-throughput single-cell transcriptome profiling of plant cell types. Cell Rep 27 (7):2241–2247.e2244. https://doi.org/10.1016/j.celrep.2019.04.054
162. Hui T, Cao Q, Wegrzyn-Woltosz J, O'Neill K, Hammond CA, Knapp DJHF, Laks E, Moksa M, Aparicio S, Eaves CJ, Karsan A, Hirst M (2018) High-resolution single-cell DNA methylation measurements reveal epigenetically distinct hematopoietic stem cell subpopulations. Stem Cell Rep 11 (2):578–592. https://doi.org/10.1016/j.stemcr.2018.07.003
163. Lee HJ, Smallwood SA (2018) Genome-wide analysis of DNA methylation in single cells using a post-bisulfite adapter tagging

approach. Springer, New York, pp 87–95. https://doi.org/10.1007/978-1-4939-7514-3_7

164. Linker SM, Urban L, Clark SJ, Chhatriwala M, Amatya S, McCarthy DJ, Ebersberger I, Vallier L, Reik W, Stegle O, Bonder MJ (2019) Combined single-cell profiling of expression and DNA methylation reveals splicing regulation and heterogeneity. Genome Biol 20(1):30. https://doi.org/10.1186/s13059-019-1644-0

165. Fujii T, Matsuda S, Tejedor ML, Esaki T, Sakane I, Mizuno H, Tsuyama N, Masujima T (2015) Direct metabolomics for plant cells by live single-cell mass spectrometry. Nat Protoc 10(9):1445–1456. https://doi.org/10.1038/nprot.2015.084

166. Masujima T (2009) Live single-cell mass spectrometry. Anal Sci 25(8):953–960. https://doi.org/10.2116/analsci.25.953

167. Barker R, Lombardino J, Rasmussen K, Gilroy S (2020) Test of *Arabidopsis* space transcriptome: a discovery environment to explore multiple plant biology spaceflight experiments. Front Plant Sci 11:147. https://doi.org/10.3389/fpls.2020.00147

168. Schneider HM, Lynch JP (2020) Should root plasticity be a crop breeding target? Front Plant Sci 11:546. https://doi.org/10.3389/fpls.2020.00546/full

Chapter 2

Evaluation of Gravitropism in Non-seed Plants

Yuzhou Zhang, Lanxin Li, and Jiří Friml

Abstract

Tropisms are among the most important growth responses for plant adaptation to the surrounding environment. One of the most common tropisms is root gravitropism. Root gravitropism enables the plant to anchor securely to the soil enabling the absorption of water and nutrients. Most of the knowledge related to the plant gravitropism has been acquired from the flowering plants, due to limited research in non-seed plants. Limited research on non-seed plants is due in large part to the lack of standard research methods. Here, we describe the experimental methods to evaluate gravitropism in representative non-seed plant species, including the non-vascular plant moss *Physcomitrium patens*, the early diverging extant vascular plant lycophyte *Selaginella moellendorffii* and fern *Ceratopteris richardii*. In addition, we introduce the methods used for statistical analysis of the root gravitropism in non-seed plant species.

Key words Gravitropism, Non-seed plants, *Physcomitrium patens*, Lycophytes, Ferns, Vertical growth index (VGI), Root bending

1 Introduction

Gravitropic growth is a crucial biological feature ubiquitously present in nearly all plant species, which facilitates plant adaptation to the Earth environment [1, 2]. For seed plants, their shoots grow upward to secure sunlight for photosynthesis [3, 4], while their roots bend toward the gravity vector for anchorage in the soil, in order to absorb water and nutrients [5]. Even though plant gravitropism has been observed and described hundreds of years ago, mechanisms underlying this process have only started to be revealed in recent decades [6]. The mechanistic insights gained regarding plant gravitropism has come mainly from extensive studies on seed plants, especially model flowering plants [7]. However, non-seed plants are seldom used to study gravitropism, as a result of numerous limiting factors, such as incompletely revealed genome and insufficient research methods [8]. Given that many gravitropic tissues/organs such as roots might evolve independently in different plant lineages [9, 10], the machinery of gravitropism in the non-seed plants is likely to differ from that in seed plants.

Elison B. Blancaflor (ed.), *Plant Gravitropism: Methods and Protocols*, Methods in Molecular Biology, vol. 2368, https://doi.org/10.1007/978-1-0716-1677-2_2,

Therefore, understanding non-seed plant gravitropism not only fills knowledge gaps about gravitropism for the whole plant kingdom, but also helps to reveal the evolution and origin of gravitropism along the plant evolutionary trajectory [1].

To better understand root gravitropism in non-seed plants, we describe protocols on how to culture representative non-seed plants (i.e., moss *P. patens*, the basal vascular plant lycophyte *S. moellendorffii*, and fern *C. richardii.*) for the root gravitropic analysis. We then introduce methods developed for the measurement and statistical analysis of non-seed plant gravitropism and further show how to take advantage of these methods to compare the gravitropism between the non-seed and seed plants.

2 Materials

2.1 Preparation of BCD Media

1. Solution B: 25 g/L $MgSO_4.7H_2O$.
2. Solution C: 25 g/L KH_2PO_4 with the pH adjusted to 6.5.
3. Solution D: 101 g/L KNO_3, 1.25 g/L $FeSO_4.7H_2O$.
4. $CaCl_2.2H_2O$:14.7 g/L $CaCl_2.2H_2O$.
5. Alternative trace element solution: 55 mg/L $CuSO_4.5H_2O$, 614 mg/L H_3BO_3, 55 mg/L $CoCl_2.6H_2O$, 25 mg/L $Na_2MoO_4.2H_2O$, 55 mg/L $ZnSO_4.7H_2O$, 389 mg/L $MnCl_2.4H_2O$, 28 mg/L KI.
6. BCD media containing 0.8% agar (*see* **Note 1**): 960 mL Distilled H_2O, 10 mL Solution B, 10 mL Solution C, 10 mL Solution D, 10 mL $CaCl_2.2H_2O$, 1 mL Alternative trace element solution, and 8 g agar. Mixed and autoclaved (max 121 °C). Pour 25–30 mL BCD media to 90 mm Petri dish.

2.2 Preparation of ½ MS Media and Materials for Tissue Sterilization

1. ½ MS medium: Half Murashige and Skoog Basal Salts supplemented with 1% sucrose and 0.8% agar (plant cell tested), and adjust pH to 5.9 by KOH.
2. Sodium hypochlorite solution (NaClO with 2% chlorine).
3. Razor blade.
4. Forceps.
5. Autoclaved distilled H_2O.

2.3 Root Scanning

1. Epson Perfection V370 Photo flatbed scanner.
2. Home-made T-shape plastic stand.
3. A plastic holder, with several empty areas for the insertion of the 90 mm Petri dishes.
4. 90 mm Petri dishes.

3 Methods

Prepare the BCD media for the cultivation of moss *P. patens*, and ½ MS media for the cultivation of lycophyte *S. moellendorffii* and fern *C. richardii* ahead of time (*see* Subheading 2) [11] (*see* **Note 1**). Prior to cultivating plants, place a commercial scanner vertically and connect it to a laptop with AutoIt program. This set up will allow the acquisition of time series images of gravistimulated growing tissues automatically.

3.1 Cultivation of Physcomitrium patens

1. Put a sheet of autoclaved cellophane on the surface of the BCD media with a sterile forcep (*see* **Note 2**), and then transfer the protonemal tissues on the cellophane-covered plates. Place the plates horizontally and set the growth conditions as follows: 24 °C in a long-day light regime (16 h light and 8 h dark), light intensity 55 μmol per m^{-2} s^{-1}.
2. After culturing for 1–2 weeks (Fig. 1a, b), position plates vertically and allow plants to continue growing for 1 or 2 weeks until the rhizoid tissues are readily seen. Rhizoid typically grows downward along the gravity vector (Fig. 1c).
3. Rotate the plates 90° to observe the bending degree of these tissues (*see* **Note 3**). After gravistimulation (Fig. 1d), scan the plates continuously for up to 48 h with an interval of 30 min. Use the captured images for downstream statistical analysis.

3.2 Cultivation of Lycophyte Selaginella Moellendorffii for Gravitropic Analysis

1. Propagate *Selaginella moellendorffii* on sterile ½ MS medium supplemented with 1% sucrose and 0.8% (w/v) agar, pH 5.9 (*see* **Note 4**).
2. Cut the shoot apical segments (around 0.5–1.0 cm) of *Selaginella* with an autoclaved razor blade (Fig. 2a). Transfer the cut segments to a Petri dish with ½ MS plus 0.8% agar (Fig. 2b). To generate root tissue, place these plates vertically in a Percival chamber, with the growth conditions set as follows: 24 °C with light intensity 30 μmol per m^{-2} s^{-1} (cool white fluorescent lamps) and regime with 16 h light and 8 h dark [12].
3. After 7–10 days of growth, the roots emerge (Fig. 2c). Reorient the plates to make the roots perpendicular to gravity vector (Fig. 2d). After gravistimulation, continuously scan the plates for up to 48 h with an interval of 30 min to capture a series of images for downstream statistical analysis of gravitropism.

3.3 Cultivation of Fern Ceratopteris richardii for Gravitropic Analysis

1. Sterilize *Ceratopteris richardii* spores with a mixture of NaClO (2% chlorine) and Tween-20 (0.1% v/v) for 15 min, and then wash three times with distilled H_2O (5 min per time) (*see* **Note 5**).

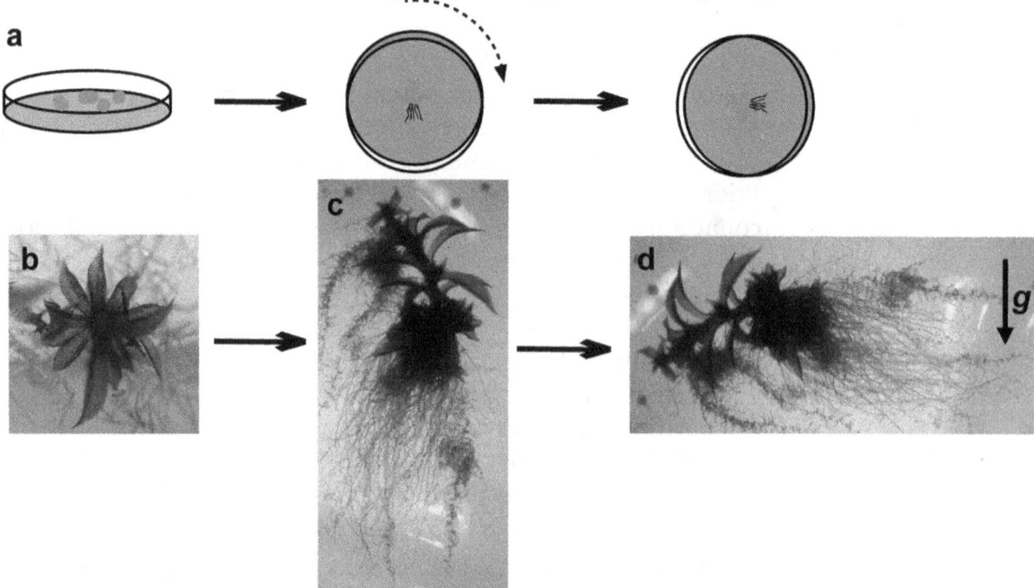

Fig. 1 Procedure of culturing moss *P. patens* for the gravitropism assay. Schematic drawing (**a**) and panels (**b–d**) showing the sample preparation for gravitropism assay. Arrow in d shows the direction of the gravity vector

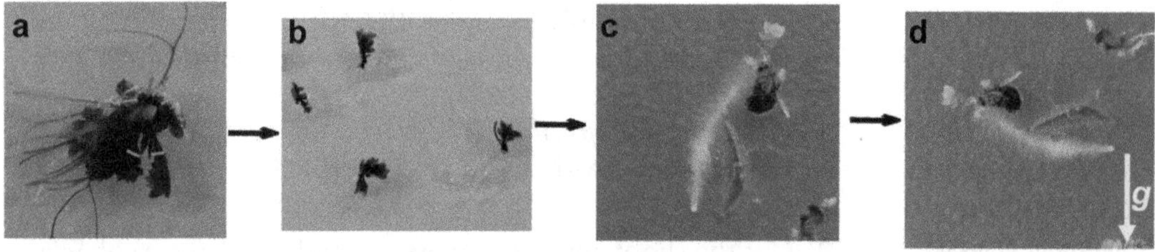

Fig. 2 Procedure of culturing lycophyte *S. moellendorffii* for gravitropism assays. (**a**, **b**) Acquiring the shoot apical segments (explants) with a razor blade. Panels (**b–d**) show the procedures for culturing *S. moellendorffii* for gravitropic evaluation. Yellow arrow in (**d**) shows the direction of the gravity vector

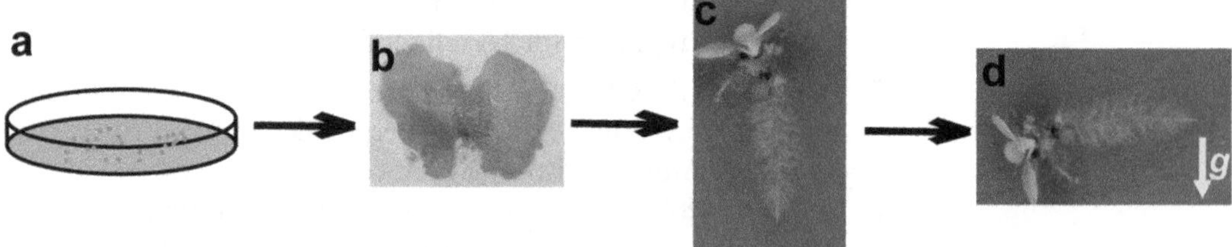

Fig. 3 Procedures of culturing fern *C. richardii* for gravitropism assays. The panels (**a–c**) show the acquisition of *C. richardii* sporophytes for gravistimulation. (**d**) A 90° reorientation of the sporophytes to evaluate gravistimulation

2. Transfer the spores to the ½ MS plates supplemented with 0.8% (w/v) agar using 1 mL pipettes (*see* **Note 6**) (Fig. 3a). Then place the plates horizontally in a Percival chamber with the setting as follows: 28 °C with light intensity 30 μmol per m^{-2} s^{-1} and regime with 16 h light and 8 h dark [13].
3. After growing for 1–2 weeks and after the spores have germinated, the gametophytes can be clearly seen (Fig. 3b). Transfer the gametophytes to new ½ MS plates with 0.8% agar. Spray sterilized H_2O to the plates and seal them tightly to maintain the high humidity, and then vertically place them in the Percival chamber.
4. Spray H_2O frequently to the plates if the agar starts to shrink. After growing for 2–3 weeks, new sporophytes generate and the roots emerge (Fig. 3c).
5. Reorient the plates to position the roots perpendicular to gravity vector. After gravistimulation (Fig. 3d), continuously scan the plates for up to 48 h with an interval of 30 min. Use the captured images for statistical gravitropic analysis.

3.4 Set up of the Vertical Scanner System Used for Gravitropism Analysis

1. Embed an Epson Perfection V370 Photo flatbed scanner into a Home-made T-shape plastic stand (Fig. 4).
2. Clip the plastic holder by magnetics into the T-shaped stand where the scanner is embedded.
3. Make sure that the surface of the scanner is in direct contact with the surface of the 90 mm Petri dishes with samples, so that the sample is scanned through a layer of media.
4. Attach a black paper on the lid of the scanner, served as a contrast-improving background.
5. Enclose the lid tightly with tapes during the scanning period.
6. Connect the scanner to a laptop (Fig. 4). Scan the sample with Epson Scan program. Set 1200 dpi and specify the saving path, and then crop the regions of interest. Turn on the AutoIt program (https://www.autoitscript.com/site/) to automatically scan with a specified time interval. The script used in AutoIt program was described previously [14], but modified the time interval to 1,800,000 ms with 96 repetitions.

3.5 Statistical Analysis of the Gravitropism of Non-seed Plant Species

In general, there are two ways to evaluate plant gravitropism: (1) Measure the bending angle after the gravistimulation, and (2) Statistical analysis of the vertical growth index (VGI).

1. In model flowering plant *Arabidopsis*, the bending angles are measured to compare the gravitropism between mutants and wild type. Apply the same strategy to measure the gravitropism of non-seed plant species and compare the gravitropism between non-seed and seed plants (Fig. 5a). Bending angles

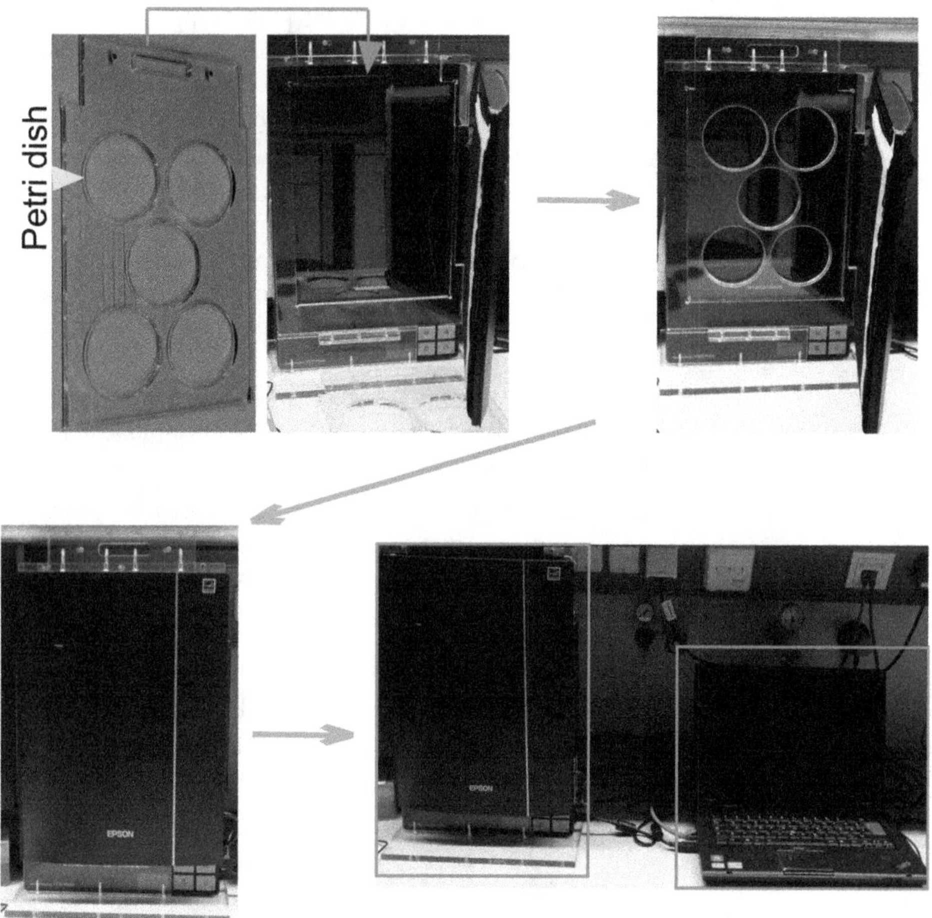

Fig. 4 Set up of the vertical scanner system used for gravitropism analysis. The transparent plastic sample holder has holes for inserting 90 mm Petri dishes carrying gravistimulated samples. The holder is clipped into the embedded vertical scanner. The lid of the scanner is enclosed (can be fixed with tapes) during scanning process. The scanner is connected to a laptop with AutoIt program installed for an automatic control

(denoted by θ) can be measured by ImageJ software (NIH; http://rsb.info.nih.gov/ij) (*see* **Notes 7 and 8**).

2. In *Arabidopsis*, statistical analysis of vertical growth index (VGI) revealed that VGI is a sensitive morphometric parameter, which enables us to detect the weak gravitropic defects [15]. As shown in Fig. 5b, the VGI is defined by dividing the full length of root (L) by the vertical travelled distance (L_y) (VGI = L_y/L). The closer the value of VGI is to 1, the more the root grows toward gravity vector. Generally, the VGI is also applicable in measuring gravitropism of non-seed plants (*see* **Note 9**).

3. Based on the observation that growth rates of non-seed plants are prominently different from that of seed plants, use a

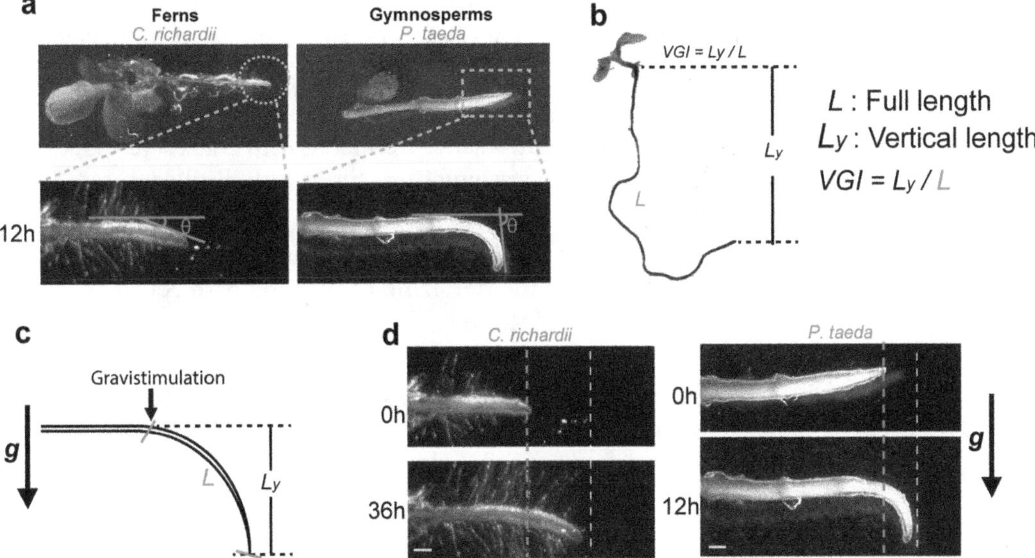

Fig. 5 Quantification of the gravitropism of non-seed plant species. (**a**) Measuring the bending root angles of non-seed plant *C. richardii* and seed plant *P. taeda* after gravistimulation for 12 h. (**b**) Schematic of vertical growth index (VGI) used in the measurement of root gravitropism, which is defined by dividing the full length of root (L) by the vertical travelled distance(Ly). (**c**) Diagram showing the modified VGI used to measure the gravitropism of different plant species with disparate growth rates. (**d**) To exclude the effect of the growth rate on the evaluation of root gravitropism between non-seed plants (such as *C. richardii*) and seed plants (such as *P. taeda*), gravistimulated roots with similar lengths are used to conduct the gravitropic analysis with modified VGI, as indicated in (**c**)

modified VGI to evaluate the plant gravitropism by ruling out the interference of different growth rates between non-seed and seed plants (Fig. 5c, d). To be specific, first measure the elongated lengths of the tissue (e.g., root) at different time points after gravistimulation. Then images with the tissue showing almost the same elongated length (L) are chosen to perform the statistical analysis with the modified VGI (Fig. 5d). The larger the value, the stronger the root gravitropism.

4 Notes

1. All stock solutions for BCD media are stored at room temperature (RT) and should be used within 3 months. The prepared BCD media is stored at 4 °C and could be kept for up to half a year.
2. To ensure the cellophanes are in close contact with the surface of BCD media, apply one or two drops of water between the contacted surfaces.

3. Patches of moss growing on the same cellophane could be separated by a razor blade and then transferred to new plates.
4. If *Selaginella moellendorffii* is contaminated by microbes, the explants could be transferred to the ½ MS media with 0.8% agar for propagation after sterilizing with NaClO (2% chlorine) for 15 min, and washing three times with distilled water (5 min each time).
5. Because spores of *C. richardii* are tiny. They get suspended in the liquid after the sterilizing or washing with NaClO solution or ddH_2O, respectively. To remove the excess liquids, spin down for a few minutes and discard the disinfectant or H_2O.
6. To ensure that the *C. richardii* grows under high humidity, frequently check the plates and spray the distilled H_2O to the plate surface every 3–5 days.
7. The samples used for quantification should be of similar size. For example, only roots with comparable lengths are taken to perform the gravitropic analysis. Occasionally, few samples show abnormal development in growth. Do not use these for gravitropism analysis.
8. The seedlings should grow under the darkness in order to avoid phototropism-induced bending, which may interfere with the interpretation of gravitropism results.
9. If the VGI is used to quantify root gravitropism of non-seed plants, we recommend to use it in species with comparable growth rates.

Acknowledgments

The *Ceratopteris richardii* spores were obtained from the lab of Jo Ann Banks at Purdue University. This work was supported by funding from the European Union's Horizon 2020 research and innovation program (ERC grant agreement number 742985), Austrian Science Fund (FWF, grant number I 3630-B25), IST Fellow program and DOC Fellowship of the Austrian Academy of Sciences.

References

1. Zhang Y, Xiao G, Wang X et al (2019) Evolution of fast root gravitropism in seed plants. Nat Commun 10:3480
2. Chen R, Rosen E, Masson PH (1999) Gravitropism in higher plants. Plant Physiol 120:343–350
3. Rakusová H, Abbas M, Han H et al (2016) Termination of shoot Gravitropic responses by auxin feedback on PIN3 polarity. Curr Biol 26:3026–3032
4. Bastien R, Douady S, Moulia B (2014) A unifying modeling of plant shoot gravitropism with an explicit account of the effects of growth. Front Plant Sci 5:136

5. Zhang Y, He P, Ma X et al (2019) Auxin-mediated statolith production for root gravitropism. New Phytol 224:761–774
6. Su SH, Gibbs NM, Jancewicz AL, Masson PH (2017) Molecular mechanisms of root gravitropism. Curr Biol 27:964–972
7. Masson PH, Tasaka M, Morita MT et al (2002) Arabidopsis thaliana: a model for the study of root and shoot gravitropism. Arab B 1:e0043
8. Chang C, Bowman JL, Meyerowitz EM (2016) Field guide to plant model systems. Cell 167:325–339
9. Qiu YL, Li L, Wang B et al (2006) The deepest divergences in land plants inferred from phylogenomic evidence. Proc Natl Acad Sci U S A 103:15511–15516
10. Menand B, Yi K, Jouannic S et al (2007) An ancient mechanism controls the development of cells with a rooting function in land plants. Science 316:1477–1480
11. Cove DJ, Perroud PF, Charron AJ et al (2009) Culturing the moss Physcomitrella patens. Cold Spring Harb Protoc 2009(2):pdb.prot5136
12. Fang T, Motte H, Parizot B, Beeckman T (2019) Root branching is not induced by auxins in Selaginella moellendorffii. Front Plant Sci 10:154
13. Plackett ARG, Rabbinowitsch EH, Langdale JA (2015) Protocol: genetic transformation of the fern ceratopteris richardii through microparticle bombardment. Plant Methods 11:37
14. Li L, Krens SFG, Fendrych M, Friml J (2018) Real-time analysis of auxin response, cell wall pH and elongation in Arabidopsis thaliana hypocotyls. Bio Protoc 8:e2685
15. Grabov A, Ashley MK, Rigas S et al (2005) Morphometric analysis of root shape. New Phytol 165:641–651

Chapter 3

Spore Preparation and Protoplast Isolation to Study Gravity Perception and Response in *Ceratopteris richardii*

Ashley E. Cannon, Tanya Sabharwal, and Stanley J. Roux

Abstract

Early studies revealed a highly predictable pattern of gravity-directed growth and development in *Ceratopteris richardii* spores. This makes the spore a valuable model system for the study of how a single-cell senses and responds to the force of gravity. Gravity regulates both the direction and magnitude of a trans-cell calcium current in germinating spores, and the orientation of this current predicts the polarization of spore development. In order to make *Ceratopteris richardii* cells easier to transform and image during this developmental process, a procedure for isolating protoplasts from *Ceratopteris richardii* gametophytes has been developed and optimized. These protoplasts follow the same developmental pattern as *Ceratopteris richardii* spores and can be used to monitor the molecular and developmental processes during single-cell polarization. Here, we describe this optimized procedure, along with protocols for sterilizing the spores, sowing them in solid or liquid growth media, and evaluating germination and polarization.

Key words *Ceratopteris richardii*, Single cell, Gravity-directed development, Polarization, Calcium, Protoplast isolation

1 Introduction

The single-celled *Ceratopteris richardii* spore is a valuable system for studying plant gravitropism. This fern spore has a highly predictable gravity-directed developmental pattern that makes it an ideal model for studying how plant cells sense and respond to gravity [1–5]. Shortly after a spore is exposed to light, germination begins, and a trans-cell calcium current can be measured [6–8]. During the next 72 h, there is a nuclear migration and an asymmetric cell division. These two events prepare the spore for the emergence of a polarly growing rhizoid. The primary rhizoid serves as a visual representation of polarity fixation because it grows in the direction of gravity as perceived by the spore in the first 24 h.

Ashley E. Cannon and Tanya Sabharwal contributed equally with all other contributors.

Elison B. Blancaflor (ed.), *Plant Gravitropism: Methods and Protocols*, Methods in Molecular Biology, vol. 2368, https://doi.org/10.1007/978-1-0716-1677-2_3, © Springer Science+Business Media, LLC, part of Springer Nature 2022

A protocol has been developed for isolating protoplasts from *Ceratopteris richardii* gametophytes. This protocol was adapted from the original procedure developed by Edwards and Roux [5]. When allowed to regenerate, *Ceratopteris richardii* protoplasts follow the same developmental pattern as *Ceratopteris richardii* spores, making them a suitable substitute for studying molecular changes during the gravity response of single cells. Plant protoplasts are valuable experimental tools because they can be reliably transformed with multiple DNA constructs [9, 10] and they can be imaged more easily than cells in intact tissues [11]. *Ceratopteris richardii* protoplasts be used molecular and developmental changes, identify the subcellular localization of proteins, and examine the influence of stress and environmental factors during the process of polarization in single cells.

In this chapter, we describe procedures for sterilizing, synchronizing, and planting spores in preparation for studies of their gravity-directed development and isolation of protoplasts from gametophyte tissue. Although we focus on the single-cell *Ceratopteris richardii* spore, these techniques could be used for other single-cell plant systems. We begin by describing the basic preparation of spores that can be used in many different assays. We then describe how to grow *Ceratopteris richardii* gametophytes and how to isolate protoplasts from these tissues. The protoplasts derived from this procedure can be used in many different assays and provide an experimental tool that is easily transformed and imaged during the process of polarization.

2 Materials

2.1 Sterilization and Synchronization of Spores

1. *Ceratopteris richardii* fern spores.
2. Spore Sterilizing Solution: Mix 1-part bleach (8.25% sodium hypochlorite) to four parts sterilized deionized water to make 1.65% sodium hypochlorite.
3. Sterile transfer pipettes.
4. 15 mL conical tubes.
5. Timer.
6. Parafilm M™.
7. Aluminum foil.
8. 28–29 °C lighted growth chamber.

2.2 Spore Culture Preparation

1. Spore Growth Media: Weigh 0.215 g of Murashige and Skoog basal salt mixture and transfer to a 250 mL Erlenmeyer flask. Add 90 mL of ddH_2O, completely mix the solution, and adjust the pH to 6.3 with 1 M NaOH and a pH meter. After the pH is adjusted, transfer the solution to a 100 mL graduated cylinder

and bring the volume up to 100 mL using ddH_2O. Pour the solution back into the 250 mL Erlenmeyer flask and add 1% (w/v) Bacto agar. Autoclave this solution before use.

2. pH meter.
3. Thermometer.
4. Hot plate.
5. 15 mL conical tubes.
6. 35 mm × 10 mm petri dishes.
7. Parafilm M™.
8. Fixed orientation rack.
9. 28–29 °C lighted growth chamber.

2.3 Imaging Spores for Rhizoid Orientation and Analysis

1. Light microscope.
2. Digital imaging software (e.g., Image J [12]).

2.4 Protoplast Isolation from *Ceratopteris* Gametophytes

1. Sterilized, pre-soaked spores grown for 16–18 days on solid Spore Growth Media.
2. Sterile, flat spatula.
3. 0.2 μm filter for media sterilization.
4. Vacuum pump for sterilization.
5. Enzyme Incubation Media: Weigh 0.215 g of Murashige and Skoog basal salt mixture and transfer to a 250 mL Erlenmeyer flask. Add 90 mL of ddH_2O, completely mix the solution, and adjust the pH to 6.3 with 1 M NaOH and a pH meter. After the pH is adjusted, transfer the solution to a 100 mL graduated cylinder, and bring the volume up to 100 mL using ddH_2O.- Pour the solution back into the 250 mL Erlenmeyer flask and add 9.1 g (0.5 M) mannitol. Autoclave and store at room temperature. Just before use, add 0.36% cellulase and 0.25% pectinase (*see* **Note 1**) to the media and filter sterilize the combined solution.
6. 100 mm × 15 mm petri dishes.
7. Parafilm M™.
8. Orbital shaker large enough for 100 mm × 15 mm petri dishes and capable of 30–50 rpm.
9. 28–29 °C lighted growth chamber.
10. Sterile transfer pipettes.
11. 15 mL conical tubes.
12. Centrifuge capable of 1302 × *g* with 15 mL tube capacity.
13. Protoplast Rinsing Solution: Weigh 0.215 g of Murashige and Skoog basal salt mixture and transfer to a 250 mL Erlenmeyer

flask. Add 90 mL of ddH_2O, completely mix the solution and adjust the pH to 6.3 with 1 M NaOH and a pH meter. After the pH is adjusted, transfer the solution to a 100 mL graduated cylinder and bring the volume up to 100 mL using ddH_2O.-Pour the solution back into the 250 mL Erlenmeyer flask and add 9.1 g (0.5 M) mannitol. Autoclave, cool to room temperature, and filter sterilize this solution prior to use.

14. 0.5 M Mannitol Solution: Add 9.1 g of mannitol to 100 mL of ddH_2O in a 250 mL Erlenmeyer flask. Autoclave and cool this solution to room temperature prior to use.
15. Aluminum foil.

3 Methods

3.1 Spore Preparation

1. Surface-sterilize spores (*see* **Note 2**) by soaking for 90 s in Spore Sterilization Solution with gentle agitation in a 15 mL conical tube.
2. Remove the bleach with a transfer pipette by squeezing the pipette bulb to expel all of the air and pushing it against the bottom of the tube. As you release the pressure on the bulb, you should see the liquid accumulating in the pipette and spores aggregating around the bottom tip of the pipette.
3. Rinse the spores by adding 3–5 mL of sterile water to the tube for 60 s with gentle agitation. Remove the water.
4. Repeat **step 3** twice so that the spores are rinsed three times.
5. Add 3–5 mL of sterile water to the 15 mL tube. Seal the lid with Parafilm and wrap the tube in two layers of foil.
6. Place the tube in complete darkness at 29 °C for 4–7 days in order to increase the synchronization of germination.

3.2 Spore Culture Preparation

1. Prepare the Spore Growth Media (*see* **Note 3**) and autoclave for sterilization. If the spores need to be grown in liquid media, leave out Bacto agar.
2. Use the transfer pipette technique from the sterilization procedure to remove the pre-soaking water from sterilized spores without remove spores.
3. Cool the media to 45–55 °C prior to adding it into the 15 mL tube of spores to reach the optimal spore density for the experiment (*see* **Notes 2 and 4**).
4. Add 1–2 mL of media with spores to each 35 mm × 10 mm petri dish or add 5 mL of media with spores to each 100 mm × 15 mm petri dish using a transfer pipette.
5. After the media has solidified, wrap the plates in one layer of Parafilm and place them in the growth chamber (*see* **Note 5**).

3.3 Evaluating Rhizoid Orientation to Determine Direction of Polarization

After at least 72 h of light-induced germination, microscopy can be used to evaluate rhizoid orientation and determine the direction of polarization.

1. Use a light microscope with a 4× or 10× objective to image the spores, taking note of the fixed orientation during germination.
2. The direction of rhizoid growth should be determined by evaluating whether the rhizoids are growing below or above the midline of the spore. Growth below the midline of the spore can be considered "down" while growth above or at the horizontal line can be considered "up" or "not down." The direction of rhizoid growth should not be assessed until the length of the rhizoid is at least one spore diameter, approximately 100 μm. When rhizoids are shorter than 100 μm, their final direction of growth may not be the same as their initial orientation.
3. Digital imaging software can be used to draw a horizontal line through the midline of each spore to allow a consistent assessment of "up" and "down" (*see* **Note 6**).

3.4 Protoplast Isolation from Ceratopteris Gametophytes

1. This protocol was modified from Edwards and Roux [5]. Plate *Ceratopteris richardii* spores using the methods described in Spore Culture Preparation (Subheading 3.2) and grow for 16–18 days prior to protoplast isolation (*see* **Note 4**).
2. Scrape the *Ceratopteris* gametophytes growing on the agar medium using a sterile flat spatula (*see* **Note 7** and Fig. 1) and add them to a sterile petri dish containing 25–30 mL of freshly prepared and filter-sterilized Enzyme Incubation Media. The media should be autoclaved without the enzymes, then the enzymes should be added after they are filter sterilized, then the media with the enzymes should be filter sterilized (*see* **Note 1**). Seal the petri dish with Parafilm and wrap the plates in two layers of aluminum foil. Incubate the gametophytes overnight (at least 16 h) in complete darkness at room temperature on an orbital shaker with gentle shaking (30 rpm).
3. Thirty minutes prior to tissue collection, turn up the speed of the orbital shaker to 50 rpm. Using sterile transfer pipettes, move the tissue and media to 15 mL conical tubes and centrifuge the samples at 2600 rpm for 4 min.
4. Wash the pellet by removing the media and resuspending the cells in Protoplast Rinsing Solution that has been autoclaved, cooled, and filter sterilized. Centrifuge the samples at 2600 rpm for 4 min. Repeat this process until the cells have been washed twice. Resuspend the washed cells in 0.5 M mannitol that has been autoclaved, cooled, and filter sterilized (*see* **Note 8**) (Fig. 2).

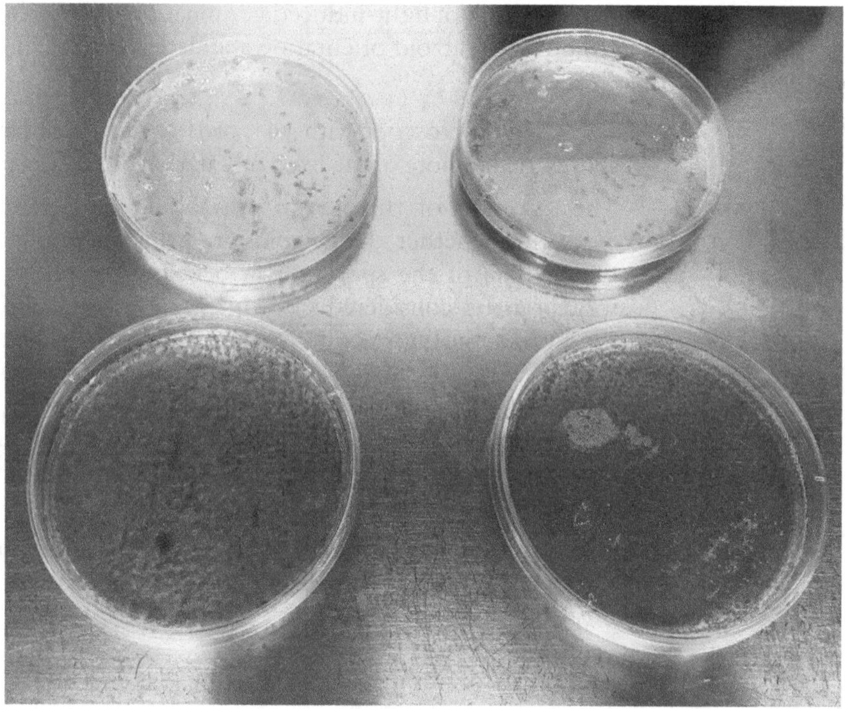

Fig. 1 Gametophytes on the surface of half-strength Murashige and Skoog basal salt mixture and Bactor agar plates before and after collection for isolating protoplasts

4 Notes

1. Filter-sterilized enzymes should be added to the media right before it is used.
2. Large volumes of liquid-grown spores encounter issues with settling and high local concentrations of spores. High spore concentrations can inhibit germination and alter development. Solid media allows for an even distribution of spores in a fixed orientation. Larger volumes of spores in liquid culture would likely need shaking or other methods to distribute spores, but this would also randomize the orientation of the spores relative to the gravity vector.
3. The percentage of Bacto agar used can vary depending on the experiment between 0.5 and 1% with no observed effect on germination.
4. If spores are being sown for protoplast isolation, the half-strength Murashige and Skoog basal salt mixture and Bacto agar media should be added directly to the petri dish rather than the spores. The spores should be rinsed once with sterile ddH_2O. Once the media has solidified, the spores should be sown on top of the agar with >500 μL of sterile ddH_2O and

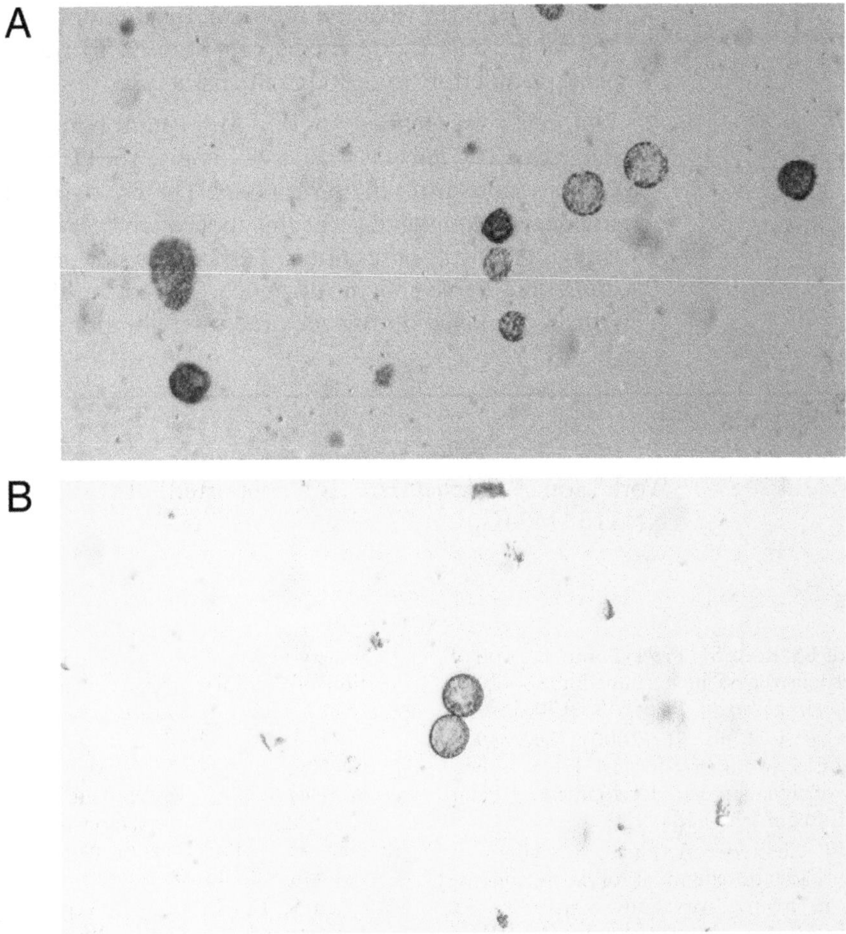

Fig. 2 After protoplasts have been isolated, you can check the material using a light microscope. (**a**, **b**) These images show what you would see using ×20 magnification if intact protoplasts are isolated

the plates should be wrapped in Parafilm. These plates should be laid flat in the growth chamber during the 16–18 days of growth.

5. If the plates are going to be used for polarization studies, place them at a fixed orientation with an arrow drawn on the bottom of the plates representing the direction of gravity during germination.
6. ImageJ or other digital imaging software can be used to measure the angle of rhizoid growth with respect to the vector of gravity or other source of polarization. The angle of rhizoid growth can be a more quantitative description of the gravity (or lack of gravity) response.
7. While removing the gametophytes from the agar surface, use a flat spatula to gently dislodge the tissue. If too much pressure is applied, the agar can dislodge and mix with the tissue and

interfere with the isolation protocol. In addition, half-strength Murashige and Skoog media can be used to wash the agar plates in an effort to dislodge the tissue.

8. The cells resuspended in 0.5 M mannitol may be passed through a 100 μm cell strainer (Biologix; 15–1100) to remove cellular debris from the protoplasts. The Biologix 100 μm cell strainer is recommended for this application because it does not clog as readily as other filters. Filtration is slow and should be done on a stable and steady bench. The set-up should not be shaken or moved during the process of filtration.

Acknowledgments

Work on *Ceratopteris* is supported by NASA grant NNX13AM54G.

References

1. Edwards ES, Roux SJ (1994) Limited period of graviresponsiveness in germinating spores of *Ceratopteris richardii*. Planta 195:150–152
2. Chatterjee A, Roux SJ (2000) *Ceratopteris richardii*: a productive model for revealing secrets of signaling and development. J Plant Growth Regul 19(3):284–289
3. Roux SJ, Chatterjee A, Hillier S, Cannon T (2003) Early development of fern gametophytes in microgravity. Adv Space Res 31 (1):215–220. https://doi.org/10.1016/s0273-1177(02)00749-4
4. Edwards ES, Roux SJ (1997) The influence of gravity and light on developmental polarity of single cells of *Ceratopteris richardii* gametophytes. Biol Bull 192(1):139–140. https://doi.org/10.2307/1542588
5. Edwards ES, Roux SJ (1998) Gravity and light control of the developmental polarity of regenerating protoplasts isolated from prothallial cells of the fern *Ceratopteris richardii*. Plant Cell Rep 17(9):711–716. https://doi.org/10.1007/s002990050470
6. Chatterjee A, Porterfield DM, Smith PS, Roux SJ (2000) Gravity-directed calcium current in germinating spores of Ceratopteris richardii. Planta 210(4):607–610. https://doi.org/10.1007/s004250050050
7. Salmi ML, ul Haque A, Bushart TJ, Stout SC, Roux SJ, Porterfield DM (2011) Changes in gravity rapidly alter the magnitude and direction of a cellular calcium current. Planta 233 (5):911–920. https://doi.org/10.1007/s00425-010-1343-2
8. ul Haque A, Rokkam M, De Carlo AR, Wereley S, Roux SJ, Irazoqui PP et al (2007) A MEMS fabricated cell electrophysiology biochip for in silico calcium measurements. Sens Actuators B 123:391–399
9. Chen S, Tao L, Zeng L, Vega-Sanchez ME, Umemura K, Wang GL (2006) A highly efficient transient protoplast system for analyzing defence gene expression and protein-protein interactions in rice. Mol Plant Pathol 7 (5):417–427. https://doi.org/10.1111/j.1364-3703.2006.00346.x
10. Walter M, Chaban C, Schutze K, Batistic O, Weckermann K, Nake C et al (2004) Visualization of protein interactions in living plant cells using bimolecular fluorescence complementation. Plant J 40(3):428–438. https://doi.org/10.1111/j.1365-313X.2004.02219.x
11. Faraco M, Di Sansebastiano GP, Spelt K, Koes RE, Quattrocchio FM (2011) One protoplast is not the other! Plant Physiol:474–478
12. Schneider CA, Rasband WS, Eliceiri KW (2012) NIH Image to ImageJ: 25 years of image analysis. Nat Methods 9(7):671–675. https://doi.org/10.1038/nmeth.2089

Chapter 4

A Multiplexed, Time-Resolved Assay of Root Gravitropic Bending on Agar Plates

Takehiko Ogura, Christian Goeschl, and Wolfgang Busch

Abstract

The ability of roots to orient their growth relative to the vector of gravity, root gravitropism (positive gravitropism), is observed in root systems of higher plants and is an essential part of plant growth and development. While there are various methods for quantifying root gravitropism, many methods that can efficiently measure gravitropism at a reasonable throughput do not yield temporal resolution of the process, while methods that allow for high-temporal resolution are often not suitable for an efficient measurement of multiple roots. Here, we describe a method to analyze the root gravitropism activity at an increased throughput with a fine time-resolution using *Arabidopsis thaliana* plants.

Key words Root gravitropism, Time-dependent change, *Arabidopsis thaliana*, Automated scanning

1 Introduction

Although different root orders of root systems have different growth angles, the angles are not random and are determined relative to the vector of gravity. These growth properties are determined by the root gravitropism of each root tip. Root gravitropism is a key determinant of root system architecture, which in turn shapes the ability of plants to access water and nutrients, as well as to anchor the plant in the soil [1]. Due to its importance and ease of observation, root gravitropism has been studied for a long time and various methods to quantify root gravitropism have been reported [2–7]. Most of these methods use the angle between root tip and the vector of gravity or the curvature of roots as phenotypes at end points or at larger time intervals. However, the dynamics of root tip re-orientation are an important feature not only for functionally dissecting gravitropism, but also to identify interesting mutants and genotypes affecting root system architecture. For instance, we have recently found that mutations in an exocytosis factor *EXO70A3* gene resulted in a delay of gravitropic bending during 6 h of gravistimulus treatment compared to Col-0 wild type. However, a

Elison B. Blancaflor (ed.), *Plant Gravitropism: Methods and Protocols*, Methods in Molecular Biology, vol. 2368, https://doi.org/10.1007/978-1-0716-1677-2_4, © Springer Science+Business Media, LLC, part of Springer Nature 2022

root angle or curvature phenotype was not observed at 24 h. Importantly, the mutants displayed a profound change in the root system architecture in the soil [8]. This underscores the importance of observing gravitropism at a high time-resolution and at a throughput that allows for robust statistical analysis. Here, we describe a method for analyzing time-dependent changes of root tip angles on agar medium plates using an automated plate image scanning system. In particular, vertically set conventional scanners are run by a program that allows continuous scanning at minimum each 4 min for 24 h, enabling us to observe root bending processes with a high time-resolution. This method allows for quantifying such dynamics as a trait for genetic studies.

2 Materials

Prepare all stock solutions using ultrapure water and at room temperature freshly.

2.1 Plant Growth Medium

1. Murashige and Skoog medium incl. MES buffer (premixed powder of nutrients; 0.52 mg $CoCl_2.6H_2O$, 0.52 mg $CuSO_4.5H_2O$, 764 mg FeNaEDTA, 129 mg H_3BO_3, 17.2 mg KI, 351 mg $MnSO_4.H_2O$, 5.21 mg $Na_2MoO_4.2H_2O$, 179 mg $ZnSO_4.7H_2O$, 6.91 g $CaCl_2$, 3.54 g KH_2PO_4, 39.6 g KNO_3, 3.76 g $MgSO_4$, 34.4 g NH_4NO_3, and 10.4 g MES/100 g powder) (*see* **Note 1**).
2. Ethylenediaminetetraacetic acid iron(III) stock solution (FeNaEDTA stock solution): Weigh 18.4 g FeNaEDTA, resolve in water and adjust the volume by adding water up to 100 mL to prepare 0.5 M (1000×) FeNaEDTA stock solution. Filter sterilize the solution and store it at room temperature (*see* **Note 2**).
3. Sucrose stock solution: Weigh 300 g sucrose, resolve in water and adjust the volume by adding water up to 1 L to prepare 30% (w/v) sucrose stock solution (30×, 0.88 M). Filter sterilize the solution and store it at room temperature (*see* **Note 2**).
4. Plant agar.
5. 12 cm × 12 cm square plates.
6. Leucopore tape (1.25 cm width).
7. OSRAM Fluora L36W/77 (light bulb, OSRAM Licht AG, Munich, Germany).
8. Fine forceps.

2.2 Automatic Scanning System

1. Scanners (EPSON Perfection V600 Photo): Except for two areas that are each the size of a plate (scanning windows), the scanning surface is covered by a black paper to minimize

unnecessary light from scanner, which could cause the light reflection on plates that increases background noises (*see* **Note 3**).

2. Desktop computer: A computer which runs the Linux operating system and to which the open source scanning framework "SANE" is installed (*see* **Note 4**).
3. Scanner operating software tool: "GravitropicScan" (download link: https://busch.salk.edu/tools/).

2.3 Data Analysis

1. Fiji: Image analysis tool.
2. Excel (Microsoft Co., Redmond, Washington, USA): Data analysis application.
3. R language (http://www.R-project.org/): Statistical analysis tool.

3 Methods

Carry out all procedures at room temperature unless otherwise specified. Prepare all media using ultrapure water and use it on the same day.

3.1 Plant Growth Media Preparation

1. Resolve 4.8 g Murashige and Skoog medium incl. MES buffer in water and adjust pH to 5.7 using 10 M KOH solution (pH-adjusted MS solution, *see* **Note 5**).
2. Add 10 g Plant agar to pH-adjusted MS solution, add water up to 966 mL, mix well and sterilize by autoclaving. Air-cool autoclaved solution until it reaches 60 °C (*see* **Note 6**).
3. In a safety cabinet, add 1 mL of 1000 × FeNaEDTA solution and 33 mL of 30 × sucrose solution to autoclaved solution, mix well. Complete this process quickly to avoid the temperature of the autoclaved solution that becomes too low as it will start to become solid.
4. Pour the solution into 12 cm × 12 cm square plates (50 mL/ plate). Chill at room temperature until the solution becomes solid (*see* **Note 7**). The final concentration of components in a plate is 1 × Murashige and Skoog medium, 0.05% w/v MES, 1.0% w/v sucrose, 1.0% w/v agar, pH = 5.7. The plate is hereafter called 1 × MS plate.

3.2 Preparation of Seedlings

1. Transfer seeds to micro tubes and sterilize by chlorine gas (*see* **Note 8**).
2. Add 400 μL sterilized water to sterilized seeds in a safety cabinet and then keep these tubes at 4 °C in the dark for 3 days to stratify seeds.

3. In a safety cabinet, put stratified seeds on 1 × MS plates and seal plates with Leucopore tape. Set up plates vertically and grow plants at 21 °C, 16 h light/16 °C, 8 h dark conditions for 5 days after germination (DAG) (*see* **Notes 9 and 10**).

3.3 Transfer Seedlings to Assay Plates

1. Select a sufficient number of seedlings that have similar sizes. When multiple genotypes are used, the size should be similar between genotypes. The sufficient number depends on the purpose of experiments and genotypes of plants, we generally use 24 plants/plate and test two genotypes in a single plate together. In this case, 12 plants/plate/genotype are necessary. We recommend repeating one experimental condition at least three times (*see* **Note 9**).
2. Sterilize fine forceps. In a safety cabinet, using fine forceps, carefully transfer selected seedlings to freshly prepared 1 × MS plates. We generally use a setup with 24 seedlings (See Fig. 1). To set transferred roots straight, first put a root on the plate surface and move the plant straightly in the direction of aerial parts so that the root slide on the surface of the medium plate into a straightened conformation (Fig. 2). Position the transferred seedlings along two lines (if the 24 seedlings setup is used, see Fig. 1). Seal plates with Leucopore tape (*see* **Notes 10** and **11**).
3. Set up plates vertically and grow plants at 21 °C, light conditions for 1 h to let plants reach at uniform state.

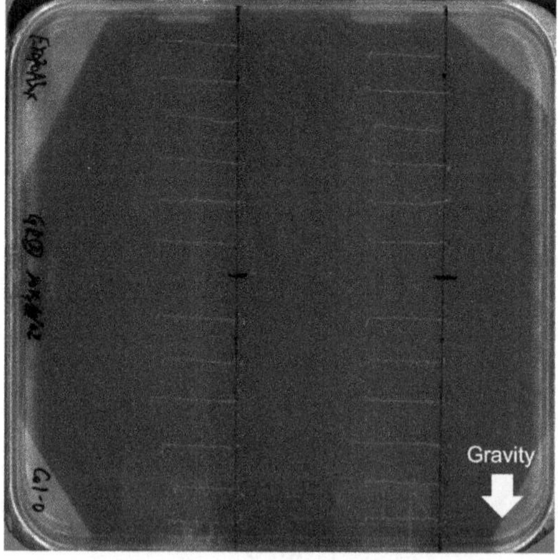

Fig. 1 Root gravitropism assay setup with 24 seedlings. A representative set up for the outlined root gravitropism assay using 24 seedlings. In this example plate, the upper half contains 12 *EXO70A3ox* mutant plants and the lower half contains 12 Col-0 wild type plants 10 h after plate re-orientation

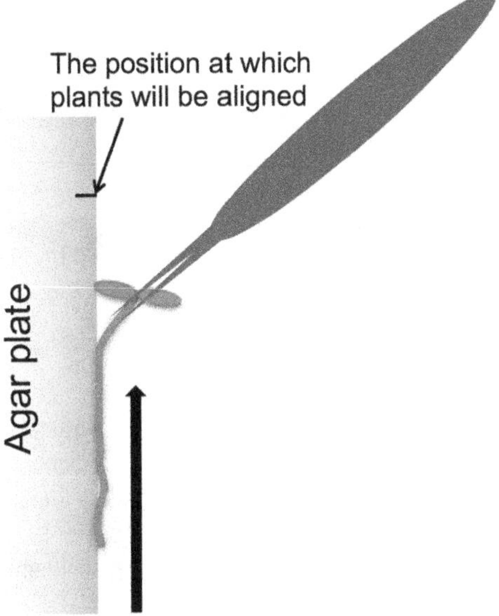

Fig. 2 Plant transfer to assay plates. Illustration for transferring plants to assay plates. A pair of fine tweezers is used to transfer the seedling to the assay plate. Only cotyledons should be touched by tweezers during the whole process. The entire seedling is set down on the plate below the assay position (marked by a line). Once most of the root adheres to the surface of the agar plate, the seedling is carefully pulled to the upper side in a straight line and aligned to the assay position. During this process, roots should stay adhered to the agar surface of a plate

3.4 Scanning Root Bending Using an Automated Scan Program

1. Set up scanners vertically in a dark room and position light sources directly above the locations of the plants (other unnecessary light sources generally lead to light reflections on plates and should not be used) to allow for sufficient image contrast between roots and the background. Set the light conditions and the room temperature to conditions used in Subheadings 3.2, **step 3**.
2. Rotate plates for 90° so that roots will be perpendicular to the vector of gravity. Fix plates firmly to their positions on the scanning surface. Attach black paper strips onto the lids of the plates to increase contrast on resulting images (Fig. 3).
3. Start the automated scanning application such as GravitropicScan to control image acquisition. Specify in the main window which scanners should be used (check box "Use:") and if images of one or both plate positions should be acquired (check box "Row 0" for bottom and "Row 1" for base scanning position in relation to the connection of a scanner and its lid.) (*see* Fig. 4).
4. Select the folder path to save the image data by clicking on the folder selection button (labeled with "...") and specifying the

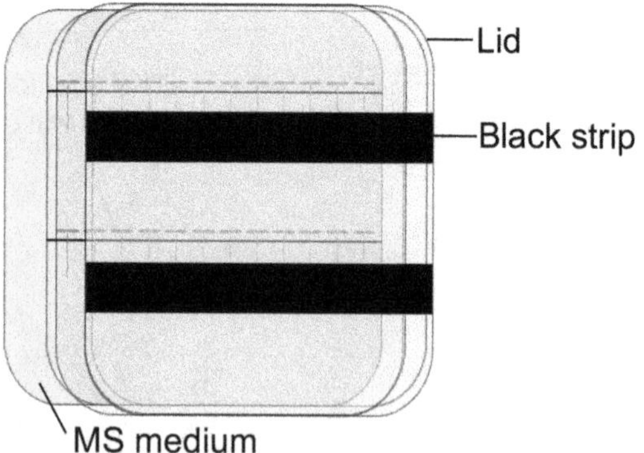

Fig. 3 Ensuring high-contrast of images via providing a black background for roots. Providing a black background for roots will lead to a superior contrast of the pictures and facilitate the analysis. This background is provided via attaching black paper strips (2 cm × 12 cm) to the lid of assay plates, on the back of roots at position where they will center on the tips of the roots

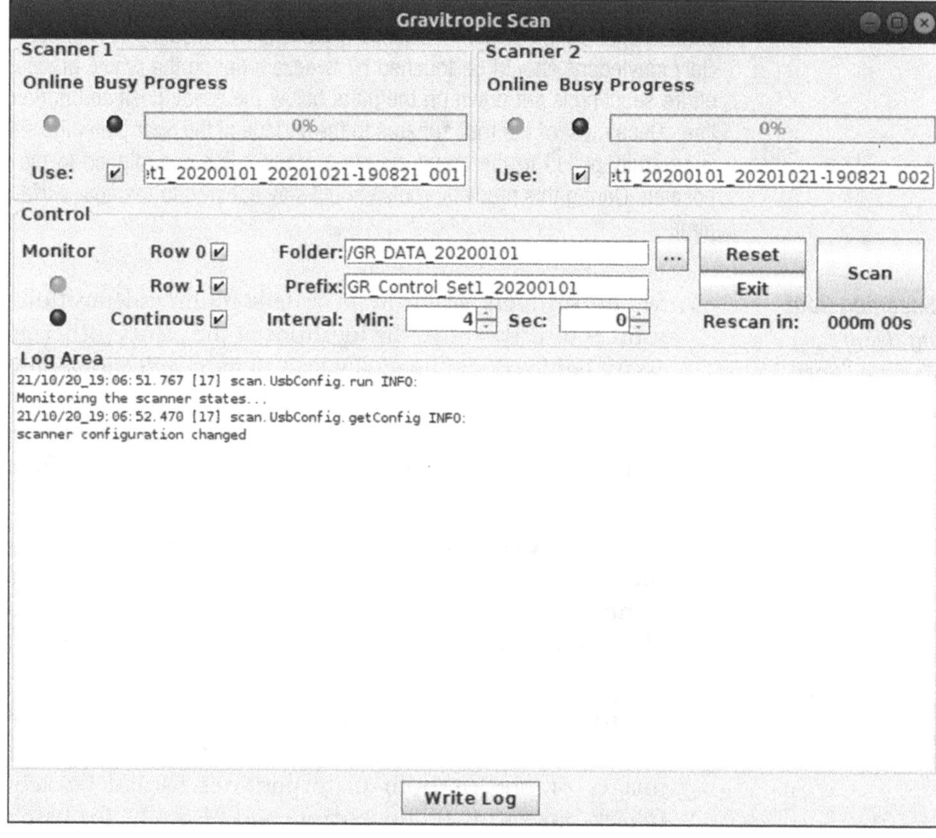

Fig. 4 The main window of the scanner operating software "GravitropicScan." A representative visual of the scanning setup for GravitropicScan is shown. Check box "Use:", check box "Row 0", and "Row 1", the folder selection button (labeled with ". . ."), the box to input prefix, scan time boxes ("Interval: Min" and "Sec"), check box "Continous", and "Scan" button to initiate scanning, that are explained in Subheading 3.4, are shown

experiment prefix that will be used for naming the files that are automatically stored.

5. Set scan time to 4 min or a desired duration which is longer than 4 min and select continuous scan mode. Click on "Scan" to initiate scanning. The images are stored as 1200 dpi 24-bit RGB TIFF files and automatically named based on the initial user input into the image acquisition tool.
6. Once scanning is completed for the time planned, stop the program.

3.5 Root Angle Analysis

1. Open Fiji and import all images of a time-series of an assay plate as Image Sequence (*see* **Notes 12** and **13**).
2. Measure the angle of each root tip at each time point. For this, use the "Angle tool" from the tool bar in Fiji and draw a straight line along the center of a root tip. Then click "Measure" in "Analyze" menu or, alternatively, press "control (command) + M". Start drawing the line from the center of the root at around 0.1–0.2 mm of the front edge of the root tip. Since the growth rate of *Arabidopsis* roots in our conditions is around the reported range 0.1 mm/h [9] and roots continuously change their angles during gravistimuli, root tips longer than this length reflect the history of bending of longer time periods than 1 h, thereby decreasing the time-resolution (*see* **Note 14**). The most efficient way to obtain measurements is by measuring each root tip throughout all time points. For this, start from time point 1 for a given root tip and go to the next time point after completion of the measurement until the given root tip has been measured at all time points. Repeat this process for all roots in a scanned image.
3. Angle measurement results are shown in a table that is automatically created by Fiji. Copy the angle data from the table and paste it into Excel. Then, using Excel, calculate averages and standard deviations of angles for each genotype at all time points and plot time course curves of angle changes for tested genotypes.
4. For statistical tests, including t-test and ANOVA, save the angle data as .csv file and import into R. Various types of plots including box plots and bean plots can be generated easily in R.

4 Notes

1. To maintain the uniformity of medium plates used in the assay particularly in terms of pH, MES buffer should be included. If MS basal salt mix, which contains MES buffer is not available, prepare 100 × (250 mM) MES solution and add 10 mL for 1 L of 1 × MS medium in advance to adjust pH.

2. We recommend to filter sterilize FeNaEDTA stock solution and sucrose solution and add these sterile solutions after autoclaving the medium. If FeNaEDTA is added to the MS medium solution in advance to the autoclave step, iron precipitation occurs during the autoclaving. Likewise, pre-inclusion of sucrose solution in the MS medium solution results in the hydrolysis of sucrose during the autoclave.
3. Any conventional scanners that are equivalent to EPSON Perfection V600 Photo and are supported by the SANE framework could potentially be utilized. However, it cannot be guaranteed that the automated scanning program will work with these.
4. Any desktop computer running Linux (we tested on the LTS Ubuntu versions 16.04 and 18.04 (https://www.ubuntu.com)) and provides enough USB ports to connect the desired number of scanners should be suitable. The SANE scanning framework (https://sane-project.org) needs to be installed for the scanning software to work.
5. To provide uniform assay conditions, a very precise adjustment of pH of 1 × MS medium is important. Do calibration of pH meter on a regular basis.
6. To avoid aggregation of agar during autoclave, the solution including plant agar should be mixed well before it is autoclaved.
7. When 1 × MS medium solution is poured into square plates, ensure that the table on which square plates are poured is completely leveled; otherwise, the thickness of agar medium will not be uniform and potentially sloped.
8. Wear gloves and conduct sterilization by chlorine gas in a chemical hood. Chlorine gas is highly toxic to the human body, and chemicals used to create chlorine gas are also toxic or harmful to the human body.
9. To obtain sufficient numbers of seedlings that have similar sizes, it is recommended to sow three times more seeds than the number of seedlings required in the assay.
10. Prepare 1 × MS plates freshly on the day of the experiment to obtain reproducible results. Old plates can result in a variability of results between experimental replicates.
11. Carefully transfer plants without damaging them. Repeated contact of roots to the surface of agar plates for pulling roots straight or direct pinching of roots will damage roots and alter root.gravitropic bending.
12. Depending on the image size and the number of images, decreasing the image size via resizing them can facilitate image analysis. High resolution images will require more

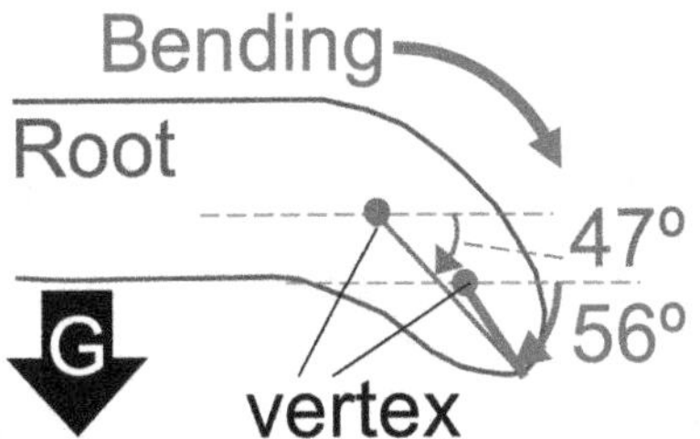

Fig. 5 The difference between angles measured with higher- and lower time-resolutions. Angle changes of a root tips are compared between a shorter root tip (higher time-resolution, blue) and a longer root tip (lower time-resolution, two folds longer, gray). A shorter root tip represents larger angle change than a longer root tip as indicated

computing resources to analyze. However, we recommend to use a higher resolution setup during scanning to avoid missing moderate changes of root angles.

13. Since at the given resolution and these assay conditions, no significant changes of root tip angles are observed during time periods such as 4–16 min, a frequency of imaging at one image per 16 min or 32 min is usually sufficient.

14. Considering the growth rate of *Arabidopsis* roots (around 0.1 mm/h, [9]) and the continuous bending of roots upon gravistimulation, taking into account a portion of the root tip that exceeds 0.1 mm will result in a decreased time-resolution as the measurement will reflect the change of the angle during a longer time period. Moreover, the angle change of a longer part of a root tip will be smaller compared to that measured on a shorter part of the root tip because of the relative position of the root tip and the vertex of the angle measurement (Fig. 5). Therefore, for a high time-resolution of the analysis, a shorter line from vertex to distal root tip for the angle measurements (0.1–0.2 mm) will provide more accuracy. However, the best method for this root tip angle measurement might depend on the genotype of plants, root growth rate and the purpose of the experiment. For example, slow bending genotypes might be analyzed taking into account longer root tip segments, or finding a genotype with stronger abnormality might not require a high time-resolution. In any case, use the same method for the angle measurement for all roots in a set of experiments.

References

1. Su S-H, Gibbs NM, Jancewicz AL, Masson PH (2017) Molecular mechanisms of root gravitropism. Curr Biol 27:R964–R972
2. Yamamoto M, Yamamoto KT (1998) Differential effects of 1-Naphthaleneacetic acid, Indole-3-acetic acid and 2,4-Dichlorophenoxyacetic acid on the Gravitropic response of roots in an auxin-resistant mutant of Arabidopsis, auxl. Plant Cell Physiol 39:660–664

3. Marchant A, Kargul J, May ST, Muller P, Delbarre A, Perrot-Rechenmann C, Bennett MJ (1999) AUX1 regulates root gravitropism in Arabidopsis by facilitating auxin uptake within root apical tissues. EMBO J 18:2066–2073
4. Grabov A, Ashley MK, Rigas S, Hatzopoulos P, Dolan L, Vicente-Agullo F (2005) Morphometric analysis of root shape. New Phytol 165:641–652
5. Mochizuki S, Harada A, Inada S, Sugimoto-Shirasu K, Stacey N, Wada T, Ishiguro S, Okada K, Sakai T (2005) The Arabidopsis WAVY GROWTH 2 protein modulates root bending in response to environmental stimuli. Plant Cell 17:537–547
6. Fortunati A, Piconese S, Tassone P, Ferrari S, Migliaccio F (2008) A new mutant of Arabidopsis disturbed in its roots, right-handed slanting, and gravitropism defines a gene that encodes a heat-shock factor. J Exp Bot 59:1363–1374
7. Wang HZ, Yang KZ, Zou JJ, Zhu LL, Xie ZD, Morita MT, Tasaka M, Friml J, Grotewold E, Beeckman T et al (2015) Transcriptional regulation of PIN genes by four lips and MYB88 during Arabidopsis root gravitropism. Nat Commun 6:9
8. Ogura T, Goeschl C, Filiault D, Mirea M, Slovak R, Wolhrab B, Satbhai SB, Busch W (2019) Root system depth in Arabidopsis is shaped by *EXOCYST70A3* via the dynamic modulation of auxin transport. Cell 178:400–412
9. Yazdanbakhsh N, Fisahn J (2010) Analysis of Arabidopsis thaliana root growth kinetics with high temporal and spatial resolution. Ann Bot 105:783–791

Chapter 5

Methods to Quantify Cell Division and Hormone Gradients During Root Tropisms

Jinke Chang and Jia Li

Abstract

Tropisms are growth-based plant directional movements, allowing plants to respond to their living environments. Plant roots have developed various tropic responses, including gravitropism, hydrotropism, chemotropism, and halotropism, in response to the gravity, moisture gradient, nutrient gradient, and salinity gradient, respectively. Revealed mechanisms of several tropic responses suggested that plant hormone gradient and cell division activity play key roles in determining these responses. Approaches to measure cell division and hormone gradients, however, have rarely been applied in root tropic analyses. Here, we describe a number of methods to quantify cell division and hormone gradients during root tropic analysis. These approaches are mainly based on our previous researches on root hydrotropism.

Key words Gravitropism, Hydrotropism, Auxin gradient, Cytokinin gradient, Cell division

1 Introduction

Plant roots evolved a series of strategies to adapt to their living environments. Among them, root tropic responses, including gravitropism, hydrotropism, and halotropism are the most effective growth-based approaches to evade stresses and obtain various needed nutrients and water for survival. Cellular and molecular mechanisms of root gravitropism, hydrotropism, and halotropism have been partially elucidated [1–5]. It was reported that gravitropism and halotropism are mainly controlled by the polar transportation of auxin, while root hydrotropism is predominantly regulated by the asymmetric distribution of cytokinins [2, 3, 6–11]. For a horizontally placed root, polar auxin transport leads to asymmetric auxin distribution and uneven cell elongation between two sides of the root tip, resulting in a root gravitropic response [1, 6]. By contrast, recent studies indicated that asymmetric distribution of cytokinins is likely the determinant to control root hydrotropism via the differential cell division in the root meristem zone [3]. Under a hydrostimulating condition, the lower water potential

Elison B. Blancaflor (ed.), *Plant Gravitropism: Methods and Protocols*, Methods in Molecular Biology, vol. 2368,
https://doi.org/10.1007/978-1-0716-1677-2_5,

side of a root tip showed enhanced cell division activity and more cells were found to accumulate within a 200-μm region from the quiescent center in the meristem zone compared to the higher water potential side of the root tip. As a result, more cells at the lower water potential side of the meristem zone can enter into the elongation zone, causing the root to grow towards higher water potential area. In addition, it was also reported that gibberellic acid (GA) showed asymmetric distribution in root tips after gravistimulation [12]. All these previously reported results suggest that cell division or cell elongation and hormone gradients play key roles in root tropic responses.

In this chapter, we summarize a number of protocols for quantifying cell division and hormone gradients during root tropisms. Root hydrotropism was used as an example to show how these approaches can be implemented. Here, we introduce three different methods to evaluate cell division activity in the root meristem region, including a 5-ethynyl-2′-deoxyuridine (EdU) staining method to test the DNA replication activity, a method to test the expression of a cell cycle marker gene *CYCB1;1*, and an approach to calculate cell numbers within a certain length of the meristem region. A few scientific software can be adopted to indirectly compare hormone gradients in root tips.

2 Materials

2.1 Quantifying Cell Division

2.1.1 EdU Staining

1. EdU imaging kit.
2. 1/2 MS liquid medium (or other liquid growth media).
3. Phosphate-buffer saline (PBS with pH 7.4).
4. PBST (PBS plus 0.5%(v/v) triton X-100).

2.1.2 GUS Staining, Cell Counting, and Hormone Gradient Determination

1. 90% acetone.
2. Deionized water.
3. 100 mM $K_3Fe(CN)_6$ (*see* **Note 1**).
4. 100 mM $K_4Fe(CN)_6$ (*see* **Note 1**).
5. 1 M Na_2HPO_4.
6. 1 M NaH_2PO_4.
7. 100 mM X-Gluc (*see* **Note 2**).
8. Ethanol.
9. Rinse solution: 50 mM sodium phosphate buffer (pH 7.2), 0.5 mM $K_3Fe(CN)_6$, 0.5 mM mM $K_4Fe(CN)_6$, 34.2 mM Na_2HPO_4, 15.8 mM NaH_2PO_4. The solution should be freshly prepared just before use.
10. Stain solution: for 100 mL rinse solution, add 2 mL 100 mM X-Gluc stock, freshly prepared before use.

11. Chloral hydrate solution: 80% chloral hydrate solution (w/v), 20% glycerin (v/v), and 20% purified water (v/v).
12. Cell wall staining solution such as propidium iodide solution (25 ng/μL), or cell membrane staining solution such as FM4-64 staining solution (*see* **Note 3**).
13. Transgenic plants containing hormone indictor genes, such as *DR5::GFP* for auxin and *TCSn::GFP* for cytokinins (*see* **Note 4**).

3 Methods

3.1 Orientation Determination During Imaging of Tropic-Treated Roots

For all tropic response analyses, it is important to ensure the orientations of tropic-treated roots remain unchanged. The following three methods can be applied.

1. Make the tropic treatments on slides and take live images without moving the seedlings. Pour the tropic medium on the slides and move 4-day-old seedlings to the tropic medium. Root tips should be tropically treated in an airtight box before imaging.
2. The orientation can remain unchanged during transfer of the seedlings from tropic medium to slides by moving root tips with some medium near the root tip, operated with a pair of flat twisters under a dissecting microscope.
3. After tropic treatment, the orientation of most root tips can be distinguished under a dissecting microscope and can be easily retrieved based on the unique morphology of the roots.

3.2 EdU Staining and Fluorescence Intensity Measurements

The fluorescence intensity from images can be measured by any confocal microscope software and Image J. Each software has its own operating system, but the principle and obtained results should be similar. Here, we take a Leica confocal software as an example.

1. Transfer 4-day-old seedlings to hydrotropic medium supplemented with 10 μM EdU, incubate for 60 min.
2. Wash the hydrotropically treated seedlings with 1/2 MS liquid medium (or other liquid growth media), 3 × 5 min, to remove excessive EdU (*see* **Note 5**).
3. Wash the seedlings in PBST (pH 7.4), 2 × 5 min.
4. Submerge the samples into the EdU detection cocktail for 30 min in the dark (*see* **Note 6**).
5. Wash the samples in PBS (pH 7.4), 3 × 5 min.
6. Cover the samples with PBS (pH 7.4) containing 8 μg/mL Hoechst 33342 (component of EdU imaging kit). Incubate for 30 min in the dark.

7. Wash in PBS (pH 7.4), 3 × 20 min.
8. Image the root tips using a confocal laser scanning microscope with the corresponding wavelength of excitation light.
9. The activities of DNA replication can be reflected by the fluorescence intensity, which can be measured by using the Leica quantification tools: Quantify→Tools (*see* **Note** 7).
10. Choose the Draw polygon or Draw rectangle tools to mark the meristematic region in each side of the root tip in the fluorescence channel.
11. In the Statistics window, the area of the selected regions and the mean value (fluorescence intensity) are presented.
12. We use the fluorescence intensity ratio to represent the cell division activity between two sides of the root tip. Using the right side/left side and lower water potential side/higher water potential side for control and hydrostimulation treatment, respectively (Fig. 1).

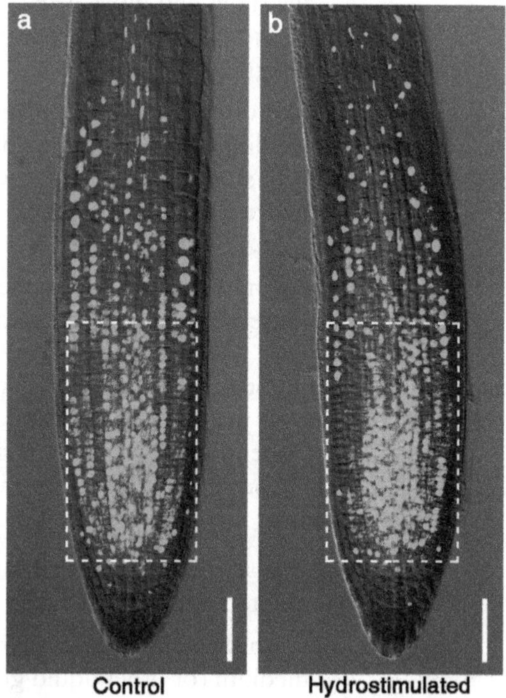

Fig. 1 Quantification of cell division using EdU staining. Alexa Fluor® 488 was used as the fluorochrome. The dashed boxes represent the regions for fluorescence intensity measurements. For control treatment (**a**), the yellow dashed box (200 μm × 50 μm) represents the right side of the meristematic region and the white dashed box represents the left side of the meristematic region. For hydrostimulation-treated roots (**b**), the yellow dashed box (200 μm × 50 μm) represents the lower water potential side and the white dashed box represents the higher water potential side. Scale bars represent 50 μm

3.3 GUS Staining and Signal Quantification

1. Transfer the tropic or control treatment seedlings to ice-incubated 90% acetone and incubate for 15 min (*see* **Note 8**).
2. Replace the acetone solution with rinse solution and keep at room temperature for 5 min.
3. Remove the rinse solution and add staining solution. Incubate at 37 °C in the dark (*see* **Note 9**).
4. Wash the samples with ethanol series, 15, 30, 50, 70, 80, 90, 100, and 80% for 30 min each at room temperature.
5. Keep the samples in 70% ethanol overnight.
6. Replace the 70% ethanol with chloral hydrate solution and keep in this solution for 48 h.
7. Image the root tips under a microscope.
8. Open the figures in Image J.
9. Convert the color mode of light microscope images from RGB to gray by using Image J software function: Image→Type→8-bit.
10. Transfer the image to bright light: Edit→Invert.
11. Set the measurements: Analyze→Set measurements→Area, Modal Gray Value.
12. Using Freehand Selection or other selections to choose the GUS staining region for measurements.
13. Press "m" to get the results.
14. To eliminate background errors, choose the background area and get the modal gray value. Each data minus to the background modal gray value will get the relative value.
15. Use the GUS signal ratio to compare the cell division activity between two sides of the root tips. Use the (right side minus background)/(left side minus background) and (lower water potential side minus background)/(higher water potential side minus background) for control and hydrostimulation-treated roots, respectively (Fig. 2).

3.4 Cell Numbers in the Root Meristem Region

1. Treat seedlings on hydrostimulating medium and then transfer them to slides for observation.
2. Add 50 μL propidium iodide solution to submerge the root tips and incubate for 5 min at room temperature.
3. Image the root tips using a confocal laser scanning microscope.
4. Cell numbers in the meristem region can be quantified by the confocal imaging software; here, we use the Leica confocal as an example. Use the quantification tools to measure the cell numbers: Quantify→Tools.

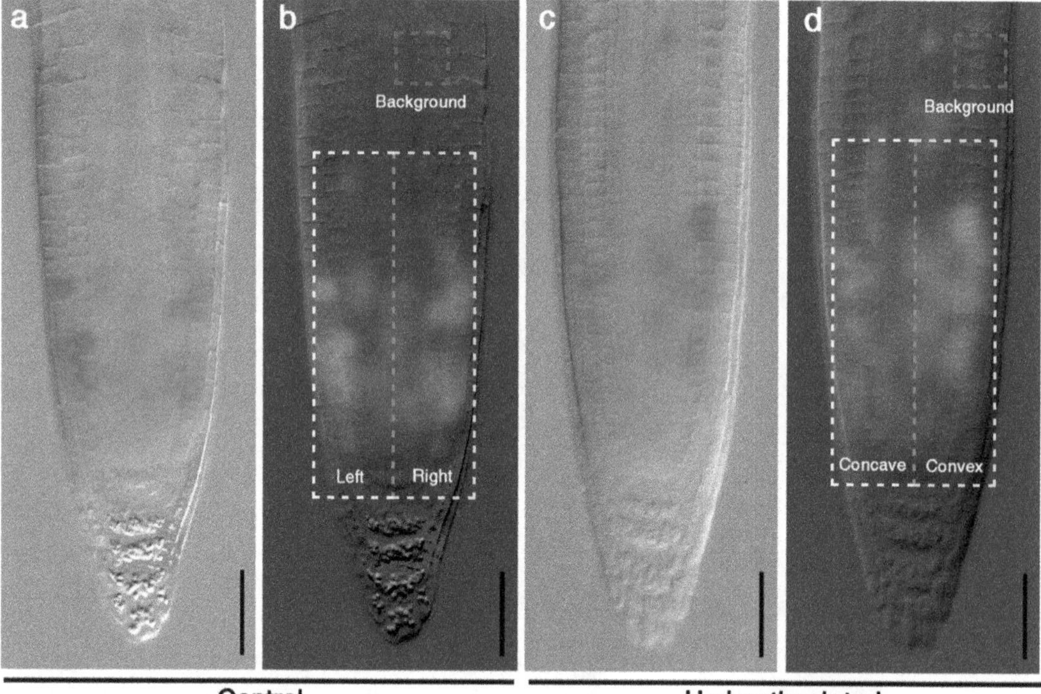

Fig. 2 Quantification of cell division by using GUS staining of transgenic plants carrying *pCYCB1;1-GUS*. (**a**, **b**) Transgenic plants without hydrostimulation treatment was used as a control. (**c**, **d**) Transgenic plants after hydrostimulation treatment. GUS stained root tips (**a**, **c**) were type-changed by using Image J software (**b**, **d**). The yellow dashed boxes (200 μm × 50 μm) represent the right side (**b**) or lower water potential side (**d**) of the meristematic region. The white boxes represent the left side (**b**) or higher water potential side (**d**) of the meristematic region. The area shown in blue boxes was used to eliminate the background errors. Scale bars represent 50 μm

5. Choose the Draw polyline tools to mark the 200 μm area from the quiescent center in the cortex cell layer at each side of a root tip.
6. In the Graphs windows, each peak represents a cell plate between two cells, so the number of peaks represents the cell number in the marked meristem region (Fig. 3).

3.5 Hormone Gradients Quantification

Cytokinins can be indirectly measured via the intensity of GFP signal using transgenic plants carrying a report construct *TCSn::GFP* [13]. *TCSn* is an artificial promoter monitoring the signaling output of cytokinins. Under a hydrostimulating condition, asymmetric distribution of cytokinins can be mainly observed in lateral root caps between lower and higher water potential sides. The following approach can be used to compare the GFP intensity between lower and higher water potential sides of the root tips, which can indirectly reveal the relative cytokinin distribution. The protocol used to quantify auxin gradient is similar to what is used

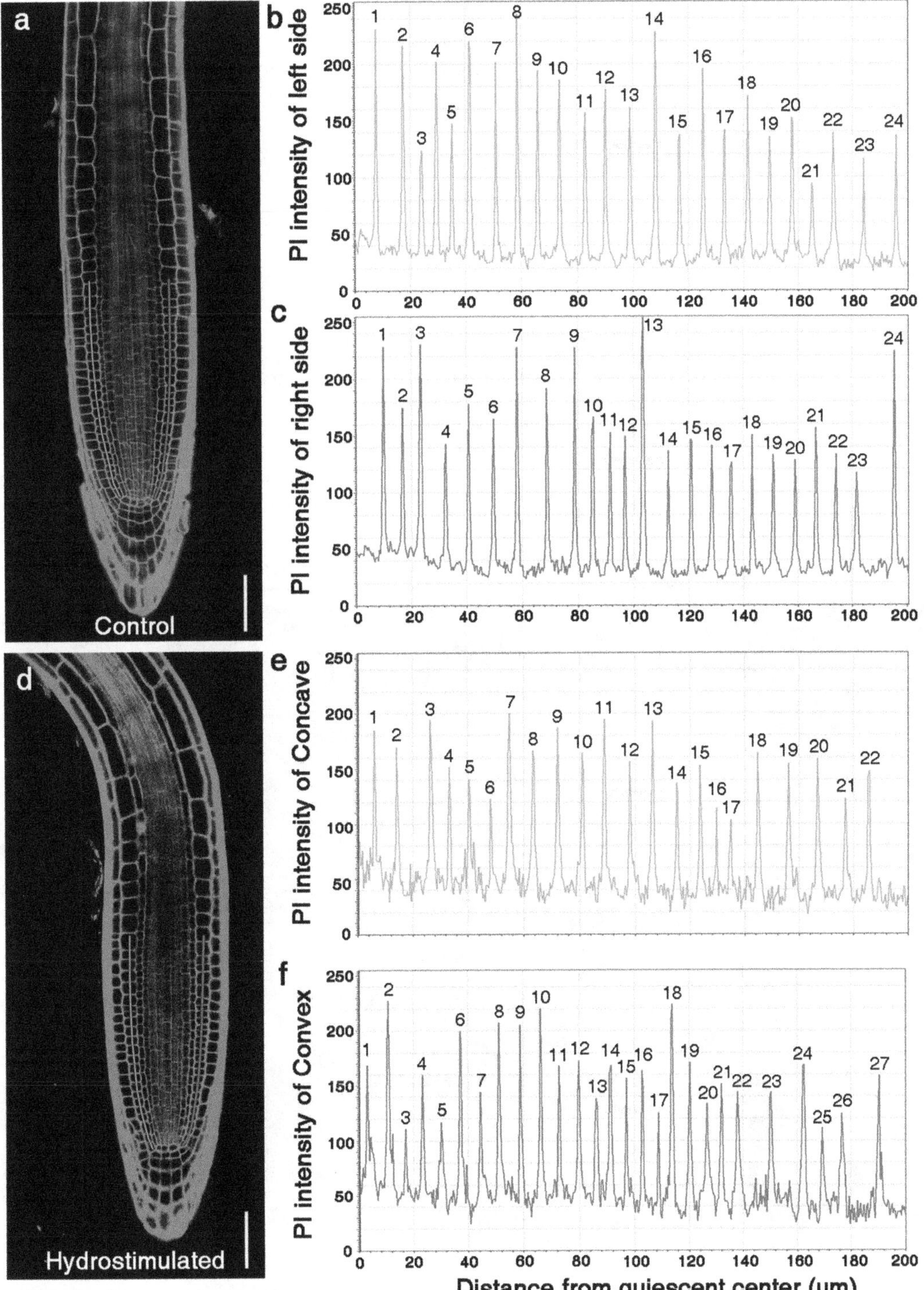

Fig. 3 Cell numbers were calculated by the peaks of cortex cell wall fluorescence within a 200-μm meristematic zone starting from the quiescent center using a Leica software after propidium iodide staining for control (**a**–**c**) and hydrostimulation treatment (**d**–**f**) roots. For control treatment, green and blue polylines represent the left and the right side cortex cell layers, respectively. The fluorescence peaks are shown in (**b**) and (**c**). For hydrostimulation treatment, green and blue polylines represent the higher water potential side and the lower water potential side, respectively. The fluorescence peak numbers are shown in (**e**) and (**f**). Scale bars represent 50 μm

for cytokinins. Transgenic plants carrying *DR5::GFP* can be used to indicate the relative concentration of auxin in various tissues [14].

1. Transferred 4-day-old seedlings were to hydrostimulating media. Treat the seedlings for 60 min by subjecting them to a water gradient.
2. Take the images of the root tips by using a confocal laser scanning microscope.
3. The GFP signal can be measured by the fluorescence intensity using quantify tools: Quantify→Tools.
4. Choose the Draw polyline tools to mark the same length at each side of the lateral root cap in the fluorescence channel (Fig. 4).
5. In the Statistics window, the length of the selected regions and the mean value (fluorescence intensity) are presented.

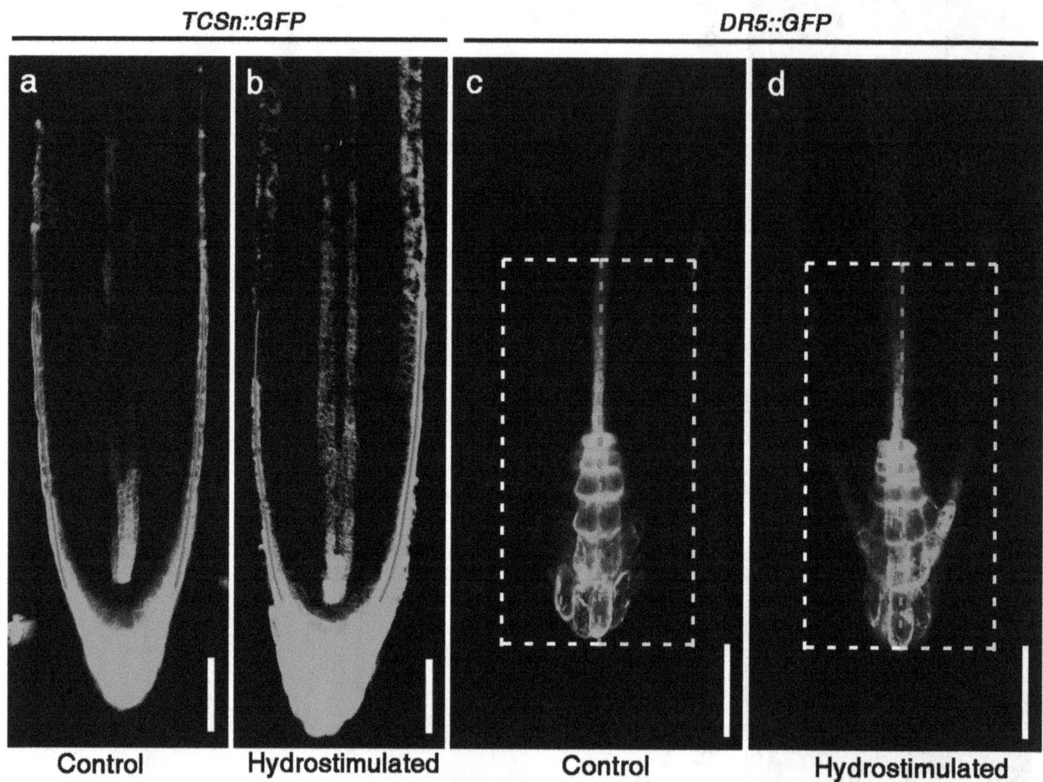

Fig. 4 Approaches used to measure the fluorescence intensity in the root tips. (**a–b**) Fluorescence intensity of *TCSn::GFP* transgenic seedlings were measured as linear average intensity within a 200-μm distance above the quiescent center after control (**a**) or hydrostimulation (**b**) treatments. The red polylines represent the right side (**a**) or lower water potential side (**b**). The purple polylines represent the left side (**a**) or higher water potential side (**b**) in the lateral root caps. (**c–d**) Fluorescence intensity of *DR5::GFP* transgenic seedlings were measured as regional average intensity within a 200 μm × 50 μm area between both sides. The yellow boxes represent the right side (**c**) or lower water potential side (**d**). The white boxes represent the left side (**c**) or higher water potential side (**d**). Scale bars represent 50 μm

4 Notes

1. The potassium ferricyanide and potassium ferrocyanide solutions should be kept in the dark at 4 °C. The solutions should be used within 2 months. If the color changes to yellow, the solutions cannot be used.
2. The X-Gluc stock solution is made up in dimethyl sulfoxide (DMSO). Keep it at −20 °C.
3. The propidium iodide and FM4-64 solutions can be kept in the dark at 4 °C for weeks.
4. There are other reporter systems for cytokinins, such as GFP driven by the promoters of type-A response regulators (*ARRs*).
5. Record the root orientation under a dissecting microscope before transferring the seedlings from hydrostimulating medium to slides for observation.
6. The EdU detection cocktail should be freshly prepared.
7. Here we use Leica confocal software as an example. Other software can be used but may differ slightly in operation.
8. Record the root orientation under a dissecting microscope before transferring the seedlings from hydrotropic medium to 90% acetone.
9. The staining time depends on the expression levels of the target genes. Pre-experiments are usually needed to determine the staining time.

Acknowledgments

Research in the laboratory of Jia Li is currently supported by the National Natural Science Foundation of China grants 31530005 and 31720103902.

References

1. Philosoph-Hadas S, Friedman H, Meir S (2005) Gravitropic bending and plant hormones. Vitam Horm 72:31–78. https://doi.org/10.1016/s0083-6729(05)72002-1
2. Muday GK (2001) Auxins and tropisms. J Plant Growth Regul 20(3):226–243. https://doi.org/10.1007/s003440010027
3. Chang J, Li X, Fu W, Wang J, Yong Y, Shi H, Ding Z, Kui H, Gou X, He K, Li J (2019) Asymmetric distribution of cytokinins determines root hydrotropism in *Arabidopsis thaliana*. Cell Res 29(12):984–993. https://doi.org/10.1038/s41422-019-0239-3
4. Galvan-Ampudia CSJM, Darwish E, Gandullo J, Korver RA, Brunoud G, Haring MA, Munnik T, Vernoux T, Testerink C (2013) Halotropism is a response of plant roots to avoid a saline environment. Curr Biol 20:2044–2050. https://doi.org/10.1016/j.cub.2013.08.020
5. Shkolnik D, Nuriel R, Bonza MC, Costa A, Fromm H (2018) MIZ1 regulates ECA1 to generate a slow, long-distance phloem-transmitted Ca^{2+} signal essential for root water tracking in Arabidopsis. Proc Natl Acad Sci U

S A 115(31):8031–8036. https://doi.org/10.1073/pnas.1804130115

6. Sato EM, Hijazi H, Bennett MJ, Vissenberg K, Swarup R (2015) New insights into root gravitropic signalling. J Exp Bot 66(8):2155–2165. https://doi.org/10.1093/jxb/eru515
7. Geisler M, Wang B, Zhu J (2014) Auxin transport during root gravitropism: transporters and techniques. Plant Biol (Stuttg) 16 Suppl 1:50–57. https://doi.org/10.1111/plb.12030
8. van den Berg T, Korver RA, Testerink C, Ten Tusscher KH (2016) Modeling halotropism: a key role for root tip architecture and reflux loop remodeling in redistributing auxin. Development 143(18):3350–3362. https://doi.org/10.1242/dev.135111
9. Shkolnik D, Fromm H (2016) The Cholodny-Went theory does not explain hydrotropism. Plant Sci 252:400–403. https://doi.org/10.1016/j.plantsci.2016.09.004
10. Shkolnik D, Krieger G, Nuriel R, Fromm H (2016) Hydrotropism: root bending does not require auxin redistribution. Mol Plant 9 (5):757–759. https://doi.org/10.1016/j.molp.2016.02.001
11. Kaneyasu T, Kobayashi A, Nakayama M, Fujii N, Takahashi H, Miyazawa Y (2007) Auxin response, but not its polar transport, plays a role in hydrotropism of Arabidopsis roots. J Exp Bot 58(5):1143–1150. https://doi.org/10.1093/jxb/erl274
12. Löfke C, Zwiewka M, Heilmann I, Van Montagu MC, Teichmann T, Friml J (2013) Asymmetric gibberellin signaling regulates vacuolar trafficking of PIN auxin transporters during root gravitropism. Proc Natl Acad Sci U S A 110(9):3627–3632. https://doi.org/10.1073/pnas.1300107110
13. Zurcher E, Tavor-Deslex D, Lituiev D, Enkerli K, Tarr PT, Muller B (2013) A robust and sensitive synthetic sensor to monitor the transcriptional output of the cytokinin signaling network in planta. Plant Physiol 161 (3):1066–1075. https://doi.org/10.1104/pp.112.211763
14. Ulmasov T, Murfett J, Hagen G, Guilfoyle TJ (1997) Aux/IAA proteins repress expression of reporter genes containing natural and highly active synthetic auxin response elements. Plant Cell 9(11):1963–1971. https://doi.org/10.1105/tpc.9.11.1963

Chapter 6

Using the Automated Botanical Contact Device (ABCD) to Deliver Reproducible, Intermittent Touch Stimulation to Plants

Caleb Fitzgerald, Cullen S. Vens, Nathan Miller, Richard Barker, Matthew Westphall, Johnathan Lombardino, Jerry Miao, Sarah J. Swanson, and Simon Gilroy

Abstract

Despite mechanical stimulation having profound effects on plant growth and development and modulating responses to many other stimuli, including to gravity, much of the molecular machinery triggering plant mechanical responses remains unknown. This gap in our knowledge arises in part from difficulties in applying reproducible, long-term touch stimulation to plants. We describe the design and implementation of the Automated Botanical Contact Device (ABCD) that applies intermittent, controlled, and highly reproducible mechanical stimulation by drawing a plastic sheet across experimental plants. The device uses a computer numerical control platform and continuously monitors plant growth and development using automated computer vision and image analysis. The system is designed around an open-source architecture to help promote the generation of comparable datasets between laboratories. The ABCD also offers a scalable system that could be deployed in the controlled environment setting, such as a greenhouse, to manipulate plant growth and development through controlled, repetitive mechanostimulation.

Key words *Arabidopsis thaliana*, Automated botanical contact device, Image analysis, Mechanostimulation, Touch

1 Introduction

Plants are exquisitely sensitive to touch stimulation that modulates a suite of responses such as leaf size and petiole length, deposition of strengthening polymers in the wall, flowering time and susceptibility to a wide range of biotic and abiotic stressors (reviewed in: [1–3]). Mechanostimulation also changes plant responses to myriad other stimuli, such as modulating the plant's gravitropic response [4]. Although plant mechanoreceptors have yet to be clearly

Caleb Fitzgerald and Cullen S. Vens contributed equally to this work.

Elison B. Blancaflor (ed.), *Plant Gravitropism: Methods and Protocols*, Methods in Molecular Biology, vol. 2368, https://doi.org/10.1007/978-1-0716-1677-2_6, © Springer Science+Business Media, LLC, part of Springer Nature 2022

molecularly defined, downstream elements ranging from altered cytoplasmic [Ca^{2+}], receptor kinase action and mitochondrial retrograde signaling, to changes in the organization of the cytoskeleton and jasmonate- and gibberellin-related events have all been linked to touch signaling (reviewed in: [1, 2, 5–9]). In addition, a host of touch-induced changes in gene expression have been reported, with estimates of up to ~8% of the Arabidopsis genome being responsive to mechanical stimulation (e.g., [10, 11]. However, these studies have also highlighted the problem of how to apply reproducible mechanical stimulation to plants in the research setting. Current approaches to this problem are divided into two major themes: (1) allow the plant to generate its own mechanical force through, e.g., growing the root into a glass or agar barrier (e.g., [4, 12–14]), or (2) directly applying touch or mechanical stimulation. This latter strategy has been applied a number of ways including: Loading the detached leaf surface with weights [15], poking with microneedles [16, 17], spraying with water [11, 18], applying air movement [19], growth on rotary shakers [20], manually brushing with a rod or paint brush [21], rubbing with fingers [21, 22], or manually bending the organs of the plant [10, 23]. These techniques are often extremely hard to apply in a reproducible way, especially over an extended period, such as the days to weeks of a plant growth cycle.

We therefore developed the Automatic Botanical Contact Device (ABCD) that applies repetitive, controlled mechanical stimulation, and automatically tracks morphological changes in the plants stimulated by this device. To facilitate implementation by multiple research groups, we also sought to make an open-source device that was relatively easy and inexpensive to build. Paul-Victor and Rowe [24] used a "chariot" carrying a polyethylene sheet that moved across the plants as way to deliver mechanical stimulation and Wang et al., [25] applied a CNC machine to draw brushes of human hair across samples. The ABCD extends such ideas and applies an automated image analysis pipeline. The system consists of: (1) a linear motion assembly to move a programmable stimulation arm back and forth over the plants, (2) an overhead camera array to continuously capture growth responses as the experiment progresses, and (3) software for automated image processing and growth analysis.

2 Materials

2.1 Construction of the ABCD

1. Frame: The ABCD is based around a frame built of standard T-slot extruded aluminum (80/20 Inc., Columbia City, IN, USA) that provides both stability (a critical factor for a mechanical stimulation experiment) and a mount for a programmable motorized stimulation bar to move back and forward over

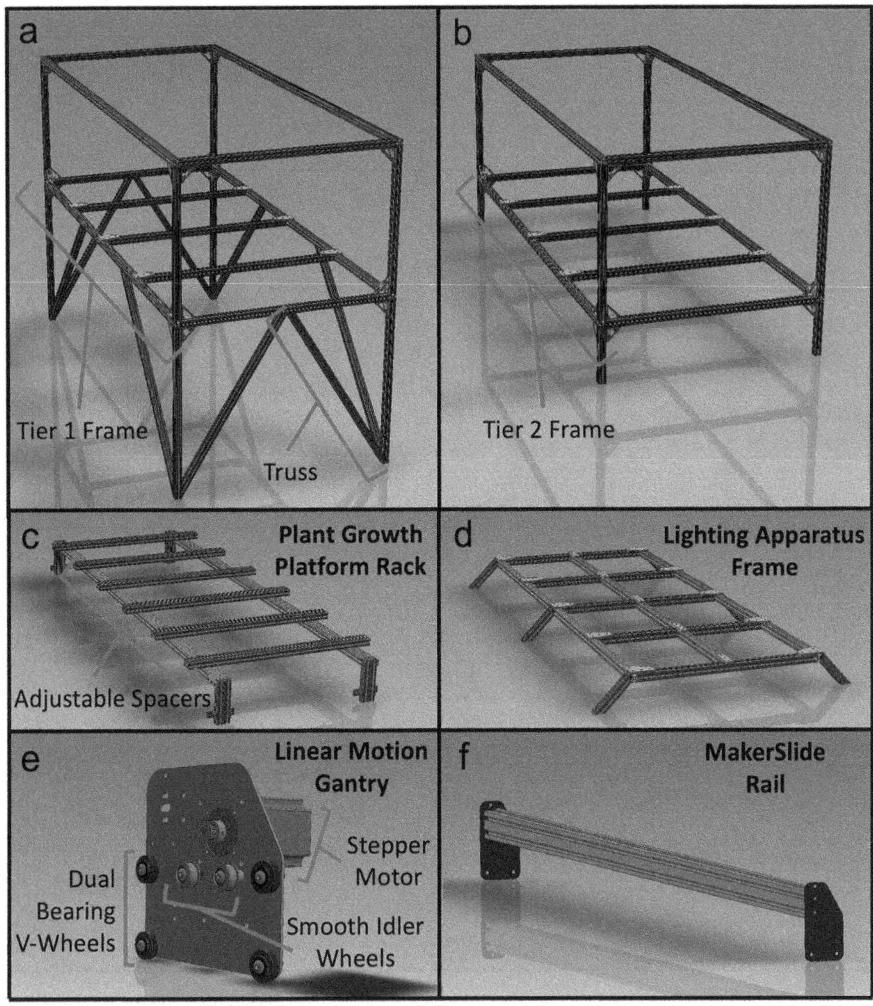

Fig. 1 Design of the ABCD frame and support structure. (**a**) Structural design of the Tier 1 frame and support trusses. (**b**) Structural design of the Tier 2 frame. The final outside dimensions of both structures are 3.05 m x 1.22 m, with a combined height of 2.44 m. (**c**) Structural design of the Plant Growth Platform Rack. Flats of plants fit between the adjustable spacers. (**d**) Structural design of the Lighting Apparatus frame. In the assembled apparatus, LED ballasts are affixed to the underside. (**e**) Linear Motion Gantry, as assembled. (**f**) MakerSlide Rail. The dual bearing V-wheels on the Linear Motion Gantry in **e** slide on the MakerSlide Rail in (**f**). Components in (**a**) to (**d**) are from 80/20 Inc. (Columbia City, IN, USA). Components of (**e**) and (**f**) are from Inventables (Chicago, IL, USA). Models rendered in Solidworks (Solidworks inc., Waltham, MA, USA)

plants (*see* **Note 1**). The frame is designed as two independent stacked experimental units, one mounted on top of the other (Figs. 1 and 2).

2. Lighting: The lighting unit is constructed of the same extruded aluminum as the frame. Five 0.61 m LED fixtures (400K, T8, Hyperikon Inc., San Diego, CA, USA), connected in series, are controlled by a commercial plug timer (Fig. 1). Affixing nylon blackout fabric (Thor labs, Newton, NJ, USA) around the

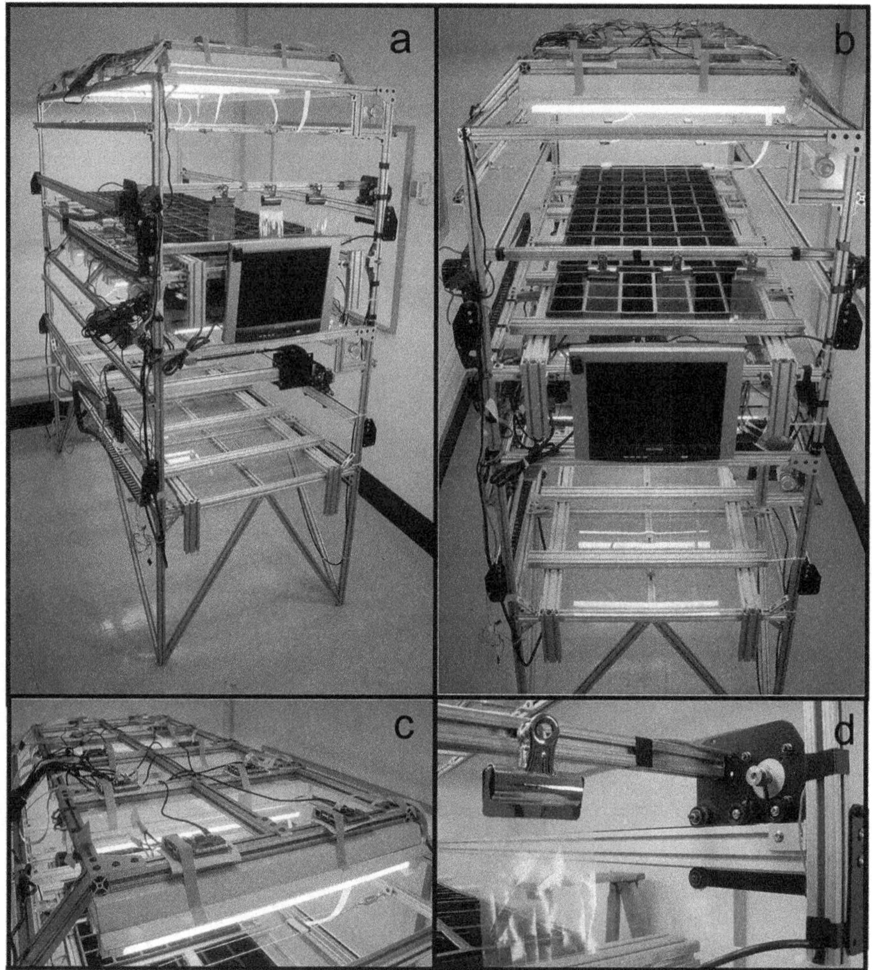

Fig. 2 Automated Botanical Contact Device, assembled. (**a**) Overview of the ABCD. An emergency stop button (top right) enables instantaneous shut down of the linear motion gantry and stimulation bar. (**b**) Front view of the ABCD. This self-contained unit can hold 10 flats (top tier shown with flats installed, bottom is empty), each with 18 inserts. The plant growth platform rack can be moved vertically to adjust light intensity. (**c**) 20 Raspberry Pi computers, each outfitted with one Pi camera, can be configured to image plants at any pre-assigned interval. (**d**) Affixed to the stimulation bar are three clips that can hold various materials for touch stimulus. The bar can be configured to apply touch stimulus to rows of plants at any pre-assigned interval

outside of the frames allows for two independent lighting zones, one for the top tier and one for the lower tier.

3. Linear Motion Assembly: The linear motion system that moves the stimulation bar is shown in Figs. 1 and 2 and consists of an off-the-shelf X-Carve linear motion gantry and MakerSlide assembly from Inventables (Chicago, IL, USA; *see* **Note 2**).
4. Touch Material: The material used to contact the plants is fixed to the Touch Bar by X-ACTO clips (Fig. 1b; Elmer's Products,

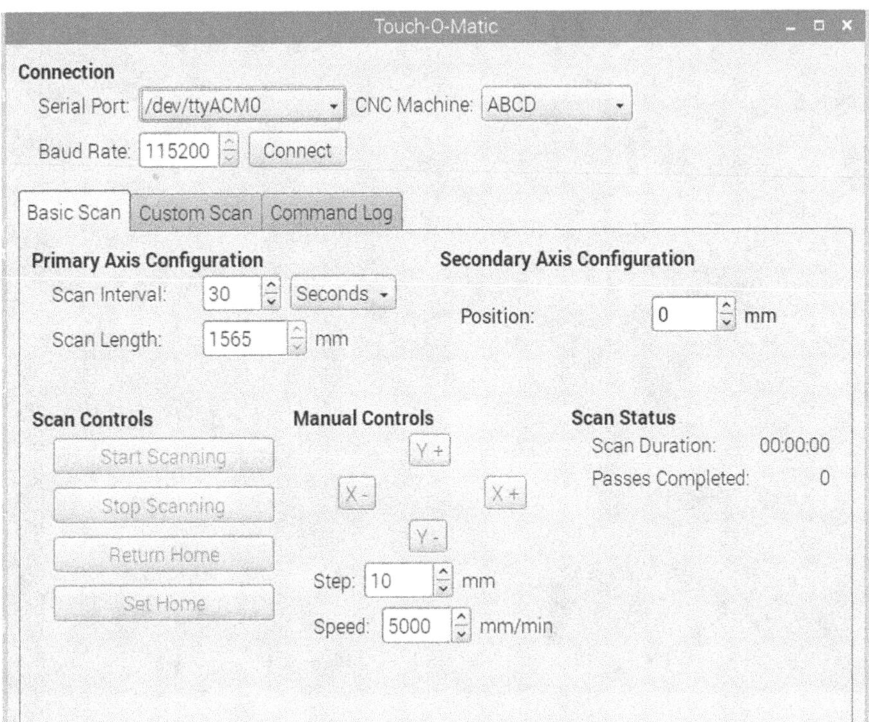

Fig. 3 ABCD software graphical user interface. Options allow parameters such as interval, speed, and extent of travel of the stimulation arm to be defined

Inc., Westerville, OH, USA) using adjustable slide-in T-nuts (80/20 Inc.; *see* **Note 3**).

5. Linear Motion Assembly Control: Timing of movement, speed and acceleration of the Linear Motion Assembly, is controlled by a combination of an Arduino Uno (Arduino, Somerville, MA, USA) and an Arduino gShield (Inventables) driving the stepper motors of the linear motion gantry. All are powered by a 24VDC 400 W enclosed power supply (Inventables).
6. Control Interface: A Raspberry Pi (Raspberry Pi Foundation, Cambridge, United Kingdom) is used to command the Arduino gShield. Master control is implemented using a Raspberry Pi running custom control software (Fig. 3; *see* **Note 4**).
7. Imaging Components: For automated image-based phenotyping, Raspberry Pi computers and Pi cameras (Fig. 4) are employed as a scalable, inexpensive option (*see* **Note 5**).
8. Data Collection and Transfer: The computers are wirelessly networked and controlled by a master Raspberry Pi via a data transfer protocol (*see* **Note 6**) running in the Python programming environment (Python Software Foundation, Wilmington, DE, USA). This interface allows users to schedule image

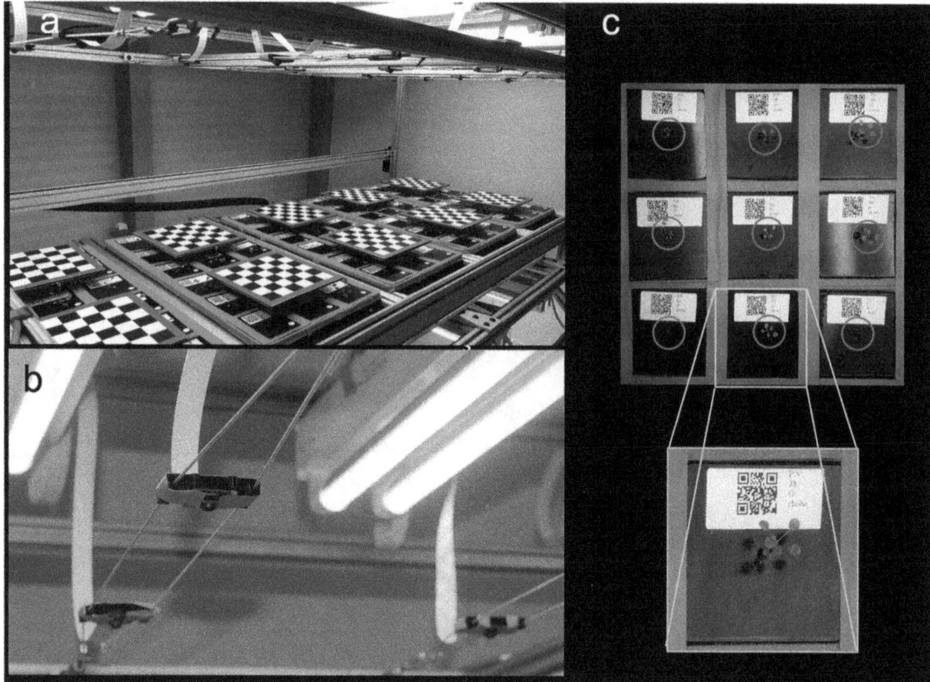

Fig. 4 ABCD camera system. (**a**) The Pi camera array is suspended over the growth trays. This image shows the ABCD at initialization of the imaging system with the flat-field checkerboard correction grids in place. (**b**) Pi cameras are suspended between vibration-isolating tensioned wire supports. (**c**) Seedlings are grown up through holes in black acrylic covers on their pots. These covers have a QR code that links to metadata such as position, plant type, and treatment

acquisition at any interval and automatically uploads to a cloud storage service, e.g., the CyVerse cyberinfrastructure site [26].

9. Data Management and Analysis: Using CyVerse's Discovery Environment for implementation, the Overhead Plant Tracker Application (OPTA) automatically tracks and analyzes plant growth from the ABCD using the PhytoMorph Image Phenomics Tool Kit (*see* **Note** 7).

2.2 Arabidopsis Media

1. Approximately 10, 9 cm round Petri plates per experimental run (using both tiers of the system).
2. Arabidopsis media: 500 mL of ½ strength Linsmaier and Skoog medium with 0.3% (w/v) sucrose, 1% (w/v) Phytagel, pH 5.7. Dissolve all components and then adjust pH using 0.1 N KOH. Make fresh and sterilize by autoclaving immediately before use.

2.3 Arabidopsis Pots, Flats, and Soil

1. Ten standard #1801 pot inserts (insert has 3×6 pots and fits a standard 1020 plant seedling flat).

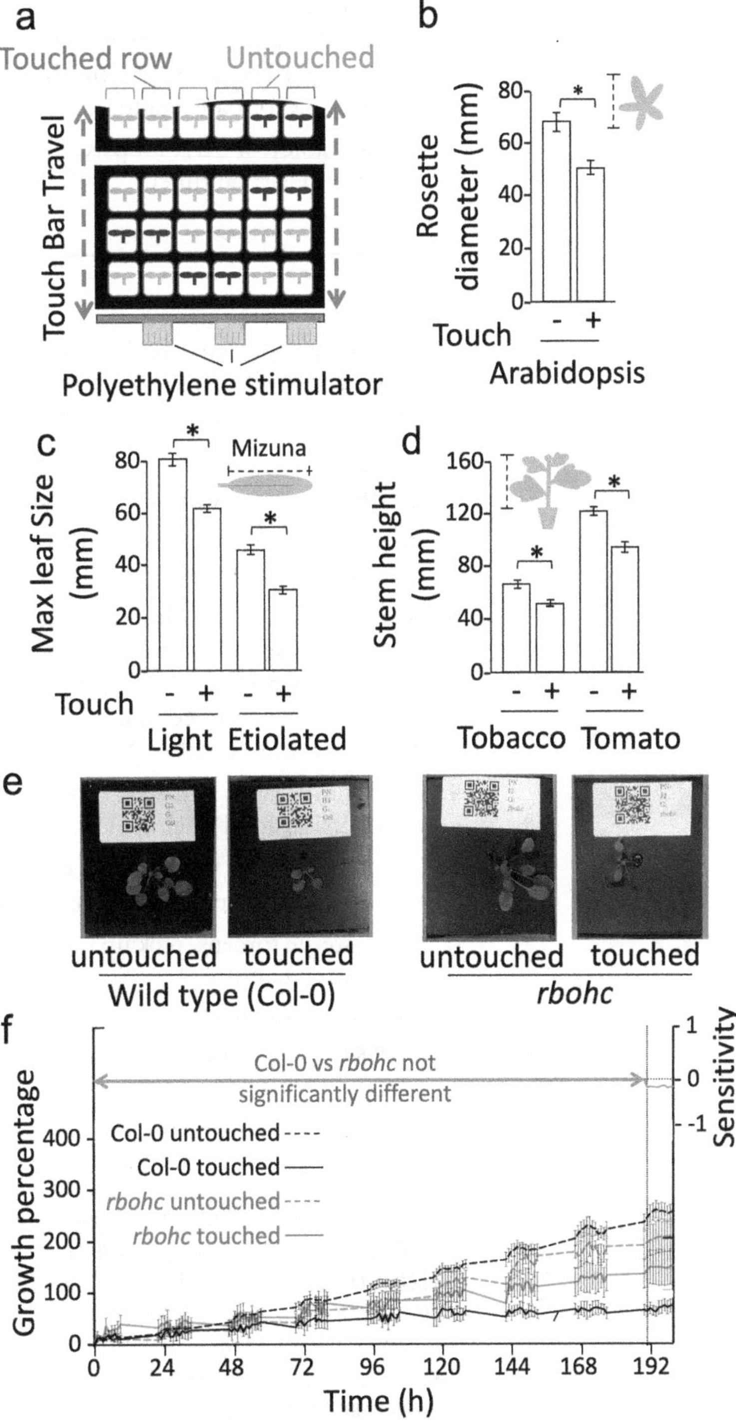

Fig. 5 Effect of ABCD on stem and leaf growth. (**a**) schematic showing the paired touched and untouched control configuration of the ABCD. (**b**) Rosette diameter of 21-day-old Arabidopsis. (**c**) Leaf length of 21-day-old mizuna seedlings (either etiolated or light grown). (**d**) Stem height in 30-day-old tobacco and tomato

2. Ten standard 1020 flats for holding the pot inserts; for Arabidopsis, the plants are maintained in flats using the 18 cell inserts described above, providing a 3×6 array of plants per flat (Fig. 5a) but any standard pot insert should be usable.
3. Soil suitable for Arabidopsis, e.g., Fafard Germination Mix (Sun Gro Horticulture).

3 Methods

The following represents the steps for setting up the ABCD for experimentation with *Arabidopsis thaliana*; however, the approach is usable with a wide variety of plant species.

1. Autoclave Arabidopsis media and allow to cool until ~70 °C and the container can be held comfortably with a gloved hand.
2. In a sterile hood, pour 15 mL of Arabidopsis media into all 10 of the sterile 9 cm Petri dishes. Allow to cool and solidify.
3. Working in a sterile hood, surface sterilize ~500 (~20 μL) Arabidopsis seeds by shaking in ~1 mL 70% (v/v) ethanol for 2 min in a 1.5 mL microfuge tube and then rinsing twice with ~1 mL sterile distilled water. Pour the seeds onto a piece of sterile (autoclaved) filter paper in a sterile Petri dish.
4. Allow filter paper and seeds to completely dry in sterile hood with Petri dish lid open.
5. Once the surface sterilized seeds are dry, close the Petri plate and wrap with Parafilm until ready to plant (*see* **Note 8**).
6. Using sterile forceps, pick up individual seeds and plant ~30 per Petri dish (*see* **Note 9**).
7. Stratify the seeds on the media in the Petri plates at 4 °C for 2 days (*see* **Note 10**).

Fig. 5 (continued) seedlings. Plants were grown for 14 days and then once established, subjected to mechanical stimulation every 5 min using the ABCD. * Significantly different from control, *t*-test, $p < 0.05$. Results are mean of $n \geq 10$. (**e**) Plant morphology in wild type and the *rbohc* mutant of Arabidopsis (*rhd2)* [16]. Representative of $n = 9$ biological replicates. (**f**) Growth percentage (rosette area/initial rosette area) extracted from images taken every 1 h during ABCD stimulation (every 5 min for 8 days). Results are mean ± s.d., $n = 9$. Sensitivity (green trace) represents the divergence of the inhibition of growth by touch at each time point of the mutant vs the wild type. For *rbohc*, the untouched growth was not significantly different to the wild type control but the degree of reduction in growth in response to ABCD stimulation was significantly lower as reflected by the negative sensitivity score after 191 h of stimulation

8. Place planted Petri plates vertically under lights (100 μmol/m^2/s), at room temperature for 7 days.
9. Fill the required number of #1801 pot inserts with soil so that each cell is completely filled. Compress the soil by ~1 cm and fill the pots to the top with more soil. It is important that all the pots are equally filled to the top. Five flats can be installed per tier of the ABCD, for a total of 90 plants per platform, or 180 plants on the two-tier system (Fig. 4; *see* **Note 11**).
10. Transplant 7-day-old seedlings, with one seedling in the center of each cell in the pot insert. Select for seedlings showing equivalent development from the Petri plates and cover each cell with a QR-coded black acrylic lid (*see* **Note 12**).
11. Using the OPTA interface, program the metadata to be associated with the QR codes present on each acrylic lid to the position and genotype of the seedling.
12. Place the flat-field calibration checkerboard on the surface of each flat of seedlings and using the OPTA, record the reference image for each camera. This image will be used to correct for camera angle and distance. The OPTA uses the colors in the images it takes to define the calibration checkerboard, the QR code and the plants (*see* **Note 13**). Remove the checkerboards.
13. Water each tray by filling with water to ~1 cm and cover with a clear seedling flat lid to allow the plants time to recover from the transplanting.
14. After 3 days, remove the clear lids from the flats of seedlings.
15. Allow seedlings to grow for a total of ~14 days at 100 μmol/m^2/s (*see* **Note 14**); providing a long-day photoperiod of 14 h light/10 h night will support bolting in Arabidopsis during the experiment, allowing assessment of the effects of touch on both vegetative growth and time to flowering (*see* **Note 15**).
16. Water the trays as needed (approximately once per week during the course of the experiment) by filling each flat with 1 cm water.
17. After 14 days, initiate touch stimulation bar movement and set imaging system to collect data (*see* **Note 16**).
18. At the end of the experiment, the aerial parts of the plant can be harvested, frozen in liquid nitrogen and used for subsequent analyses such as qPCR or RNAseq.
19. Analysis of the touch sensitivity of the plants can be made manually from downloaded images from CyVerse using, e.g., Image J [27] (*see* **Note 17**), or by using the automated measurements of rosette area produced by the OPTA (*see* **Notes 18 and 19**).

4 Notes

1. Extruded 6105-T5 aluminum (80/20 Inc.) is used to construct the ABCD frame in a two-tier design, with two identical stacked stimulation systems (Figs. 1 and 2). A platform rack with an adjustable frame can accommodate flats of different sizes located under the path of the stimulation bar (Fig. 1). The parts list, annotated images, and data on dimensions needed to construct the ABCD is available at https://github.com/Gilroy-Lab/ABCD/blob/master/ABC%20Parts%20List-2.xlsx
2. An extruded aluminum bar (touch bar; Fig. 1b) is connected between two gantry side plates and to the outside, stepper motors are wired in series to allow the bar to be moved back and forward under user control. Dual bearing V-wheels mounted to the gantry plates provide tension against the frame and smooth idler wheels reduce friction as the drivetrain rolls along the MakerSlide rails. The stepper motors are guided by GT2 belting connected at both ends of the MakerSlide rails. Outside of the gantry, a drag chain protects power and communication cables from interfering with the drivetrain mechanisms.
3. Although multiple materials can be used for the contact material, we have found that clear polyethylene plastic sheets (Fig. 1b) do not readily foul with soil, plant materials, or water during prolonged stimulation. Encrusting with foreign matter is a problem regularly seen when using other materials such as fabrics or other natural products for the stimulation sheeting.
4. A user interface based on the PyQt5-based graphical tool generates G-Code sequences for the Arduino to run the motors of the stimulation bar. The PyQt5 communicates with the Arduino via pySerial. Source code is available at https://github.com/EnSpec/Plant_CNC_Controller.git
5. Ten Pi computers and Pi cameras are installed on each tier. Two cameras provide an imaging area entirely covering one of the five flats per tier and so the ten cameras together image all of the experimental area. The cameras are suspended from tensioned wires that run across the width of the lighting hood (Fig. 4b), providing insulation against any minor vibrations that might distort the images during the acquisition process.
6. Code for automated data transfer is available at https://github.com/CoSE-Jerry/ABCD.
7. The OPTA is initialized with a checkerboard image to provide automatic calibration of magnification for each of the cameras

(Fig. 4a, c). A Quick Response barcode (QR code) on each plant station encodes plant and experiment information. The OPTA is freely accessible at CyVerse in the phytoMorph Image Phenomics Tool Kit module "Overhead Plant Tracker."

8. When stored dry at 4 °C, Arabidopsis seeds that have been surface sterilized in this way have good germination for at least 1 month.
9. Starting the seedlings on Petri plates allows for pre-selection prior to transplantation to soil, providing a uniform initial starting population of plants for the experimental materials.
10. Cold treatment synchronizes germination in Arabidopsis.
11. This arrangement allows for three parallel sets of paired samples with a touch-stimulated plant and its immediately adjacent untreated controls (Fig. 5a).
12. The central holes in each pot cover allows the seedling to grow up through the opening. This means that once grown, the plant's leaves can be visualized against a high contrast uniform background of the acrylic cover (75 mm L × 75 mm W × 4.5 mm H; Fig. 4d, e). This approach removes the color complexities of the soil surface from the images and provides a high contrast image of the leaf system of the plants, facilitating the automated extraction of plant leaf data by the OPTA.
13. The image scene to be analyzed with the OPTA was designed with five colors encoding different parameters of the setup, and so enabling a straightforward machine vision solution. Red (tape) is used as a boundary to segment the pot regions, blue is used as the identifying color for the periphery of the calibration checkerboard (Fig. 4a, c), white is the QR code and black, the background. Green pixels in each individual pot are identified by clustering via a Gaussian Mixture Model within the OPTA. Plant digital biomass is reported as green pixels in each pot after removal of background (Fig. 4d, e).
14. Seedlings need to be well-established prior to starting the touch stimulation to ensure they are not dislodged from the soil by the travel of the stimulation sheet.
15. Experiments can be run for several weeks providing the plants are kept well-watered and that they do not grow taller than the height of the stimulation bar.
16. For Arabidopsis, with the bar moving once every 5 min, a significant reduction in vegetative growth is detectable within ~2 days. Analysis of other plants, or when eliciting different responses from Arabidopsis, such as investigating touch-induced alteration in flowering, will require experimentation for the optimal frequency of stimulation.

17. For manual quantitative growth analysis, images are downloaded from CyVerse and features such as rosette diameter extracted using ImageJ's [27] built in measurement features. Examples of manual data extraction from images for a range of plants are shown in Fig. 5b–d. Note, all show a reduction in vegetative growth in response to stimulation by the ABCD.
18. For automated analysis using the OPTA, quantitative data on rosette size is automatically extracted by OPTA analysis in CyVerse is exported in a file accessible to, e.g., Microsoft Excel. Multiple *t*-tests are then performed to compare the mean percentage growth (i.e., the average increase in growth when compared to the initial size of the plant) at each time point in the untouched and touched samples. For comparisons between, e.g., cultivars or wild type and mutant lines, a relative measure of touch sensitivity is required to normalize for potential differences in absolute growth rates between the plants being compared. This normalization can be accomplished by applying a mechano-sensitivity index. This sensitivity index monitors the deviation of response from the mean of the all untouched samples relative to all the touched samples. For example, when comparing *Arabidopsis* wild type to a mutant, for those time points where the difference between the means of untouched and/or touched wild type and mutant plant growth percentage were significant, the difference between mean growth percentage of untouched Col-0 (U_{WT}) and mean untouched mutant (U_{mutant}) and the difference between mean touched Col-0 (T_{WT}) and mean touched mutant, (T_{mutant}) at each time point is calculated. The expression (U_{WT}-U_{mutant} + T_{WT}-T_{mutant})/2 then calculates a measure of the divergence of mean growth of the mutants from wild type for both untouched and touched plants, providing a measure of the "sensitivity" to touch that is normalized to the growth rates of the plants. Negative "sensitivity" values indicate mutants that are less sensitive to this mechanostimulation than wild type. Positive "sensitivity" values indicate mutants that are more sensitive than wild type.
19. The results shown in Fig. 5e, f are the output of the automated imaging system and a sensitivity analysis for the wild type Arabidopsis Col-0 ecotype and a mutant in the respiratory burst oxidase homolog isoform C (*RBOHC*, also in the Col-0 ecotype). RBOHC is an NADPH oxidase, an enzyme that produces reactive oxygen species, and that has been linked to mechanical responses in roots (e.g., [16]). The ABCD analysis reveals that the *rbohc* mutant also shows reduced touch sensitivity in its rosette growth, but this is only detectable when touch stimulation is regularly applied over many days.

Acknowledgments

This research was supported by grants from the National Science Foundation (IOS1557899) and NASA (NNX13AM50G, NNX17AD52G, and 80NSSC19K0132).

References

1. Chehab EW, Eich E, Braam J (2009) Thigmomorphogenesis: a complex plant response to mechano-stimulation. J Exp Bot 60:43–56
2. Toyota M, Gilroy S (2013) Gravitropism and mechanical signaling in plants. Am J Bot 100:111–125
3. Braam J (2005) In touch: plant responses to mechanical stimuli. New Phytol 165:373–389
4. Massa GD, Gilroy S (2003) Touch modulates gravity sensing to regulate the growth of primary roots of *Arabidopsis thaliana*. Plant J 33:435–445
5. Xu Y, Berkowitz O, Narsai R et al (2019) Mitochondrial function modulates touch signalling in *Arabidopsis thaliana*. Plant J 97:623–645
6. Monshausen GB, Haswell ES (2013) A force of nature: molecular mechanisms of mechanoperception in plants. J Exp Bot 64:4663–4680
7. Lange MJP, Lange T (2015) Touch-induced changes in Arabidopsis morphology dependent on gibberellin breakdown. Nat Plants 1:14025
8. Hamant O, Haswell ES (2017) Life behind the wall: sensing mechanical cues in plants. BMC Biol 15:59
9. Shih HW, Miller ND, Dai C et al (2014) The receptor-like kinase FERONIA is required for mechanical signal transduction in Arabidopsis seedlings. Curr Biol 24:1887–1892
10. Lee D, Polisensky DH, Braam J (2005) Genome-wide identification of touch- and darkness-regulated Arabidopsis genes: a focus on calmodulin-like and XTH genes. New Phytol 165:429–444
11. van Moerkercke A, Duncan O, Zander M et al (2019) A MYC2/MYC3/MYC4-dependent transcription factor network regulates water spray-responsive gene expression and jasmonate levels. Proc Natl Acad Sci U S A 116:23345–23356
12. Nakagawa Y, Katagiri T, Shinozaki K et al (2007) Arabidopsis plasma membrane protein crucial for Ca^{2+} influx and touch sensing in roots. Proc Natl Acad Sci U S A 104:3639–3644
13. Wang Y, Wang B, Gilroy S et al (2011) CML24 is involved in root mechanoresponses and cortical microtubule orientation in Arabidopsis. J Plant Growth Regul 30:467–479
14. Zha G, Wang B, Liu J et al (2016) Mechanical touch responses of Arabidopsis *TCH1-3* mutant roots on inclined hard-agar surface. Int Agrophysics 30:105–111
15. Jacques E, Verbelen JP, Vissenberg K (2013) Mechanical stress in Arabidopsis leaves orients microtubules in a "continuous" supracellular pattern. BMC Plant Biol 13:163
16. Monshausen GB, Bibikova TN, Weisenseel MH et al (2009) Ca^{2+} regulates reactive oxygen species production and pH during mechanosensing in Arabidopsis roots. Plant Cell 21:2341–2356
17. Gus-Mayer S, Naton B, Hahlbrock K et al (1998) Local mechanical stimulation induces components of the pathogen defense response in parsley. Proc Natl Acad Sci U S A 95:8398–8403
18. Braam J, Davis RW (1990) Rain-, wind-, and touch-induced expression of calmodulin and calmodulin-related genes in Arabidopsis. Cell 60:357–364
19. Knight MR, Smith SM, Trewavas AJ (2006) Wind-induced plant motion immediately increases cytosolic calcium. Proc Natl Acad Sci U S A 89:4967–4971
20. Der Loughian C, Tadrist L, Allain J-M et al (2014) Measuring local and global vibration modes in model plants. Comptes Rendus Mécanique 342:1–7
21. Jensen GS, Fal K, Hamant O et al (2017) The RNA polymerase-associated factor 1 complex is required for plant touch responses. J Exp Bot 68:499–511
22. Benikhlef L, L'Haridon F, Abou-Mansour E et al (2013) Perception of soft mechanical stress in Arabidopsis leaves activates disease resistance. BMC Plant Biol 13:133
23. Richter GL, Monshausen GB, Krol A et al (2009) Mechanical stimuli modulate lateral root organogenesis. Plant Physiol 151:1855–1866
24. Paul-Victor C, Rowe N (2011) Effect of mechanical perturbation on the biomechanics, primary growth and secondary tissue

development of inflorescence stems of *Arabidopsis thaliana*. Ann Bot 107:209–218

25. Wang K, Law K, Leung M et al (2019) A labor-saving and repeatable touch-force signaling mutant screen protocol for the study of thigmomorphogenesis of a model plant Arabidopsis thaliana. J Vis Exp 150:e59392
26. Merchant N, Lyons E, Goff S et al (2016) The iPlant collaborative: cyberinfrastructure for enabling data to discovery for the life sciences. PLoS Biol 14:e1002342
27. Schindelin J, Rueden CT, Hiner MC et al (2015) The ImageJ ecosystem: an open platform for biomedical image analysis. Mol Reprod Dev 82:518–529

Chapter 7

High-Resolution Kinematic Analysis of Root Gravitropic Bending Using RootPlot

Aditi Bhat, Cody L. DePew, and Gabriele B. Monshausen

Abstract

Root gravitropic bending is a complex growth process resulting from differential expansion of cells on the upper and lower sides of a gravistimulated root. In order to genetically dissect the molecular machinery underlying root bending, a thorough understanding of the kinetics and spatial distribution of the growth process is required. We have developed an experimental workflow that enables us to image growing roots at high spatiotemporal resolution and then convert XY-coordinates of root cellular markers into 3D representations of root growth profiles. Here, we present a detailed description of the setup for monitoring vertically oriented roots before and after gravistimulation. We also introduce our newly developed custom R-based program RootPlot, which calculates root velocity profiles from root XY-coordinate data obtained using a previously published image processing software. The raw velocity and derived relative elemental growth rate (REGR) curves are then fitted via LOWESS regression for assumption-free data analysis. The resulting smoothed growth profiles are plotted as heatmaps to visualize how different regions of the root contribute to the growth response over time. Additionally, RootPlot provides analysis of overall growth and bending rates based on root XY-coordinates.

Key words Gravitropism, Kinematic analysis of growth, Relative elemental growth rate, Root growth

1 Introduction

Plants respond to many environmental and endogenous cues by altering their rate of growth. How these growth rate changes are distributed along the zone of cell expansion determines the shape of the growth response of the plant organ. Symmetrical changes along organ flanks can lead to rapid acceleration or retardation of organ elongation (e.g., [1, 2]) while asymmetrical changes across opposing flanks produce tropic (e.g., [3–9]), nastic (e.g., [10–12]) or nutational organ bending [13, 14]. Even seemingly simple

Aditi Bhat and Cody L. DePew contributed equally to this work.

Elison B. Blancaflor (ed.), *Plant Gravitropism: Methods and Protocols*, Methods in Molecular Biology, vol. 2368, https://doi.org/10.1007/978-1-0716-1677-2_7,

growth responses may be realized by growth rate changes that are heterogeneous at the tissue level when neighboring cells exhibit differential sensitivity or acclimation to a given cue (e.g., [15, 16]). In order to advance exploration of the mechanisms underlying growth responses, it is therefore helpful to first develop a detailed understanding of the spatial and temporal characteristics of a growth process.

Approaches to study root growth profiles at high spatial resolution were introduced in the 1940s and 1950s ([17] and references therein) when images of roots growing in front of vertically oriented microscopes were captured using film-based photography, and photographic prints were manually analyzed by tracing the displacement of epidermal marker structures relative to the root apex. While such studies yielded important insights into the growth responses of roots to the environment, they were prohibitively labor-intensive. Today, thanks to recent advances in automated digital image acquisition, -processing and -analysis, growth patterns can be monitored with comparatively minimal effort, allowing investigation of growth dynamics at high spatial as well as temporal resolution [18–22]. In investigations of root gravitropism, for example, computer-vision assisted analyses have helped to resolve subtle alterations in root bending responses of *Arabidopsis* mutants with impaired auxin signal transduction. Unlike *Arabidopsis* wild-type roots, where gravitropic bending is initiated in the apical elongation zone and then propagates into the central elongation zone, auxin-transport defective *pin3* mutants develop only a single, apical curvature zone after gravistimulation; *pin3* roots thus initially bend at the same rate but then slow down relative to wild-type roots [19]. *Arabidopsis cngc14* mutants, on the other hand, are impaired in transducing auxin accumulation into cytosolic Ca^{2+} signals in the root [23, 24]. These mutants exhibit a brief (~7 min), but distinct delay in the onset of gravitropic bending, confirming a key role for Ca^{2+} in the earliest phase of the growth response [23]. Investigating organ bending using a combination of complementary high-resolution approaches thus can provide information about where in the signal transduction cascade candidate proteins are positioned and what specific effects they have on the downstream growth response.

Here, we present a detailed description of the setup required to image roots exposed to a gravitropic stimulus; we then provide a step-by-step protocol for computation of root relative elemental growth rate (REGR) patterns and bending rates using the newly developed R-based program RootPlot_v1.

2 Materials

1. Agar for germination of *Arabidopsis thaliana* Col-0 seedlings (granulated agar).
2. Murashige and Skoog medium (basal salts) (MS medium).
3. Sucrose.
4. Polystyrene sterile petri dishes.
5. Parafilm M.
6. Pipettor and sterilized 20–1000 μL pipette tips.
7. 50% bleach (v/v) in dH_2O.
8. Sterilized dH_2O.
9. Eppendorf microcentrifuge tubes (e.g., 1.5 mL tubes).
10. 40 × 24 × 0.13 ~ 0.17 mm coverslips.
11. Custom-made acrylic glass (Perspex) experimental chambers.
12. Paraffin (Gulf Wax Household Paraffin Wax), camel hair paint brush #6, hotplate, glass petri dish.
13. Glass knife strip (histology grade; 6 mm × 25.4 mm × 203 mm).
14. Magenta boxes GA-7.
15. Fine curved forceps (e.g., Technik Tweezers 5B-SA).
16. Razor blades.
17. Compound microscope with camera mount for vertical stage microscopy (e.g., ZEISS Axioplan positioned on its back).
18. High-resolution camera (e.g., Stingray model F504B; Allied Vision Technologies) and computer.

3 Methods

3.1 *Arabidopsis* Seed Germination

1. To prepare the nutrient agar medium, autoclave ¼ strength (1.1 g/L) MS medium supplemented with 1% (w/v) sucrose and 1% (w/v) agar, pH 5.8. Once the nutrient medium has cooled down to 50–55 °C, pour into sterile petri dishes under a laminar flow hood to ensure sterile conditions. Allow the agar to solidify completely before closing the petri dishes to avoid excessive accumulation of condensation (*see* **Note 1**).
2. To surface sterilize the seeds, completely suspend approximately 10 μL of seeds in ~1 mL of 50% bleach in a 1.5 mL Eppendorf microcentrifuge tube, and constantly agitate the tube for 10 min. Remove the bleach solution using a sterile pipette and rinse the seeds thrice with sterilized dH_2O.

3. Allow the seeds to settle to the bottom of the tube by leaving the tube on the laminar flow hood bench for a few minutes (or by spinning in a mini centrifuge), and remove all but ~100 μL dH_2O from the tube (some liquid is required for transfer of the seeds to agar plates). Individually transfer the seeds to the agar plate by pipetting approximately 2–3 μL of liquid plus seeds at a time and placing each seed separately on the agar surface. Once all the excess transferred liquid has dried off around the seeds, close the lid of the petri dish and seal it using parafilm. Stratify the seeds by placing the agar plate in a refrigerator/cold room at 4 °C for 2 days; this promotes uniform germination (*see* **Note 2**).
4. Once the seeds have been stratified, position the agar plate in vertical orientation under fluorescent or LED grow lights for continuous illumination at 22 ± 1 °C.

3.2 Transfer of Arabidopsis *seedlings into Experimental Chambers*

1. To prepare the experimental chambers, place a 40 × 24 mm coverslip into the recessed acrylic glass chamber frame and seal the coverslip in place using hot, melted paraffin wax. Melt the paraffin wax in a glass petri dish placed on a hotplate preheated to ~90–110 °C. Use a paint brush to spread the melted paraffin onto the edges of the coverslip to secure it in place (leakproof) (Fig. 1a, b).
2. Prepare the experimental nutrient agar medium by autoclaving ¼ strength MS medium supplemented with 1% (w/v) sucrose and 1% (w/v) agar, titrated to pH 5.8 (*see* **Note 3**). Pipette 700 μL of warm nutrient agar medium onto a clean glass knife strip to make rectangular agar "cushions" (Fig. 1c; *see* **Note 4**). Upon cooling, trim the sides of the agar "cushion" to fit the interiors of the experimental chamber (Fig. 1d).
3. Pipette ~10 μL droplet of ¼ MS medium with 1% sucrose, pH 5.8 in the center of the experimental chamber. Take a 4.5-day-old *Arabidopsis* seedling from the germination agar plate by hooking curved forceps under cotyledons. Gently lift the seedling up from the plate and quickly place it into droplet in the center of the experimental chamber (*see* **Note 5**).
4. Using a razor blade, carefully slide the trimmed agar cushion on top of the seedling to cover the hypocotyl and root, while ensuring that cotyledons remain exposed to air (Fig. 1e). It is important that the root is positioned as straight as possible; if additional straightening is required, hook the forceps under the cotyledons and gently pull the seedling up to help straighten it.
5. To prevent sliding of the roots, seal the agar cushion in place by pipetting warm (~40 °C) nutrient agar medium around the cushion along the edges of the experimental chamber. Observe the roots under a stereo microscope and discard those which show signs of heat damage.

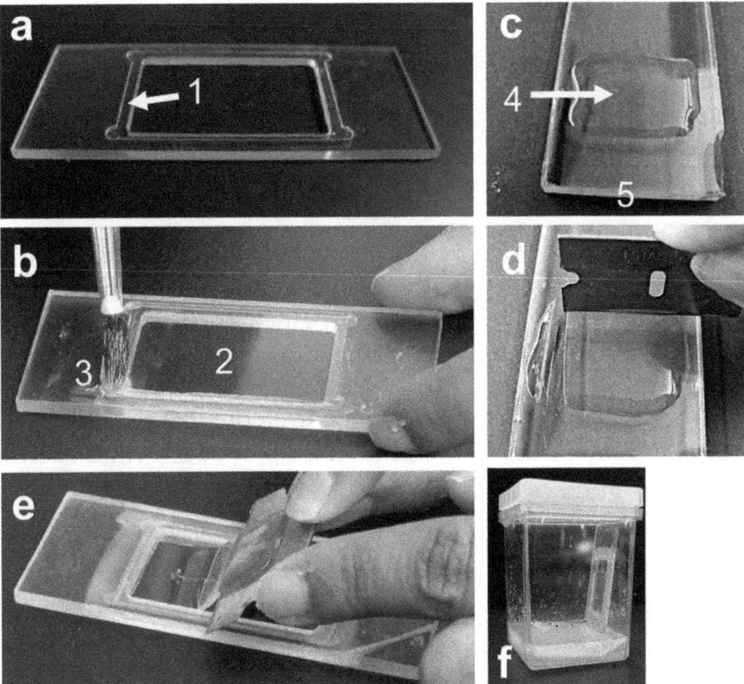

Fig. 1 Preparing experimental chambers. (**a**) Acrylic (Perspex) chamber frames for gravitropism experiments. Note that frames are recessed (1) for lipless mounting of coverglass. (**b**) A glass coverslip (2) is placed into the recess of the chamber frame and sealed in place by "painting" along the edges with hot melted paraffin (3) using a natural hair paintbrush. (**c**) Rectangular agar "cushions" (4) are formed by pipetting warm nutrient agar medium onto a clean, thick glass bar (5). (**d**) After solidifying, the cushion is trimmed using a razor blade to fit the interior of the chamber. (**e**) A seedling is quickly placed into a drop of nutrient medium in the center of the coverglass bottom of the chamber and then the trimmed agar "cushion" is gently slid on top of the seedling root, keeping the upper hypocotyl and cotyledons exposed to air. (**f**) Experimental chambers are transferred to Magenta boxes to let seedlings recover overnight from the stress of transfer. A shallow pool of tapwater in the box serves to maintain a high humidity environment and a square of paper towel renders the bottom of the box slip-proof. Chambers are leaned against a Magenta box wall at a ~10° angle to ensure that the seedling root grows along the cover glass surface

6. Fill a magenta box with ~20 mL of tap water to create a humid environment. Transfer the experimental chambers to the magenta box and let the seedlings grow overnight for recovery from the stress of transfer and to acclimate to the new growth conditions. Lean chambers against walls of Magenta boxes at a ~10° angle to ensure downward root growth with minimal root waving along the surface of the cover glass bottom of the experimental chambers (Fig. 1f).

3.3 Vertical Stage Microscopy of Root Growth and Gravitropism

1. Place a compound microscope on its back to convert it into a makeshift vertical stage microscope. Ensure that the microscope stage remains in position during an experiment and that its weight does not cause it to slide downward: tighten translational control knobs or place jack screw underneath stage (Fig. 2; if available, use circular rotating stage) (*see* **Note 6**).
2. Before starting the experiment, observe the roots under a stereo microscope to confirm that the seedlings are developing normally. Pay particular attention to the distance of the root hair initiation site from the root tip. In 4- to 5-day-old wild-type seedlings, root hair initiation should become detectable at >600 μm from the root tip (Fig. 3a); if root hairs initiate closer to the root tip, root growth is likely impaired and the seedling should be discarded.
3. After selecting a well-growing seedling, close the experimental chamber by sealing another 40 × 24 mm cover glass to the acrylic glass frame with melted paraffin wax. Leave an opening gap of about 1 mm at the top of the chamber to aid gas exchange during longer term growth experiments. Transfer the experimental chamber to the room where experiments will be performed, and let the seedling acclimate for at least 1 h.
4. Place the experimental chamber onto the microscope stage in vertical orientation to monitor growth prior to gravistimulation (Fig. 2b). Position the root in the field of view of the 10× objective and focus on the edges of the root (Fig. 3a).
5. After monitoring vertical growth for 15–30 min, rotate the microscope stage to gravistimulate the seedling (Fig. 2c; *see* **Note 7**). All subsequent steps have to be completed quickly (ideally in less than 2 min) to track the earliest phase of gravitropic response. Position the root in the field of view of the 10× objective and focus on the edges of root. Ensure that there is sufficient space for the root to grow downward without leaving the field of view over the course of the experiment.

3.4 Tracking Root Growth

Processing of imaging data may be performed by any program of choice. The key prerequisite for subsequent determination of growth rates, growth patterns, and root angles using RootPlot_vl is that the output of image processing is saved as a comma-separated values (.csv) file listing the XY-coordinates of all tracked points for all frames of the image series [pairs of columns:= XY-coordinates of all points starting with reference point; rows:= frames (or time points)]. We perform high-throughput image processing using Image Processing Toolkit v10. This program tracks user-defined points along a root in every image of a series and saves the data as XY-coordinates to a .csv file (Figs. 3a, b and 4a). The

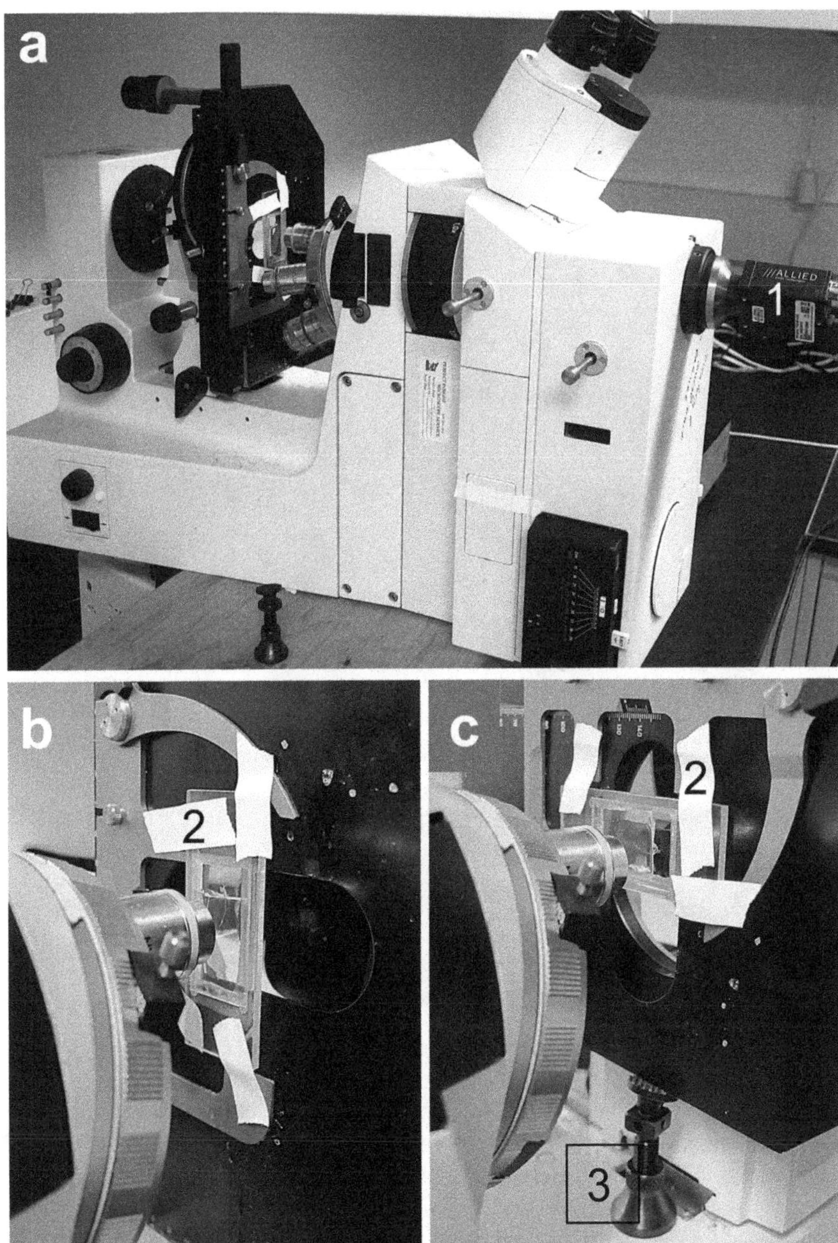

Fig. 2 Experimental setup. (**a**) Microscope turned on its back for vertical stage microscopy. High-resolution digital camera (1) mounted via photoport. (**b**, **c**) Adhesive tape (2) secures the chamber to the microscope stage. Unless the microscope is equipped with a circular, freely rotating stage, mounting a chamber for gravitropism experiments requires some improvization. The stage on the Zeiss Axioplan microscope can be turned by $>90°$, but may need to be mechanically supported (e.g., by a jack screw (3)) in the upright (**b**) or horizontal orientation (**c**)

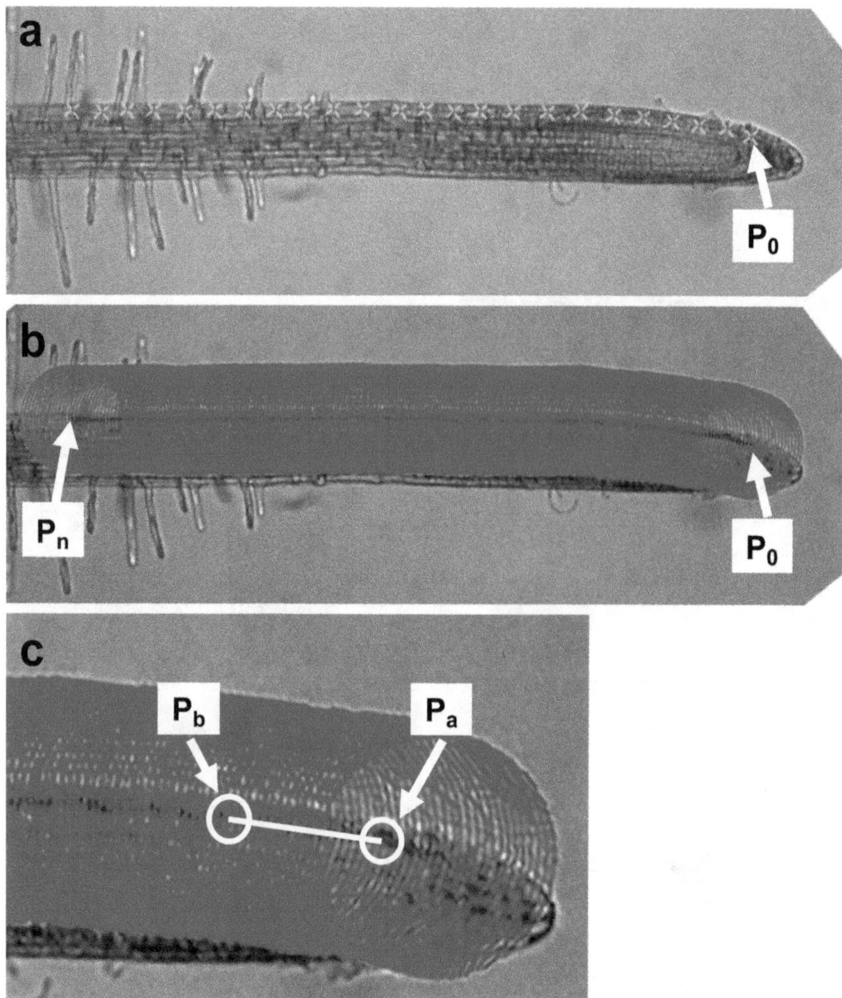

Fig. 3 Image acquisition and analysis. (**a**) During image acquisition, the root should be positioned such that the *entire* growing region of the root is included in the camera field of view and the entire length of the imaged root is in focus. Precise focus is critical if the goal is to analyze differential growth on the upper and lower sides of a gravistimulated root. Note the highly "textured" appearance of the root, which facilitates tracking of cellular features. Image Processing Toolkit v10 allows the user to mark points (*blue-purple asterisks*) along the root length that will define the position and boundaries of a "midline" (or "edgeline"). We typically place the first marker (reference point P_0) close to the first tier of the columella and the last marker well within the root hair zone to ensure that the midline/edgeline spans the entire growing region. (**b**) Image Processing Toolkit v10 then connects these markers into a virtual midline/edgeline and positions points P_i along this line at user-defined distances. The points are subsequently tracked across the entire image series and their XY-coordinates are saved to a comma-separated values file (xy_coordinates.csv). *Red dots*, program-positioned tracked points; *blue circles*, tracking disks. (**c**) Points P_a and P_b (here corresponding to P_{11} and P_{25}, respectively) selected for angle calculation

program was developed by and is available upon request from Drs. Nathan Miller and Edgar Spalding (Department of Botany, University of Wisconsin, Madison) (see also [21]).

3.5 Kinematic Analysis of Root Growth Using RootPlot_v1

RootPlot is an R script designed for robust and assumption-free data analysis and visualization of high spatiotemporal resolution root growth data; this growth data must be provided as a .csv file listing the XY-coordinates of all tracked points as described above. Instructions for running RootPlot_v1 are provided in the Supplemental File. Along with the software package, instructions can also be downloaded as RootPlot_v1 README.txt file from (https://github.com/depewcod/rootplot).

RootPlot_v1 produces the following output files ("Prefix" refers to the component of the output file name specified by the user in "user-defined-parameters.csv" in the RootPlot_v1 home folder on Github; see also Supplemental File):

3.5.1 Root Growth Rate

Root growth rates are calculated as changes in root "midline" (or "edgeline" for analysis of differential growth on opposite root flanks) length per frame. For midline-based root growth rates to correctly reflect total growth rate, tracked points must encompass the *entire* root elongation zone (Fig. 3a, b).

1. *Prefix-midlines.csv*: Length of root midline in successive frames (Fig. 4b). Each column corresponds to the midline in a particular frame (time point); each row lists the cumulative length of the root midline (in pixels of the original image) with increasing distance from the reference point P_0; the midline length is calculated as the sum of distances between neighboring tracked points P_i by applying the Pythagorean equation to their XY-coordinates (Fig. 4a). The last cell in each column thus reflects the entire length L_n of the measured midline in a frame.

$$L_n = \sum_{i=1}^{n} \sqrt{\left(P_{i,X} - P_{i-1,X}\right)^2 + \left(P_{i,Y} - P_{i-1,Y}\right)^2}$$

2. *Prefix-midline growth rate.csv*: Root growth rates [pixels frame^{-1}] (Fig. 4c) calculated as differences in total midline length between frames ($L_{n;\ t\ +\ 1} - L_{n;\ t}$) taken from *Prefix-midlines.csv*.
3. *Prefix-midline growth rate.png:* Graphs of root growth rates [pixels frame^{-1}].

3.5.2 Root Angle

Root angles are calculated as angles of a straight line connecting two user-defined points P_a and P_b (specified in "user-defined-parameters.csv"). These points are selected from the list of tracked points making up the root midline/edgeline (Fig. 3c).

1. *Prefix-angles.csv*: Angles and changes in angle (bending rate) of line connecting two user-defined points $P_{a(X,Y)}$ and $P_{b(X,Y)}$ along the root midline [given as degrees and Δdegrees frame^{-1}, respectively]. Angle α is calculated for each frame t using the arctangent function:

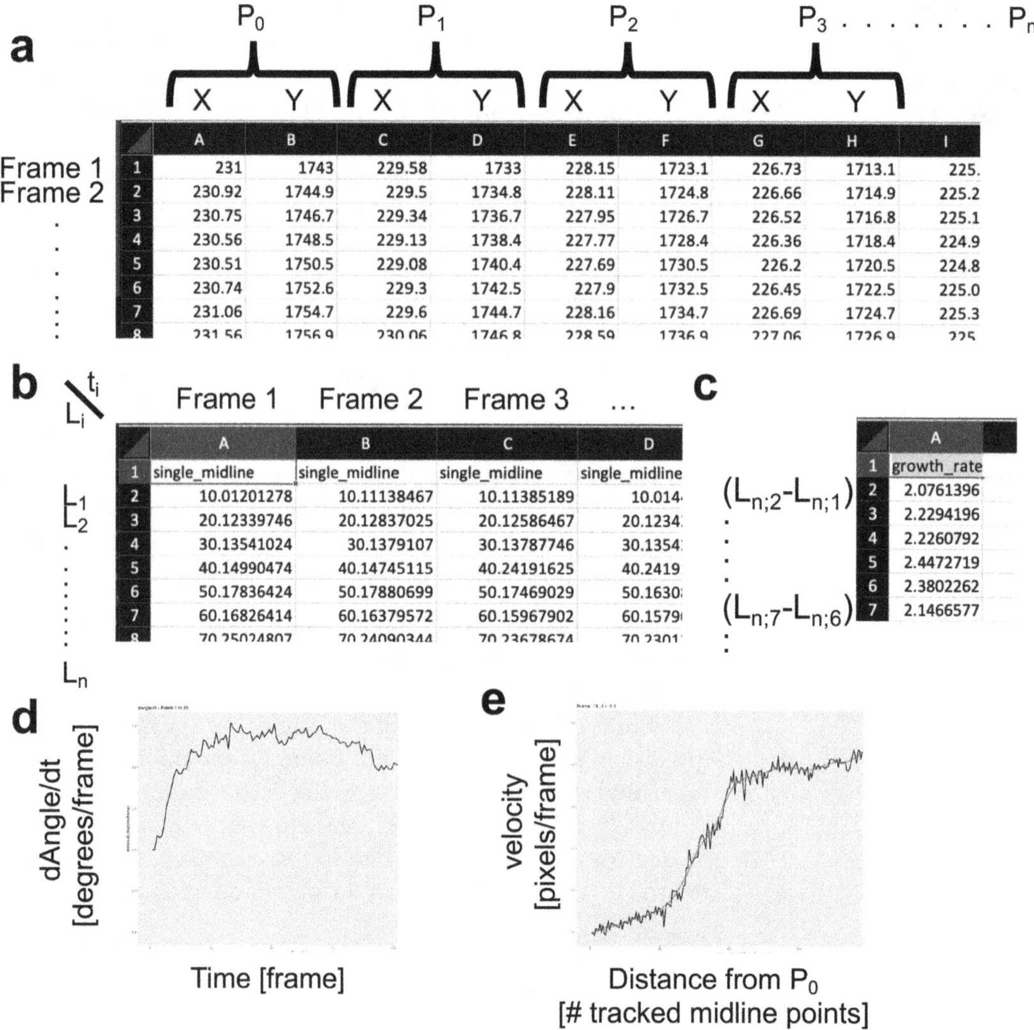

	A	B	C	D	E	F	G	H	I
1	231	1743	229.58	1733	228.15	1723.1	226.73	1713.1	225.
2	230.92	1744.9	229.5	1734.8	228.11	1724.8	226.66	1714.9	225.2
3	230.75	1746.7	229.34	1736.7	227.95	1726.7	226.52	1716.8	225.1
4	230.56	1748.5	229.13	1738.4	227.77	1728.4	226.36	1718.4	224.9
5	230.51	1750.5	229.08	1740.4	227.69	1730.5	226.2	1720.5	224.8
6	230.74	1752.6	229.3	1742.5	227.9	1732.5	226.45	1722.5	225.0
7	231.06	1754.7	229.6	1744.7	228.16	1734.7	226.69	1724.7	225.3
8	231.56	1756.9	230.06	1746.8	228.59	1736.9	227.06	1726.9	225

	A	B	C	D
1	single_midline	single_midline	single_midline	single_midline
2	10.01201278	10.11138467	10.11385189	10.014
3	20.12339746	20.12837025	20.12586467	20.1234
4	30.13541024	30.1379107	30.13787746	30.1354
5	40.14990474	40.14745115	40.24191625	40.2419
6	50.17836424	50.17880699	50.17469029	50.1630
7	60.16826414	60.16379572	60.15967902	60.1579
8	70.25024807	70.24090344	70.23678674	70.2301

	A
1	growth_rate
2	2.0761396
3	2.2294196
4	2.2260792
5	2.4472719
6	2.3802262
7	2.1466577

Fig. 4 XY-coordinate input and growth analyses output files. (**a**) Example of Image Processing Toolkit v10 root tracking data set saved to xy_coordinates.csv. (**b**) -midline.csv file lists the length (in pixels) of the root "midline/edgeline" connecting tracked points in each frame and (**c**) -midline growth rate.csv lists changes in midline/edgeline length between frames (i.e., growth rate [in pixels per frame]). (**d**) Bending rate of gravistimulated wild-type root calculated as change in angle of line connecting P_a and P_b (see Fig. 3c). (**e**) Graph of LOWESS smoothed curve (red) fitted to raw velocity profile (black) at user-selected time point

$$\alpha_t = \mathrm{Arctan}\left(\frac{\left|Y_{b;t} - Y_{a;t}\right|}{\left|X_{b;t} - X_{a;t}\right|}\right) * \frac{180}{\pi}.$$

2. *Prefix-angle.png and -dAngle-dt.png*: Graphs of root angles and bending rates (i.e., change in angle over time; Fig. 4d).

3.5.3 Root Velocity Profile

The velocity value for any point P_i along the root midline/edgeline reports how quickly the reference point P_0 at the root tip "moves" away from point P_i and thus reflects the total growth of all cells along the midline between points P_0 and P_i. Raw velocity for each point along the midline is calculated as $(L_{i;\ t\ +\ 1}-L_{i;\ t})$*frame^{-1} (Figs. 3b and 4b).

1. *Prefix-velocity-raw.csv*: Velocity data for each tracked point P_i (rows) in each frame (time t; columns).
2. *Prefix-velocity-raw-unadjusted.png*: 3D heatmap of velocity profiles for each frame and each point along the root length. X-axis represents time (in frames), Y-axis represents the position P_i along the root axis, and autoscaled color represents velocity (pixels frame^{-1}).
3. *Prefix-2D-velocity-smoothing.png*: Graph of raw velocity profile and regression curve along entire root at user-defined frame (time point) t (specified in "user-defined-parameters.csv") (Fig. 4e).
4. *Prefix-velocity-smoothed-and-midlineshiftcorrected.csv*: Smoothed velocity data; local regression is performed using user-defined Locally Weighted Scatterplot Smoothing (LOWESS) (specified in "user-defined-parameters.csv"). The position of each point along the root axis is corrected by accounting for root growth over time (*see* **Note 8**).
5. *Prefix-velocity-auto-scale.png*: 3D heatmap of smoothed velocity profiles for each frame and each (corrected) position P_i along the root length. Autoscaled color represents velocity (pixels frame^{-1}).
6. *Prefix-velocity-manual-scale.png*: 3D heatmap of smoothed velocity profiles for each frame and each (corrected) position P_i along the root length (Fig. 5a). Color represents velocity (pixels frame^{-1}) per user-defined scale (minimum, maximum values in LUT; specified in "user-defined-parameters.csv").

3.5.4 Root REGR (=Strain) Profile

REGR values along the root are calculated as the derivatives of the smoothed velocity curve. Values reflect local relative expansion as fraction frame^{-1}; to convert into units of [% h^{-1}], multiply values by 100 [%] * (60 [s]/frame interval [s]) * 60 [h].

1. *Prefix-REGR-raw.csv*: Relative elemental growth rate for each position P_i (rows) in each frame (time; columns).
2. *Prefix-REGR-raw.png*: 3D heatmap of raw (unsmoothed) REGR profiles for each frame and each position along the root length. Autoscaled color represents REGR (fractions frame^{-1}).
3. *Prefix-REGR-smoothed.csv*: Smoothed REGR data; local regression is performed using user-defined Locally Weighted

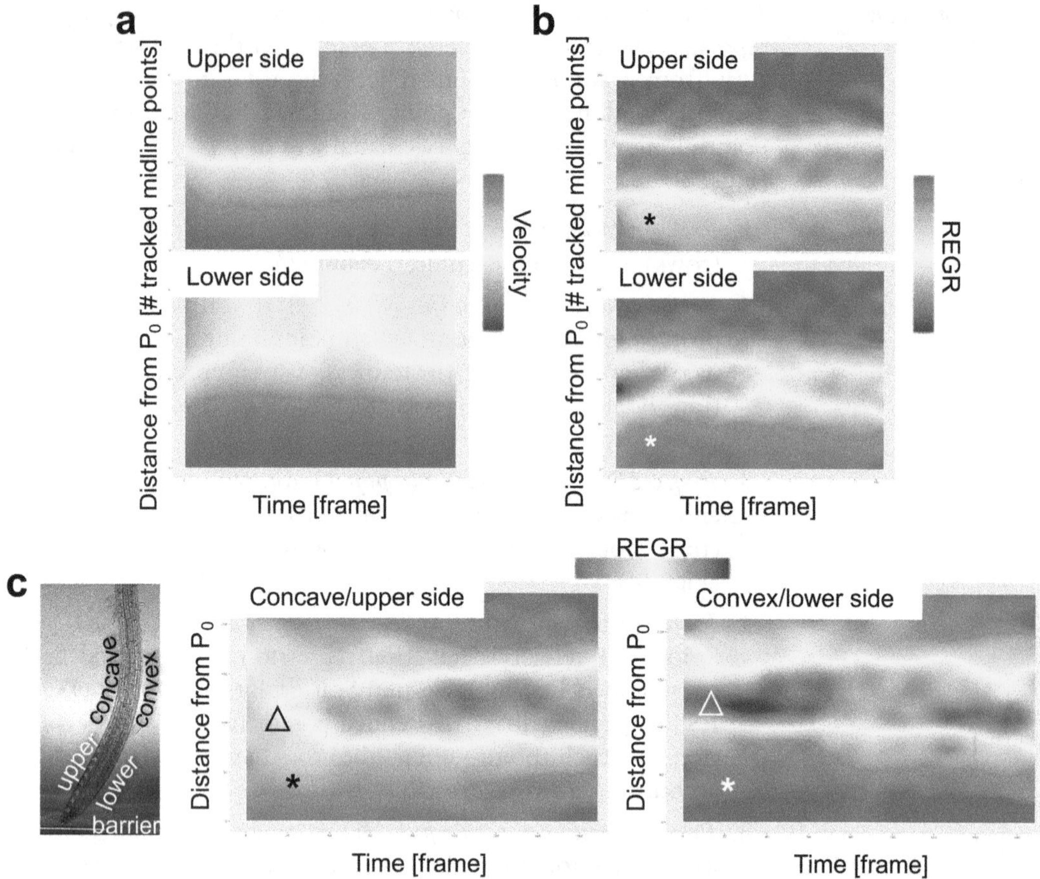

Fig. 5 3D heatmaps of root growth patterns. Manually scaled (**a**) velocity and (**b**) REGR profiles of gravistimulated *Arabidopsis* WT root 2–62 min after tilting the root by 90°. Heatmap colors reflect the (user-defined) magnitude of velocity/REGR, with warmer colors corresponding to larger magnitude. Note the asymmetry of expansion rates in the apical elongation zone, with the upper side of the root showing accelerated and the lower side reduced expansion (*). **c** RootPlot v1.0 reveals complex cell expansion patterns in Arabidopsis roots tracking along a cover glass barrier (see [6, 21]). The image series was started ~10 min after the root had encountered and started to slip along the barrier. Note the differential expansion in the primary bending region of the central/proximal elongation zone (corresponding to convex and concave regions; triangles) as well as the initiation of gravitropic bending in the apical elongation zone (asterisks)

Scatterplot Smoothing (LOWESS) (specified in "user-defined-parameters.csv").

4. *Prefix-REGR-auto-scale.png*: 3D heatmap of smoothed REGR profiles for each frame and each position along the root length. Autoscaled color represents REGR (fractions frame^{-1}).
5. *Prefix-REGR-manual-scale.png*: 3D heatmap of smoothed REGR profiles for each frame and each position along the root length (Fig. 5b, c). Color represents REGR (fractions frame^{-1}) per user-defined scale (minimum, maximum values for LUT; specified in "user-defined-parameters.csv").

4 Notes

1. Once prepared, any unused agar plates can be stored at 4 °C for up to a month. Although a longer storage period does not affect seed germination, we do see a subtle effect on root growth as the roots typically exhibit shorter elongation zones on older plates (root hairs initiate at <500 μm from the tip).
2. Stratification helps in promoting uniform germination of seeds. While there is no fixed duration specified, it is essential to employ a uniform stratification period throughout all the experiments, as it will ensure that the seedlings are not only at the same chronological age, but at the same developmental age as well.
3. Different types of gelling agents (agar, agarose, phytagel, etc.) have different gel strengths even when used at the same concentration. Here, the main objective in choosing an appropriate gel concentration is to have an agar gel which is easy to manipulate (cutting, trimming, etc.) and at the same time does not inhibit root growth.
4. The thickness of the agar "cushion" should be less than the depth of the experimental chamber, as a thick agar cushion will interfere with placing the second coverslip before starting the experiment. Autoclaved agar medium should not be reheated more than twice.
5. The transfer of the seedling should be done swiftly and gently as *Arabidopsis* seedlings are extremely delicate and dry out very quickly. Once the seedling has been carefully transferred to the experimental chamber, it is advisable to observe it under a stereo microscope to ensure that the root is not dead or damaged. If the root has dried out during the transfer or has been heat damaged while securing agar cushion with warm agar, the cytoplasm will condense and the root tip will seem dark. Only roots with the apical ~1 mm oriented perfectly vertically should be selected for an experiment.
6. It is important that the experimental chamber is stably secured to the microscope as any slight movement will hinder the tracking of points during image analysis. Once properly positioned, the chamber can be fixed to the microscope stage using adhesive tape.
7. If a rotating stage is used, root growth can be monitored before and after gravistimulation. The stage on our microscope does not fully rotate, but can be tilted by >90° in one direction. If the microscope stage does not rotate/tilt, one may have to omit growth analysis in vertical orientation. Instead, imaging starts as quickly as possible after mounting the chamber in

horizontal orientation. It is useful to first mount a test chamber with another seedling to identify the focal plane and set imaging parameters for optimal contrast and "texture" (to provide sufficient information for successful tracking of points). By optimizing settings prior to the start of an actual experiment, delays are avoided and the earliest phase of the gravitropic response can be captured.

8. At $t = 0$, any RootPlot input consists of a reference midline with equally spaced XY-coordinates. As the root grows, the distance between these points increases at a rate dependent on the rate of cell expansion in that root region. To report accurate distances from the P_0/root tip at later time points, evenly spaced values (based on reference midline point interval) are extrapolated from a smoothed velocity curve for each time point. These corrected values are then used to report velocity and REGR in all subsequent graphs with accurate distances from the P_0/root tip. It is important to note that this correction is not applied to either the velocity smoothing sample 2-D graph or the raw velocity heatmap, which both report uncorrected values by necessity.

Acknowledgments

This work was supported by NSF grant MCB-1817934 to GBM. The authors thank Drs. Edgar Spalding and Nathan Miller for many helpful discussions and the use of the custom software Image Processing Toolkit v10.

References

1. Morgan DC, O'Brien T, Smith H (1980) Rapid photomodulation of stem extension in light-grown *Sinapis alba* L.: studies on kinetics, site of perception and photoreceptor. Planta 150:95–101. https://doi.org/10.1007/BF00582351
2. Cosgrove DJ (1985) Kinetic separation of phototropism from blue-light inhibition of stem elongation. Photochem Photobiol 42:745–751. https://doi.org/10.1111/j.1751-1097.1985.tb01642.x
3. Zieschang HE, Sievers A (1991) Graviresponse and the localization of its initiating cells in roots of *Phleum pratense* L. Planta 184:468–477. https://doi.org/10.1007/BF00197894
4. Ishikawa H, Hasenstein KH, Evans ML (1991) Computer-based video digitizer analysis of surface extension in maize roots: kinetics of growth rate changes during gravitropism. Planta 183:381–390. https://doi.org/10.1007/BF00197737
5. Orbovic V, Poff KL (1993) Growth distribution during phototropism of *Arabidopsis thaliana* seedlings. Plant Physiol 103:157–163. https://doi.org/10.1104/pp.103.1.157
6. Massa GD, Gilroy S (2003) Touch modulates gravity sensing to regulate the growth of primary roots of *Arabidopsis thaliana*. Plant J 33:435–445. https://doi.org/10.1046/j.1365-313x.2003.01637.x
7. Galvan-Ampudia CS, Julkowska MM, Darwish E, Gandullo J, Korver RA, Brunoud G, Haring MA, Munnik T, Vernoux T, Testerink C (2013) Halotropism is a response of plant roots to avoid a saline environment. Curr Biol 23:2044–2050. https://doi.org/10.1016/j.cub.2013.08.042

8. Dietrich D, Pang L, Kobayashi A, Fozard JA, Boudolf V, Bhosale R, Antoni R, Nguyen T, Hiratsuka S, Fujii N, Miyazawa Y, Bae TW, Wells DM, Owen MR, Band LR, Dyson RJ, Jensen OE, King JR, Tracy SR, Sturrock CJ, Mooney SJ, Roberts JA, Bhalerao RP, Dinneny JR, Rodriguez PL, Nagatani A, Hosokawa Y, Baskin TI, Pridmore TP, De Veylder L, Takahashi H, Bennett MJ (2017) Root hydrotropism is controlled via a cortex-specific growth mechanism. Nat Plants 3:17057. https://doi.org/10.1038/nplants.2017.57
9. Bastien R, Guayasamin O, Douady S, Moulia B (2018) Coupled ultradian growth and curvature oscillations during gravitropic movement in disturbed wheat coleoptiles. PLoS One 13: e0194893. https://doi.org/10.1371/journal.pone.0194893
10. Jaffe MJ, Galston AW (1968) Physiology of tendrils. Annu Rev Plant Physiol 19:417–434
11. van Doorn WG, van Meeteren U (2003) Flower opening and closure: a review. J Exp Bot 54:1801–1812. https://doi.org/10.1093/jxb/erg213
12. Polko JK, Voesenek LA, Peeters AJ, Pierik R (2011) Petiole hyponasty: an ethylene-driven, adaptive response to changes in the environment. AoB Plants 2011:plr031. https://doi.org/10.1093/aobpla/plr031
13. Brown AH (1993) Circumnutations: from Darwin to space flights. Plant Physiol 101:345–348. https://doi.org/10.1104/pp.101.2.345
14. Bastien R, Meroz Y (2016) The kinematics of plant nutation reveals a simple relation between curvature and the orientation of differential growth. PLoS Comput Biol 12:e1005238. https://doi.org/10.1371/journal.pcbi.1005238
15. Sharp RE, Silk WK, Hsiao TC (1988) Growth of the maize primary root at low water potentials: I. Spatial distribution of expansive growth. Plant Physiol 87:50–57. https://doi.org/10.1104/pp.87.1.50
16. Ishikawa H, Evans ML (1993) The role of the distal elongation zone in the response of maize roots to auxin and gravity. Plant Physiol 102:1203–1210. https://doi.org/10.1104/pp.102.4.1203
17. Goodwin RH, Avers CJ (1956) Studies on roots. III. An analysis of root growth in *Phleum pratense* using photomicrographic records. Am J Bot 43:479–487
18. Miller ND, Parks BM, Spalding EP (2007) Computer-vision analysis of seedling responses to light and gravity. Plant J 52:374–381. https://doi.org/10.1111/j.1365-313X.2007.03237.x
19. Chavarria-Krauser A, Nagel KA, Palme K, Schurr U, Walter A, Scharr H (2008) Spatio-temporal quantification of differential growth processes in root growth zones based on a novel combination of image sequence processing and refined concepts describing curvature production. New Phytol 177:811–821. https://doi.org/10.1111/j.1469-8137.2007.02299.x
20. Brooks TL, Miller ND, Spalding EP (2010) Plasticity of *Arabidopsis* root gravitropism throughout a multidimensional condition space quantified by automated image analysis. Plant Physiol 152:206–216. https://doi.org/10.1104/pp.109.145292
21. Shih HW, Miller ND, Dai C, Spalding EP, Monshausen GB (2014) The receptor-like kinase FERONIA is required for mechanical signal transduction in *Arabidopsis* seedlings. Curr Biol 24:1887–1892. https://doi.org/10.1016/j.cub.2014.06.064
22. Bastien R, Legland D, Martin M, Fregosi L, Peaucelle A, Douady S, Moulia B, Höfte H (2016) KymoRod: a method for automated kinematic analysis of rod-shaped plant organs. Plant J 88:468–475. https://doi.org/10.1111/tpj.13255
23. Shih HW, DePew CL, Miller ND, Monshausen GB (2015) The cyclic nucleotide-gated channel CNGC14 regulates root gravitropism in *Arabidopsis thaliana*. Curr Biol 25:3119–3125. https://doi.org/10.1016/j.cub.2015.10.025
24. Dindas J, Scherzer S, Roelfsema MRG, von Meyer K, Muller HM, Al-Rasheid KAS, Palme K, Dietrich P, Becker D, Bennett MJ, Hedrich R (2018) AUX1-mediated root hair auxin influx governs $SCF^{TIR1/AFB}$-type Ca^{2+} signaling. Nat Commun 9:1174. https://doi.org/10.1038/s41467-018-03582-5

Chapter 8

Staging of Emerged Lateral Roots in *Arabidopsis thaliana*

Sascha Waidmann and Jürgen Kleine-Vehn

Abstract

The root system in plants plays a fundamental role in water and nutrient uptake. Lateral roots emerge from the primary root (PR) and its directional organ growth allows the plant to strategically explore the surrounding area. Compared to the main root, lateral roots initially display a distinct gravitropic set point angle, which is established shortly after emergence. Here, we describe a unifying protocol for the morphological description and classification of emerged, young lateral roots.

Key words Lateral roots, Imaging, Quantification, Gravitropic setpoint angle

1 Introduction

The plant root system interacts with the soil environment, supplying the plant with water and nutrients [1]. It is comprised of downwards growing primary roots (PRs) and angular growth displaying later roots (LRs), which determine the depth and overall root system size [2]. In contrast to PRs, LRs partially suppress gravitropic growth and establish a distinct gravitropic setpoint angle (GSA; [3]), allowing the radial spread of the root system. In the model plant *Arabidopsis thaliana,* the so-called stage I LR emerge in a 90° angle from the PR. Subsequently, the LRs undergo a maturation of gravity sensing columella cells [4] as well as the formation of a de novo formation of an elongation zone [5]. Temporally defined asymmetric maxima of auxin and lower cellular elongation at the lower side of stage II LRs define organ bending to gravity [5, 6]. On the other hand, cytokinin signaling at the upper root flank of stage II LRs spatially interferes with cellular elongation and expansion rates, suppressing gravitropic bending [7]. Accordingly, a spatially distinct phytohormonal crosstalk mechanism at the upper (anti-gravitropic input by cytokinin) and lower (gravitropic input by auxin) LR flanks defines the angular

Elison B. Blancaflor (ed.), *Plant Gravitropism: Methods and Protocols*, Methods in Molecular Biology, vol. 2368,
https://doi.org/10.1007/978-1-0716-1677-2_8,

growth and depth of the root system. This developmental stage II lasts between 8 and 9 h, afterwards transiting to stage III with symmetric elongation rates at the opposing organ flanks, which consequently maintains the acquired GSA. Stage IV lateral roots display alterations in GSA [5].

In previous publications, LRs where staged either by the different columella cell types [8] or based on differential growth [5]. Here, we present a protocol for the morphological classification of emerged LRs and a detailed description of the different stages in *Arabidopsis thaliana* unifying both hallmarks.

2 Materials

2.1 Plant Growth Medium and Seedling Growth

1. *Arabidopsis thaliana* Col-0 (WT) or mutants to be analyzed.
2. Solid plant growth medium: 2.3 g/L Murashige and Skoog (MS) salt, 0.5 g/L MES, 10 g/L sucrose, 8 g/L plant agar, adjust pH to 5.9, autoclave and store at RT.
3. Liquid growth medium for mounting: 2.3 g/L MS salt, 0.5 g/L MES, 10 g/L sucrose, adjust pH to 5.9, autoclave, and store at RT.
4. Sterile square petri dishes (12 × 12 × 1.7 cm) with vents.
5. Air permeable tape.
6. 70% and 100% ethanol.
7. Sterile tooth pics for sowing.

2.2 Specimen Preparation

1. Soft touch tweezers for transferring the seedlings.
2. Microscope slides.
3. Coverslips (24 × 50 mm #1.5).
4. Propidium iodide (PI) solution, 20 μL ml^{-1} in ddH_2O.

2.3 Imaging and Analysis

1. Upright confocal laser scanning microscope (CLSM) (for imaging PI) or standard light microscope equipped with 40× or 63× (water immersion) objective.
2. ImageJ 1.41 software (http://rsb.info.nih.gov/ij/) for cell length measurements.

3 Methods

3.1 Plant Material Growth

1. Grow Arabidopsis seedlings in a plant growth chamber at 22 °C in a 16-h-light/8-h-dark regime at 150 μmol/m^2/ s light intensity and 60% relative humidity.

2. In vitro growth is carried out in standard, sterile square petri dishes (see above) with vents. Pour enough media into plates to cover approximately half of the depth of the plate (*see* **Note 1**). Allow the plates to cool at room temperature for about an hour to allow the agar to solidify. If the plates are not to be used immediately, wrap them in plastic and store at 4 °C.
3. Sterilize the seeds by soaking them first in 70% and then in 100% ethanol.
4. Remove ethanol and allow the seeds to dry in a laminar flow hood.
5. Sow seeds separately using a sterile tooth pick in a straight line in the upper part of the petri dish with 0.5 cm space between the seeds.
6. Seal the plates with air-permeable paper tape to prevent desiccation while allowing slight aeration.
7. Stratify the seeds for 1–2 days by keeping the plates at 4 °C in the dark.
8. Grow the seedlings in a growth chamber for 8 days. The plates should be placed vertically into the plant incubator, allowing gravitropic growth on the surface of the agar (*see* **Note 2**).

3.2 Specimen Preparation

To prevent dehydration of the seedlings during imaging, first place a drop of liquid growth medium on the microscope slide. For PI staining, use a drop of PI. Use a pipette; the amount of water would vary according to the number of seedlings to be mounted (typically 80–200 μL for 1–3 seedlings). Place seedlings on the wet slide. Gently place a coverslip (cover glass) over the mounted seedlings (*see* **Note 3**).

3.3 Image Acquisition

1. Observe the samples with a light microscope using bright-field illumination or CLSM for PI imaging (excitation 569 nm, emission peak 593 nm) using a 40× or 63× (water immersion) objective.
2. Take pictures of the regions of interest (*see* **Note 4**).

3.4 LR Classification

The different stages of LRs can be defined by the following hallmarks:

1. Stage I (Fig. 1a): emerged LRs with only two rows of non-elongated columella (CI and C1) cells and no defined elongation zone.
2. Stage II (Fig. 1b): LRs with only two rows of columella (CI and C1) cells, C1 cells display onset of elongation, and the establishment of cellular elongation in the epidermal cell file (adjacent to the base of the main root).

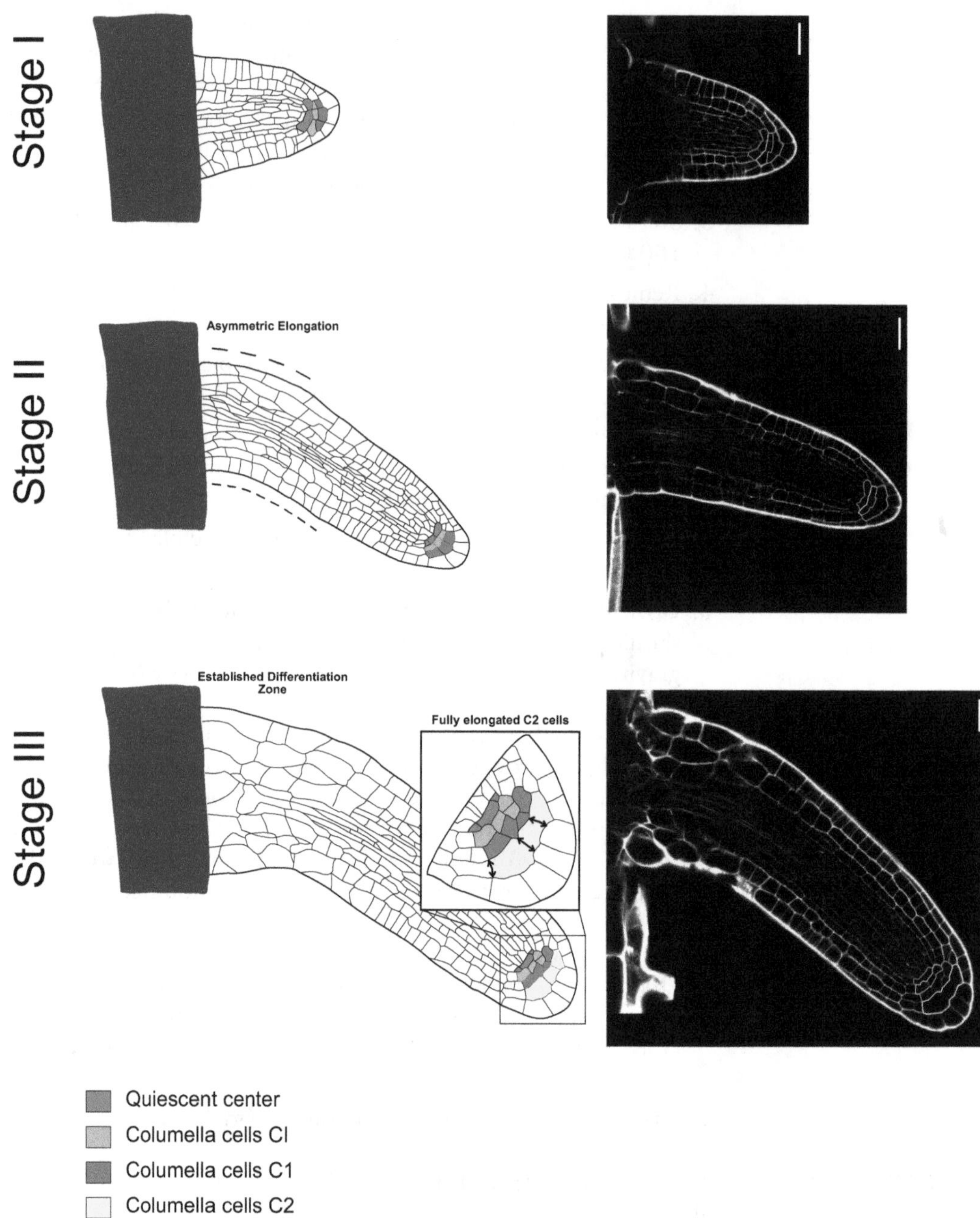

Fig. 1 Schematic representation of stage I–III lateral roots

3. Stage III (Fig. 1c): LRs with three rows of columella (CI, C1, and C2) cells, C2 cells are elongated, appearance of the differentiation zone at the base of the main root, and symmetric elongation rates in the elongation zone.
4. Stage IV (not depicted): LRs display additional bending to gravity, establishing a novel GSA.

4 Notes

1. To reduce growth variability between plates, it is advisable to pour always the same amount of medium.
2. Depending on the exact growth conditions roots may display emerged stage I-III lateral roots between 7 and 9 days.
3. Successful PI staining can be achieved by incubations as short as 30 s, however for very young, cuticular containing LRs a longer incubation might be necessary.
4. Keep the same magnification and microscope settings during the same experiment.

Acknowledgments

This work was supported by the Vienna Research Group (VRG) program of the Vienna Science and Technology Fund (WWTF) (to J.K-V.), the Austrian Science Fund (FWF) (P29754 to J.K-V., and P33497 to S.W.), the European Research Council (ERC) (Starting Grant 639478-AuxinER) (to J.K-V.).

References

1. Zürcher E, Muller B (2016) Cytokinin synthesis, signaling, and function—advances and new insights. Int Rev Cel Mol Bio 324:1–38
2. Waidmann S, Sarkel E, Kleine-Vehn J (2020) Same same, but different: growth responses of primary and lateral roots. J Exp Bot 71:2397–2411
3. Digby J, Firn RD (1995) The gravitropic set-point angle (GSA): the identification of an important developmentally controlled variable governing plant architecture. Plant Cell Environ 18:1434–1440
4. Kiss JZ, Miller KM, Ogden LA, Roth KK (2002) Phototropism and gravitropism in lateral roots of Arabidopsis. Plant Cell Physiol 43:35–43
5. Rosquete MR, von Wangenheim D, Marhavý P, Barbez E, Stelzer EHK, Benková E, Maizel A, Kleine-Vehn J (2013) An auxin transport mechanism restricts positive orthogravitropism in lateral roots. Curr Biol 23:817–822
6. Rosquete MR, Waidmann S, Kleine-Vehn J (2018) PIN7 auxin carrier has a preferential role in terminating radial root expansion in Arabidopsis thaliana. Int J Mol Sci 19:1238
7. Waidmann S, Rosquete MR, Schöller M, Sarkel E, Lindner H, LaRue T, Petřík I, Dünser K, Martopawiro S, Sasidharan R, Novák O, Wabnik K, Dinneny JR, Kleine-Vehn J (2019) Cytokinin functions as an asymmetric and anti-gravitropic signal in lateral roots. Nat Commun 10:3540
8. Taniguchi M, Furutani M, Nishimura T, Nakamura M, Fushita T, Iijima K, Baba K, Tanaka H, Toyota M, Tasaka M, Morita MT (2017) The Arabidopsis LAZY1 family plays a key role in gravity signaling within statocytes and in branch angle control of roots and shoots. Plant Cell 29:1984–1999

Chapter 9

Methods for a Quantitative Comparison of Gravitropism and Posture Control Over a Wide Range of Herbaceous and Woody Species

Félix P. Hartmann, Hugo Chauvet-Thiry, Jérôme Franchel, Stéphane Ploquin, Bruno Moulia, Nathalie Leblanc-Fournier, and Mélanie Decourteix

Abstract

Quantitative measurements of plant gravitropic response are challenging. Differences in growth rates between species and environmental conditions make it difficult to compare the intrinsic gravitropic responses of different plants. In addition, the bending movement associated with gravitropism is competing with the tendency of plants to grow straight, through a mechanism called proprioception (ability to sense its own shape). Disentangling these two tendencies is not trivial. Here, we use a combination of modeling, experiment and image analysis to estimate the intrinsic gravitropic and proprioceptive sensitivities of stems, using *Arabidopsis* as an example.

Key words Kinematics, Proprioception, Posture control, Plantlets, Image analysis

1 Introduction

Interest for how plants maintain a straight and erected growth habit led plant biologists to focus on tropisms, the orientation of growth in the direction of an environmental stimulus. Among the tropisms, major progress has been gained primarily in the study of gravitropism and phototropism.

Plant gravitropism entails slow growth movement allowing a reorientation of organs relative to the gravity field. Despite a growing interest for such movements and the recent insights into the molecular mechanisms of gravitropic sensing (see [1] for a recent review), little is known about its quantitative biological control. In

Félix P. Hartmann and Hugo Chauvet-Thiry contributed equally to this work and should be considered co-first authors. Nathalie Leblanc-Fournier and Mélanie Decourteix acted as co-PI and should be considered co-last authors.

Elison B. Blancaflor (ed.), *Plant Gravitropism: Methods and Protocols*, Methods in Molecular Biology, vol. 2368, https://doi.org/10.1007/978-1-0716-1677-2_9,

plants, gravisensing relies on small starch-filled organelles called statoliths. When a plant is reoriented, statolith sedimentation in the direction of the gravity field triggers asymmetric growth and, consequently, curvature of the organ. For a long time, this growth response has been interpreted to be a consequence of the perception of the weight of statoliths and/or of the protoplast, meaning that these structures would behave as force sensors. Our research team recently developed an original device enabling us to: (1) study the effect of the direction (i.e., the angle between the initial inclination of the plant shoot and the gravity vector) vs the intensity of the gravity vector for the study of long-term stimuli [2], or (2) realize gravity dose–response studies in the context of transient stimulations [3]. With the help of this device [2], we were able to disprove the statolith weight and protoplast pressure models in aerial organs by demonstrating that shoot gravitropism relies on the sensing of the inclination and not on the sensing of gravitational force or acceleration. This enabled us to validate the other long-standing model of gravitropic control based on the sensing of the inclination angle with respect to the gravity field. It would thus seem logical to continue to focus on measures related to gravitropism with well-proven techniques consisting of following the tip angle of organs. However, in the last decade, studies based on kinematic analyses combined with mathematical modeling highlighted that the gravitropic response cannot be explained by the sensing of the inclination angle only [4]. Such approaches revealed the importance of the straightening movement in aerial organs. To preserve their erected habit and grow straight, plant aerial organs must have the ability to sense their inclination with respect to the gravity field (graviception) as well as to read their own shape (proprioception) [4, 5]. Graviception tends to generate a bending or curving movement upward toward the vertical while proprioception tends to rectify the organ by decreasing its curvature. To understand the gravitropic uprighting process and to separate the part of the response due to graviception from the part due to proprioception, it thus appeared necessary to not only measure the inclination of the organ but also to monitor the local curvature C over the entire growth zone.

How is local curvature measured and what are the underlying biological concepts for acquiring such measurements? The curvature C can be defined as the spatial rate of change of the angle A along the organ [6, 7]:

$$C = \frac{\partial A}{\partial s}$$

s is the spatial coordinate along the organ and $\frac{\partial A}{\partial s}$ is the derivative of A versus spatial position (partial derivative). When $C = 0$ everywhere, the organ is straight. During a tropic response, the curvature changes locally with time. To properly quantify the response of an organ, it is useful to describe the growth mechanism at the origin of the curvature changes. Any change in curvature results from differential growth, i.e., a difference between the elongation rate of the lower face ($\dot{\varepsilon}_{\mathrm{low}}$) and the elongation rate of the upper face ($\dot{\varepsilon}_{\mathrm{up}}$) of the organ. For instance, if the lower face elongates faster than the upper face, the organ curves upward. Locally, curvature variation is directly driven by the difference in elongation rates:

$$\frac{\mathrm{DC}}{\mathrm{Dt}}R = \frac{\dot{\varepsilon}_{\mathrm{low}} - \dot{\varepsilon}_{\mathrm{up}}}{2}$$

where R is the radius of the organ and $\frac{\mathrm{DC}}{\mathrm{Dt}}$ is the time derivative of the curvature (curving rate) attached to each successive segment (i.e., material derivative, see [6] for a further biologist-friendly priming on the quantification of tropic response including material derivatives).

However, these relative elongation rates depend a lot on growth conditions and can vary from one species to another. The effect of such variations on the computation of curvature changes can be taken into account by considering the mean relative elongation rate $\dot{\varepsilon}_{\mathrm{mean}}$ defined as:

$$\dot{\varepsilon}_{\mathrm{mean}} = \frac{\dot{\varepsilon}_{\mathrm{low}} + \dot{\varepsilon}_{\mathrm{up}}}{2} \tag{1}$$

Then, Eq. (1) can be rewritten as:

$$\frac{\mathrm{DC}}{\mathrm{Dt}}R = \epsilon_{\mathrm{mean}}\Delta \tag{2}$$

where $\widetilde{\Delta}$ is the differential growth ratio and is defined as:

$$\widetilde{\Delta} = \frac{\dot{\varepsilon}_{\mathrm{low}} - \dot{\varepsilon}_{\mathrm{up}}}{\dot{\varepsilon}_{\mathrm{low}} + \dot{\varepsilon}_{\mathrm{up}}}$$

The interest of $\widetilde{\Delta}$ is that it measures specifically the redistribution of the mean growth toward the lower and upper faces.

Having established a method to assess curvature, it is important to understand how graviception and proprioception contribute to this curvature. To this end, it is necessary to assess the graviceptive and proprioceptive sensitivities of the organ. Bastien *et al.* [8] provided a mathematical model, the $AC\dot{E}$ model, enabling the quantification of these metrics. It assumes that the relative differential growth $\widetilde{\Delta}$ is a local response to the inclination angle A (graviception) and to the curvature C (proprioception):

$$\widetilde{\Delta} = -\widetilde{\beta}A - \widetilde{\gamma}CR$$

$\widetilde{\beta}$ and $\widetilde{\gamma}$ are, respectively, the gravitropic and proprioceptive sensitivities. The tilde symbols indicate that these parameters are dimensionless and normalized for the mean elongation rate and the radius, in contrast with the previous formulation in [4]. This makes it possible to directly compare the sensitivities of two different plants, independently of their size and mean elongation rate.

Note that A, C, and $\widetilde{\Delta}$ are functions of the position along the organ (s) and of time (t). The organ radius R is assumed constant and uniform along the organ. The mean elongation rate $\dot{\varepsilon}_{mean}$ is assumed uniform along the growth zone and constant in time. $\widetilde{\beta}$ and $\widetilde{\gamma}$ are intrinsic parameters of the plant and depend on neither space nor time.

Once the organ has reached a steady-state (i.e., its shape does no longer depend on time), the inclination angle along the growth zone is given by:

$$A_{st}(s) = A_0 \exp\left(-B\frac{s}{L_{gz}}\right)$$

A_0 is the angle at the base of the growth zone, L_{gz} is the length of the growth zone, and B is the dimensionless "balance number" expressed as:

$$B = \frac{\widetilde{\beta} L_{gz}}{\widetilde{\gamma} R}$$

B is an important integrative trait which combines the gravitropic and proprioceptive sensitivities with the geometry of the organ (aspect ratio of the growth zone L_{gz}/R). It determines both the final shape of the organ and the time required to reach this shape (see details in [7]).

In this chapter, we describe how to monitor the kinematics of the gravitropic response (inclination and curvature), and how to estimate the gravitropic and proprioceptive sensitivities through experiment and semi-automatic image analysis, with the help of a software (Interekt) which has been developed in our laboratory. We hereafter describe our method by taking Arabidopsis as an example, but it can be adapted to studies of other species.

2 Materials

2.1 Preparation and Imaging of Plants

1. Seeds (*Arabidopsis thaliana*, Columbia-0 accession) (*see* **Notes 1–3**).
2. 30 mL polypropylene cylindrical containers.
3. Potted soil supplemented with fertilizing media.
4. Growth chamber or growth room.
5. Refrigerator (4 °C) to vernalize the seeds.
6. A system to clamp pots horizontally and a dark background (*see* **Notes 2–4**).
7. NIKON D200 wide-angle lens camera (*see* **Note 4**).

2.2 Software

Interekt (Interactive Exploration of Kinematics and Tropisms) is an open-source software devoted to the quantitative analysis of one-dimensional organ growth. It is written in the programming language Python and depends on several Python libraries. It has been tested on Linux, Windows, and OS X. The version of Interekt used in this chapter, along with installation guidelines, is available online (https://forgemia.inra.fr/hugo.chauvet-thiry/rootstemextractor/-/tree/chapter_gravitropism).

3 Methods

3.1 Preparation and Imaging of Plants

1. Vernalize Arabidopsis seeds at 4 °C for at least 3 days.
2. Fill up cylindrical containers with potted soil and add appropriate amount of fertilizer. Sample containers can be pierced with holes to water the plantlets by sub-irrigation before the imaging process (*see* **Note 1**).
3. Dispense one seed per container.
4. Grow Arabidopsis plants in a growth chamber until inflorescences are 6–8 cm tall. Plants will be ready for analysis after about 6 weeks in the growth chamber under typical conditions (16 h/8 h day/night cycle; 23 °C during the day and 19 °C during the night; 50% air hygrometry). Water the plants regularly by sub-irrigation.
5. Once the plants have reached the required stage, and the day before reorienting the plant, cover the rosette leaves with a net. This step is necessary because when plants are reoriented horizontally, leaves can become hyponastic and mask the lower part of the floral stem. Alternatively, attach the rosette leaves to the cylindrical container with pieces of tape (*see* **Note 5**).
6. In order to ease the digital (automatic) tracking of the inflorescence movements, it is better to work with a single unbranched floral stem. For this purpose, the day before reorienting the plant, remove all the inflorescence ramifications and cauline leaves with a scalpel (*see* **Note 5**).
7. A few hours before reorienting the plant, place a black mark with a permanent marker at the tip of the inflorescence, at the bottom of the terminal group of flowers. This mark will be used by the *Interekt* software to track the movements of the floral stem (*see* **Note 6**).
8. Clamp the container horizontally on a dark background.

   Program the camera to acquire time-lapse: Acquire 1 photo every 6 min during the first and fastest part of the gravitropic movement (about 2 h), and a final photo after 36 h to have the final shape of the stem (this will be the steady-state image). Start picture acquisition on the camera.

3.2 Image Analysis and Semi-Automatic Estimation of Parameters

1. Import the sequence of images generated from the experiment by clicking on the "Open" button in the toolbar and then selecting images. The software automatically stores the images into a single archive file (*see* **Note 7**). This can take some time. After this operation is completed, the first image of the sequence is displayed in the main window.
2. If you wish, inspect visually the experiment by browsing through the sequence. Press the right (resp. left) arrow key to move forward (resp. backward) in the sequence. Alternatively, click on the "Image list" button in the toolbar to open a small floating window, from which you can directly access a given image by clicking on its name.
3. Select the steady-state image (i.e., the image at which you consider the organ has reached its final shape; usually, the last image of the sequence). Click on "Options" and then "Select stead-state image". A popup window appears, with the list of images (Fig. 1). Click on the image you consider at steady-state. You have the possibility to exclude this image from time series by checking the box below the list. This can be useful when the steady-state image has been taken a long time after other images.
4. The image scale can be set with the "Scale" window (Options → Scale). Click on "Measure distance". Then, choose two points of the current image separated by a known distance. Click successively on these two points. A line joining the points

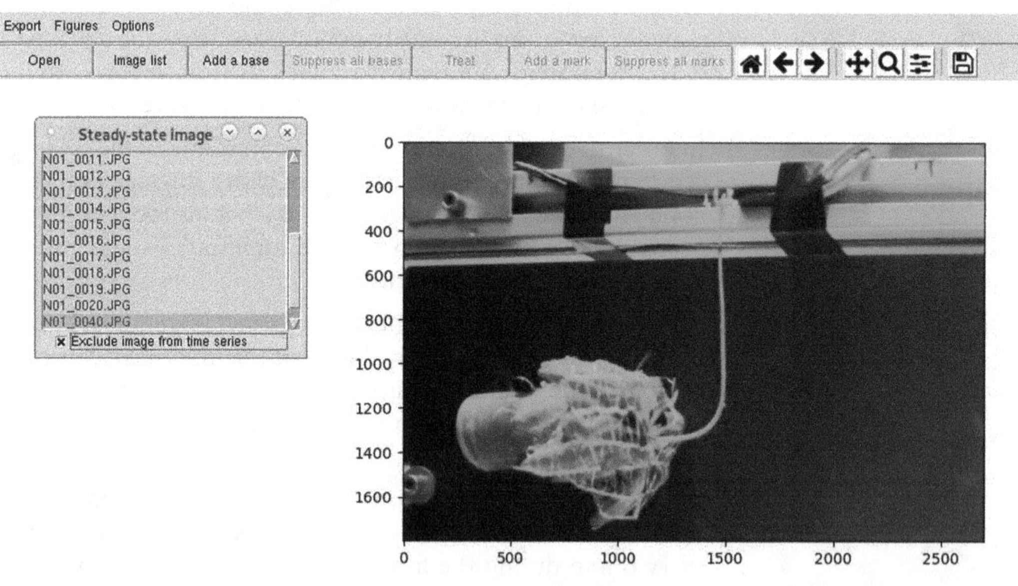

Fig. 1 Interekt interface after images have been loaded. Selection of the steady-state image

appears, and the "pixels" field of the "Scale" window is automatically filled in with the number of pixels between the points. Fill in the "cm" field with the distance in centimeters between the points. Then click on "Update the scale" and close the "Scale" window. All distances in pixels will be converted into cm using this ratio (*see* **Note 8**).

5. For extracting organ shapes, move to the first image of the sequence and click on the "Add a base" button in the toolbar. A "base" is an object, which indicates the beginning of the organ and its direction. To create the base, left-click successively on both sides of the organ (each time outside the organ), close to the anatomical base of the organ. A line segment crossing the organ appears, with a shorter line segment orthogonal to the main segment. This orthogonal line segment indicates the direction, which will be followed by the algorithm extracting the organ shape. It depends on the order in which the sides of the organ are clicked when creating the base. For an organ growing from left to right, click first above and then below the organ (and inversely for an organ growing from right to left). For an organ growing from top to bottom, click first right and then left of the organ (and inversely for an organ growing from bottom to top).
6. Once a base has been created on each organ to analyze, click on the "Treat" button in the toolbar. This triggers organ shape extraction and may take some time. At the end of the treatment, extracted organ skeletons appear as colored lines on the organs (Fig. 2). For convenience, options to provide a specific label to the organ are available (*see* **Note 9**). Errors with shape extraction can also lead to an incorrect skeleton, which can be corrected (*see* **Note 10**).
7. Geometric quantities can be computed from the skeleton of an organ for the current image. They can be visualized by right-clicking on a skeleton and selecting "Angles and curvatures" from the context menu. A new window pops up, with three plots. The topmost plot shows the profile of the organ, which is nothing more than its extracted skeleton. The middle plot shows the inclination (angle from vertical) along the organ. The bottom plot shows the curvature along the organ.
8. It is possible to get an overview of the kinematics of an organ, and thus a first insight into possible tropisms, by right-clicking on a skeleton and selecting "Time series" from the context menu (or, alternatively, simply by left-clicking on the skeleton). A new window pops up, with three plots (Fig. 3). The plot from the left (Fig. 3a) is made of the superimposition of the profiles of the organ during the experiment. The successive profiles are color-coded using a colormap ranging from purple

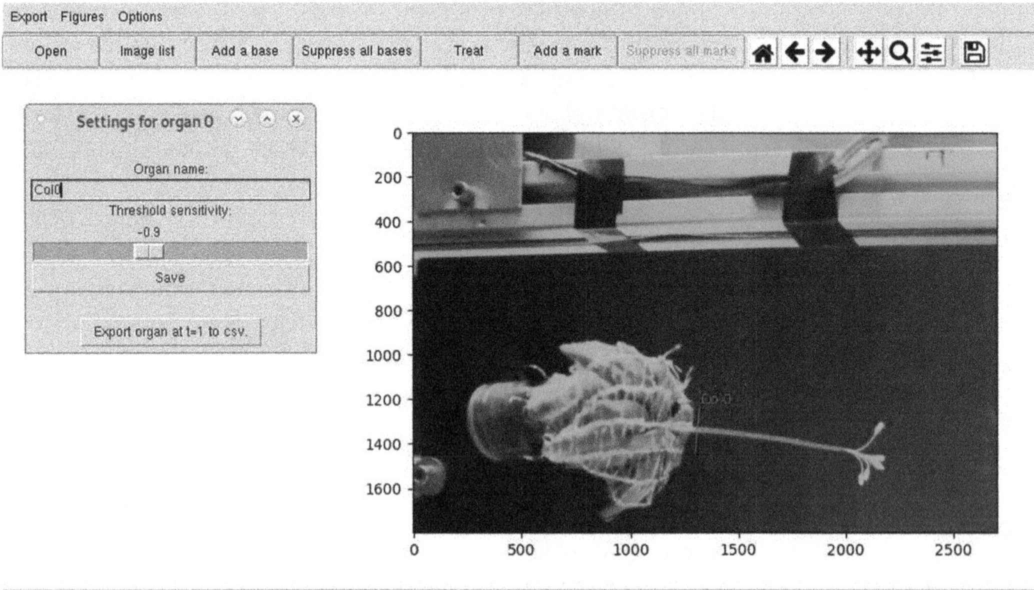

Fig. 2 Interekt interface after skeleton extraction of the stem. The threshold sensitivity has been set to −0.9

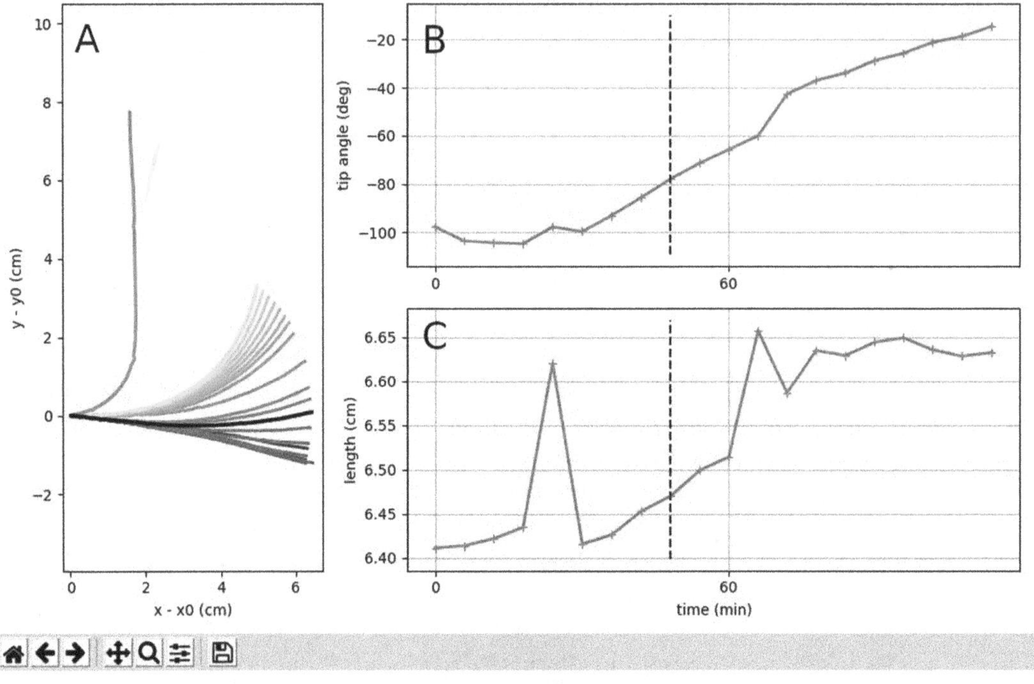

Fig. 3 Kinematic description of the tropic movement. (**a**) Successive profiles of the organ. The steady-state profile is in red and the current profile is in black. (**b**) Time series of the tip angle. (**c**) Time series of the organ length (growth curve). You can see that the growth curve is a bit irregular, due to imperfections in skeleton extraction. In both time series, the vertical dotted line represents the current image

(for the beginning of the experiment) to yellow (for the end of the experiment). The steady-state profile is colored separately in red. The top-right plot (Fig. 3b) represents the tip angle over time. The bottom-right graph (Fig. 3c) represents the length of the organ over time. If you had decided to exclude the steady-state from time series, it does not appear in these two last graphs.

9. For estimating the growth rate of an organ (dL/dt), right-click on its skeleton and select "Estimate growth rate" from the context menu. A new window pops up, with four plots. The plot from the left is the same superimposition of organ profiles as explained above ("Time series"). The top-right plot shows the growth curve of the organ (length as a function of time). The growth rate is estimated through a linearization of the growth curve, or at least a part of it. On this top-right plot, you can select the part of the curve, which is closest to linearity. Left-click on the plot and move the mouse pointer while clicking to draw a rectangle on the zone you want to select (*see* **Note 11**).
10. The middle-right plot of the "Growth rate" popup window shows the growth length of the organ, zoomed-in according to the rectangle drawn on the plot above. In addition, there are two red dots joined by a line segment. By moving the dots with the mouse (left-click), you can position the line segment such that it represents the linearization of the growth curve. The growth rate dL/dt (computed from the slope of the line) is displayed on the left plot. The bottom-right plot shows the residuals of the linearization. This gives a way to estimate the goodness of the linearization.
11. For estimating the growth length L_{gz} of an organ, right-click on its skeleton and select "Estimate growth length" from the context menu. A new window pops up, with three plots. The bottom plot represents the organ curvature across space and time using a heatmap. Space goes along the x-axis, while time goes along the y-axis. Warm colors (yellow, red, dark red) code for positive curvatures, while cold colors (blue, dark blue) code for negative curvatures. On the right edge, a solid black line represents the linearized growth as estimated above. A draggable dotted red line parallel to the linearized growth enables you to select the growth length, or more exactly the boundary between the non-growing zone and the growing zone of the organ (*see* **Note 12**).
12. During gravitropic movement, differential growth is strong, which means that the curvature is changing with time everywhere in the growth zone. This is how growth can be spotted: Where curvature does not change along the y-axis (time), there is no growth, whereas a curvature change along the y-axis signals growth. A horizontal draggable dotted line provides

an auxiliary tool for detecting the boundary of the growing zone. Dragging this line sets the current time, with an effect on both top plots. The top-left plot shows the angle along the organ, at the initial time and at the current time. Comparing the angles at two different times can help spotting where growth occurs. The top-right plot pictures the organ in its initial state, with the superimposition of the organ skeleton at the current time. The big red dot on the image represents the position of the vertical draggable line. This superimposition offers a third way to check the estimation of the growth length. The numerical values of the estimated growth length L_{gz} is displayed at the bottom of the window.

13. For estimating the parameter $\tilde{\beta}$ of an organ, right-click on its skeleton and select "Estimate beta". A new window pops up, with three plots (Fig. 4). The left plot (Fig. 4a) shows the successive profiles of the organ, as already described. The

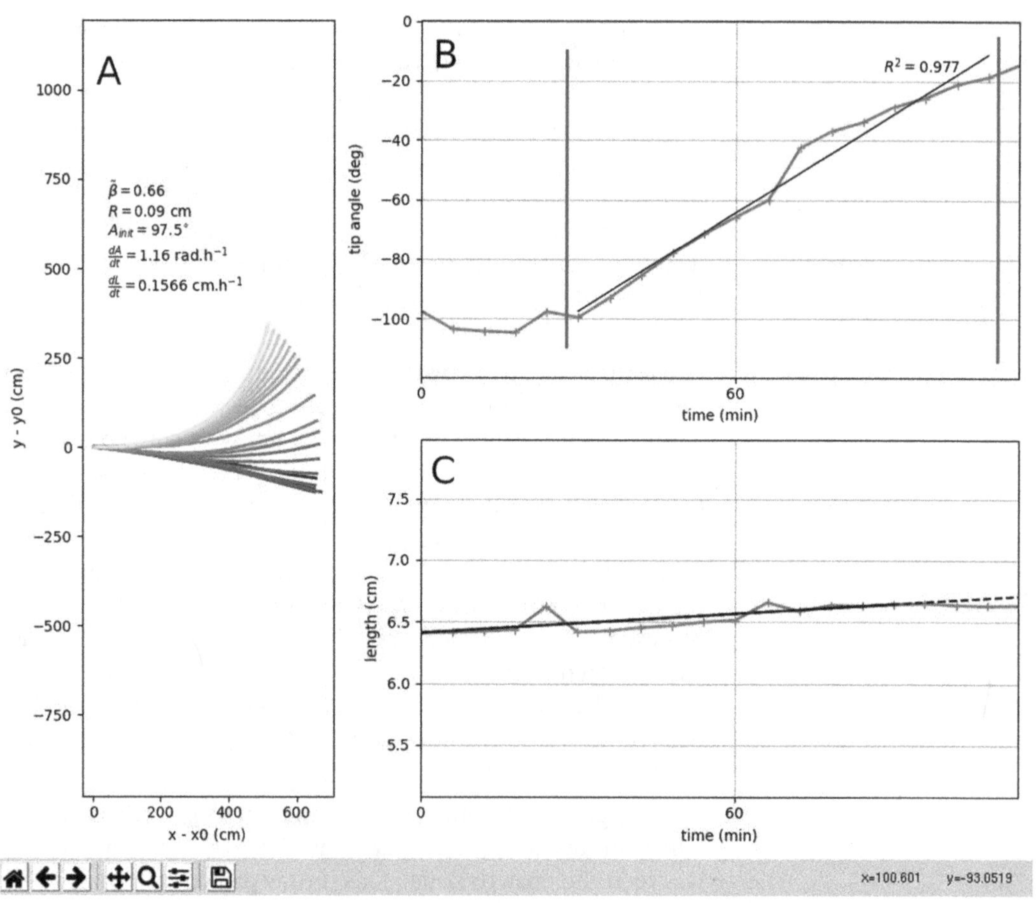

Fig. 4 Interactive estimation of $\tilde{\beta}$. (**a**) Successive profiles of the organ. (**b**) Tip angle over time, with adjustable linear regression. (**c**) Growth curve, with its linearization

top-right plot (Fig. 4b) represents the tip angle over time. You need to make a linear regression on this curve to estimate $\widetilde{\beta}$. Two vertical blue draggable lines allow you to select the regression range. The R^2 of the fit is displayed on the plot. The estimated value of $\widetilde{\beta}$ is displayed on the left plot, along with quantities involved in the estimation: The radius R, the initial tip angle A_{init}, the rate of angle change $\mathrm{d}A/\mathrm{d}t$ (obtained from the linear regression), and the growth rate $\mathrm{d}L/\mathrm{d}t$ (*see* **Note 13**). The bottom-right plot (Fig. 4c) represents the growth, with its linearization, as a reminder.

14. For estimating the balance number B of an organ, right-click on its skeleton and select "Estimate B". A new window pops up, with three plots (Fig. 5). The top-left plot (Fig. 5a) represents the inclination angle along the growth zone, for the profile selected before as steady-state. The top-right plot

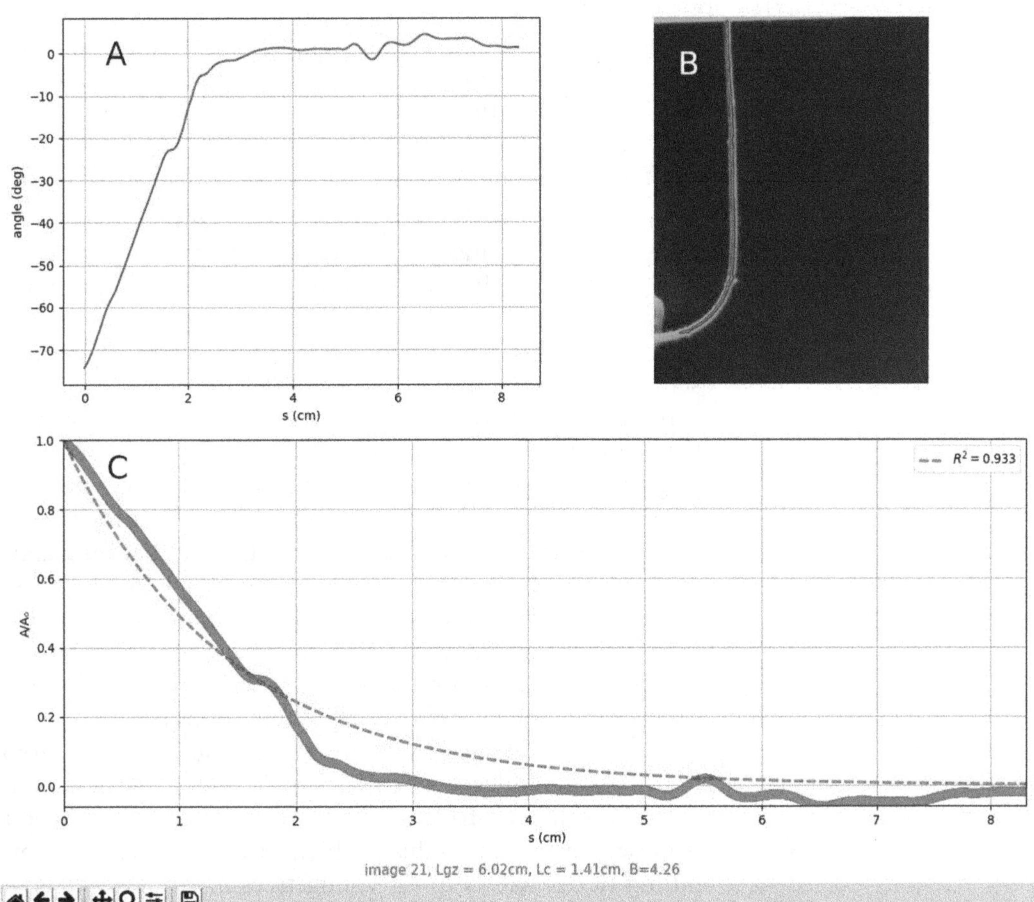

Fig. 5 Estimation of B. (**a**) Inclination angle along the growth zone at steady-state. (**b**) Steady-state image of the organ, with the extracted skeleton. (**c**) Normalized inclination angle along the growth zone, fitted with an exponential

(Fig. 5b) shows the steady-state image of the organ, with the extracted skeleton. The bottom plot (Fig. 5c) represents the normalized inclination angle A/A_0 along the growth zone (where A_0 is the angle at the base of the growth zone). This curve is fitted with an exponential (red dotted line, *see* **Note 14**). No action is required from the user; the regression is automatic. If you are satisfied with the fit, just close the popup window. Otherwise, you can try selecting another image as steady-state and/or re-estimating the growth length.

15. As a last step, export the results into an output CSV file. Click on "Export" and then "Phenotype (graviception, proprioception)". The resulting CSV file contains all previously estimated values, plus the proprioceptive parameter $\tilde{\gamma}$ (*see* **Note 15**).

4 Notes

1. In our hands, the method also works well with several plants [8] including wheat coleoptiles (*Triticum aestivum*) (between 1.5 and 2.5 cm tall-about 4 days after germination) originating from seeds that were germinated on cotton [2] and sunflower hypocotyls (*Helianthus annuus* cv Germline) (between 2 and 4 cm tall) originating from seeds that were germinated on soil.
2. The method can also be adopted for taller plants and/or plants having a long curving phase, provided that a special device called "isotropic light spheres" is used. The device, which was first developed in our lab to disentangle gravitropism and phototropism [9], has recently been modified and can now be used in studies aiming at disentangling graviperception and proprioception. In our hands, it works well with young trees (1-year-old) and we gained a good command with poplar. The uprighting and recovery movements of secondary growing plants (e.g., young trees) can be slow (several weeks for poplar) compared to those of small plants or organs displaying mainly primary growth. It then becomes necessary to maintain day-night cycles for several weeks, and it is therefore not possible to analyze these movements in the dark as is practiced in the protocol described above. However, under anisotropic lighting conditions and outside microgravity conditions (spaceflight for example), it is difficult to clearly separate gravitropism from phototropism. Indeed, when a plant is tilted, one side is lit while the other is shaded, which can induce a phototropic response. To overcome this problem, the use of an isotropic lighting device allows to tilt young trees in an isotropic light environment. It thus triggers a response devoid of its phototropic component. In our case, this device consists of two hemispheres (1.5 m diameter) mounted on two metallic and hexagonal tracks.

3. In order to increase the isotropy of the light (1) hemisphere are made of transparent polymethyl methacrylate so that lighting LED tubes can be positioned outside the hemisphere, (2) the inside of each hemisphere has been painted in white so as to obtain a good diffusion of light inside the sphere, (3) LED tubes were placed according to Den Dulk's "TURTLE" pattern [10]. The isotropic spheres were first equipped with fluorescent tubes as a lighting system [9]. However, to limit overheating inside the sphere, they were recently replaced by LED tubes. Despite the use of LEDs, the system produces a lot of heat. Thus, to maintain the temperature at approximately 25 °C during the lighted periods, a mobile cooling system (Blyss WAP 267EC) was connected to the sphere.
4. In our case, this system consists of a medium-density fiberboard, which is painted in black. Fixing collars (3 cm diameter) were attached to this board allowing to clamp Arabidopsis plants (grown in cylindrical containers) horizontally. To avoid phototropism effects, this device has to be placed in a dark room and pictures have to be taken in the dark. Thus, the flash must be filtered in the green safelight waveband.
5. Plants respond to rosette leaves manipulation and to removing the ramifications and cauline leaves of the inflorescence by transiently stopping growth (a process named thigmomorphogenesis) [11]. Growth comes back to normal several hours after these two manipulations. This is the reason why they are carried out the day before leaning the plant.
6. The black mark is placed in a growing zone. Avoid placing the mark too soon on the inflorescence unless growth will stretch the mark and reduce its intensity. This reduced intensity prevents the proper detection of the mark by the *Interekt* software. The mark isn't placed right before leaning the plant because of the thigmomorphogenetic Touch response (*see* **Note 5**).
7. Once a sequence of images from an experiment has been imported into Interekt, they are automatically archived into a single hdf5 file, saved in the same folder as the images, with the default name "interekt_data.h5". Any data obtained from subsequent processes and analyses will be automatically saved in the same archive file. Therefore, this file contains every data from the experiment. It can be opened directly from Interekt (using the "Open" button) for any future use, which is much faster than reopening the original sequence of images.
8. The scale is the same for all images of a sequence. Make sure that the photographs were taken from the same viewpoint, with same angle and settings.
9. By default, each organ is labeled with a number. You may prefer to use a more informative label, for example with the date of

the experiment and/or the name of the mutant. To do that, right-click on the skeleton, select "Settings", and then fill in the label field in the popup window.

10. If the skeleton extraction algorithm did not stop at the black mark before the inflorescence, you can try changing the threshold sensitivity. Right-click on the skeleton and select "Settings". If, for some images, the algorithm missed the mark and included the flowers, try decreasing the threshold sensitivity. If, conversely, the algorithm did not reach the mark, try increasing the sensitivity. Then close the "Settings" window and click again on "Treat". Tune iteratively the sensitivity until you are satisfied with the extracted skeleton.
11. Once a rectangle selector has been drawn, you can resize it using the square handlers at the corners and in the middle of each edge. You can also move the rectangle using the central handler.
12. The length of the growing zone (the growth length L_{gz}) is assumed constant throughout the experiment. This is why the boundary between the non-growing zone and the growing zone is oblique on the curvature heatmap and parallel to the linearized growth.
13. The exact formula is

$$\tilde{\beta} = \frac{R}{\sin\left(A_{\text{init}}\right)} \times \frac{dA/dt}{dL/dt}$$

The radius R is averaged along the organ.

14. At steady-state, the inclination angle along the growth zone is given by: $\mathrm{A_{st}(s)} = \mathrm{A_0 exp}(-B\frac{\mathrm{s}}{\mathrm{L_{gz}}})$

    Knowing the value of the growth length L_{gz} (estimated before), an exponential fit on the inclination angle gives the value of B.
15. The "balance number" B is related to the graviceptive and proprioceptive parameters through the relation: $B = \frac{\tilde{\beta} L_{gz}}{\tilde{\gamma} R}$.

Acknowledgments

This work was supported by CNES (Centre National d'Etudes Spatiales). This research (development of the *Interekt* software and experimental procedure on Arabidopsis) was financed by European Research Area Network for Coordinating Action in Plant Sciences through MURINAS project.

References

1. Nakamura M, Nishimura T, Morita MT (2019) Gravity sensing and signal conversion in plant gravitropism. J Exp Bot 70 (14):3495–3506
2. Chauvet H, Pouliquen O, Forterre Y, Legué V, Moulia B (2016) Inclination not force is sensed by plants during shoot gravitropism. Sci Rep 6:35431
3. Chauvet H, Moulia B, Legué V, Forterre Y, Pouliquen O (2019) Revealing the hierarchy of processes and time-scales that control the tropic response of shoots to gravi-stimulations. J Exp Bot 70(6):1955–1967
4. Bastien R, Bohr T, Moulia B, Douady S (2013) Unifying model of shoot gravitropism reveals proprioception as a central feature of posture control in plants. PNAS 110:755–760
5. Hamant O, Moulia B (2016) How do plants read their own shapes? New Phytol 212:333–337
6. Moulia B, Fournier M (2009) The power and control of gravitropic movements in plants: a biomechanical and systems biology view. J Exp Bot 60:461–486
7. Moulia B, Bastien R, Chauvet-Thiry H, Leblanc-Fournier N (2019) Posture control in land plants: growth, position sensing, proprioception, balance, and elasticity. J Exp Bot 70:3467–3494
8. Bastien R, Bohr T, Moulia B, Douady S (2014) A unifying modeling of plant shoot gravitropism with an explicit account of the effects of growth. Front Plant Sci 5:136
9. Coutand C, Adam B, Ploquin S, Moulia B (2019) A method for the quantification of phototropic and gravitropic sensitivities of plants combining an original experimental device with model-assisted phenotyping: exploratory test of the method on three hardwood tree species. PLoS One 14:e0209973
10. Den Dulk JA (1989) The interpretation of remote sensing, a feasibility study. PhD thesis, Wageningen University
11. Chehab EW, Eich E, Braam J (2009) Thigmomorphogenesis: a complex plant response to mechano-stimulation. J Exp Bot 60(1):43–56

Chapter 10

The Analysis of Gravitropic Setpoint Angle Control in Plants

Suruchi Roychoudhry, Marta Del Bianco, and Stefan Kepinski

Abstract

The history of research on gravitropism has been largely confined to the primary root–shoot axis and to understanding how the typically vertical orientation observed there is maintained. Many lateral organs are gravitropic too and are often held at specific non-vertical angles relative to gravity. These so-called gravitropic setpoint angles (GSAs) are intriguing because their maintenance requires that root and shoot lateral organs are able to effect tropic growth both with and against the gravity vector. This chapter describes methods and considerations relevant to the investigation of mechanisms underlying GSA control.

Key words Gravitropic setpoint angle, Gravitropism, Lateral root, Branch, GSA, Arabidopsis, Clinostat, Auxin

1 Introduction

The non-vertical growth of plant organs is one of the most important but least understood components of plant architecture. Root and shoot branches that grow out from the main root–shoot axis facilitate the capture of the resources required for plant growth and variation in the growth angle of these branches is a fundamental determinant of variety of plant forms observed throughout nature [1, 2]. In many species, lateral branches grow at specific gravitropic setpoint angles or GSAs, meaning that their angle of growth with respect to gravity is actively maintained [1]. The GSA concept, introduced by Digby and Firn, provides a neat system for investigating growth angles that are dependent on graviresponse. In this system, an organ being maintained vertically and growing downward has a GSA of 0° (e.g., a primary root), while an organ growing vertically upward would have a GSA of 180° (e.g., a primary shoot). Organs being maintained at non-vertical angles have GSA designations between these two extremes [1].

This chapter outlines simple methods for the investigation of gravity-dependent non-vertical growth. A most important first step in understanding mechanisms of growth angle regulation is to

Elison B. Blancaflor (ed.), *Plant Gravitropism: Methods and Protocols*, Methods in Molecular Biology, vol. 2368,
https://doi.org/10.1007/978-1-0716-1677-2_10,

establish whether or not the organ in question has a GSA. The test for this involves a very simple reorientation assay in which the organs are shifted to angles that are either more vertical or more horizontal than their original growth angle [1, 3, 4]. If the organ undergoes tropic growth to bring it back close to its original growth angle, then the organ has a GSA. The corollary of this is that lateral roots and shoots with a GSA must be able to effect tropic growth upward as well as downward to achieve this. There are two other points of note here. The first, we contend, is that an organ does not need to move back to precisely its original angle of growth to be considered to have a GSA. This is because in many species the regulation of GSA is dynamic in that it changes as the organ develops. In Arabidopsis for example, lateral roots become increasingly vertical over the course of several days of growth and their development has been classified into stages [1–6]. Stage I lateral roots have just emerged from the primary root axis and grow close to the horizontal. Stage II lateral roots undergo a brief period of downward growth that is associated with the development of starch statoliths in the columella and a recognizable elongation zone [3, 6]. Stage III lateral roots then undergo stable, angled growth at specific non-vertical angles, while stage IV roots grow at near-vertical orientations [6]. In our experimental conditions, stage III lateral roots range from 0.5–3 mm in length and remain at this stage for approximately 24 h. At all points tested during Stages III–IV of non-vertical growth, lateral roots have the capacity to effect a tropic response that counteracts the displacement from their GSA [1, 4]. It follows that the growth profile of the Arabidopsis lateral root represents a continuous sequence of increasingly vertical GSA states [1, 2] and hence, there may be a natural developmental shift in the GSA of the organ during the course of a reorientation experiment.

The second point is that for an organ to be considered as having a GSA then the capacity to perceive and maintain its growth angle in the gravity field must be inherent in the organ itself. A branch that is maintained at a particular growth angle because of the gravitropic response in some other part of the plant to which it is attached, would not be considered to have a GSA [1, 2].

It is with these considerations of the dynamic nature of GSA regulation throughout the development of individual organs that these methods have been written. They include basic protocols for quantifying GSA and also experimental systems that can be used to gain insights into the mechanisms underlying the maintenance of gravity-dependent non-vertical growth. Specifically, we describe the use of a simple two-dimensional clinostat to investigate the growth patterns observable in lateral root and shoot branches in the absence of a stable reference to gravity. Clinostats are devices that allow experimental material to be rotated at various angles with respect to the gravity vector. For studies of graviresponse, the most common implementation involves slowly rotating plants perpendicular to the gravity vector, such that the plant is subject to omnilateral gravitational stimulation. Using this approach, it has recently been shown that the lateral roots and shoots of *Arabidopsis thaliana* are maintained at non-vertical GSAs through the activity of an antigravitropic offset mechanism that operates in tension with underlying gravitropic response in the branch to generate stable, gravity-dependent, angled growth. The methods described here are the ones we have used to analyze GSA in Arabidopsis, bean, wheat, and rice. They can, of course, be adapted and applied to any species.

2 Materials

2.1 Lateral Root Growth Angle and Reorientation Assays

1. 100% Ethanol.
2. 10% Bleach.
3. Sterile distilled water.
4. *Arabidopsis thaliana* salts (ATS) plant growth medium: 5 mM KNO_3, 2.5 mM KH_2PO_4, 2 mM $MgSO_4$, 2 mM $Ca(NO_3)_2$, 50μM Fe-EDTA, 1 ml/L micronutrient stock solution (consisting of 70 mM H_2BO_3, 14 mM $MnCl_2$, 0.5 mM $CuSO_4$, 1 mM $ZnSO_4$, 0.02 mM $NaMoO_4$, 10 mM NaCl, 0.01 mM $CoCl_2$), 1% Sucrose, 0.8% Plant Agar.
5. Polystyrene 9 cm round and 12 cm square sterile petri dishes.
6. Sterile filter paper.
7. Sterile pointed end cocktail sticks.
8. Aluminium foil.
9. Protractor.
10. Spirit level.
11. Imaging equipment: a digital camera (e.g., Canon G9 or similar) or flatbed scanner (e.g., Hewlett Packard Scanjet G4050 Scanner or similar).
12. ImageJ analysis software (http://imagej.nih.gov/ij/).

2.2 Lateral Shoot Branch Growth Angle and Reorientation Assays

1. Soil mixture.
2. Soil trays with individual 4 × 4 cm square cells.
3. Aluminium foil.
4. Spirit level.
5. Digital camera with a hot shoe (e.g., Canon G9 or similar).
6. Hot shoe-mounted two axis spirit level (e.g., Polaroid Hot Shoe Two Axis Spirit Level).
7. Camera tripod.
8. ImageJ analysis software (http://imagej.nih.gov/ij/).

2.3 Clinorotation of Lateral Root and Shoot Branches

1. Horizontal two-dimensional clinostat with rotation speed of 1 revolution per minute (r.p.m.).
2. Horizontal two-dimensional clinostat with rotation speed of 4 revolutions per hour (r.p.h.).

2.4 Reorientation Kinetics of Lateral Roots

1. 100% Ethanol.
2. 10% commercial bleach.
3. Sterile distilled water.
4. *Arabidopsis thaliana* salts (ATS) plant growth medium.
5. Polystyrene 9 cm round and 12 cm square sterile petri dishes.
6. Sterile filter paper.
7. Sterile pointed end cocktail sticks.
8. Aluminium foil.
9. Protractor.
10. Infrared converted camera with a macro lens (*see* **Notes 1 and 2**).
11. Hot shoe-mounted two axis spirit level (e.g., Polaroid Hot Shoe Two Axis Spirit Level).
12. LED backlight panel with dimmer (Addlux, WL: 940 nm) (*see* **Note 3**).
13. Computer with software for remote imaging that is compatible with the camera and computer in use. We used the "Image Capture" software on OS El Capitan on a 2013 MacBook Pro laptop.

2.5 Lateral Root Growth Angle and Reorientation Assays in Larger Seedlings

1. Seed of species of interest.
2. Sterile polystyrene 9 cm petri dishes.
3. Sterile filter paper disks.
4. Sterile water.
5. Dropper.
6. "CYG" ™ seed germination pouches (Mega-International, USA).
7. Pouch holder.
8. Scissors.
9. Clean, large container such as a large Tupperware.
10. Sterile forceps.
11. Hoagland's No. 2 basal salt mixture (Sigma-Aldrich, Cat. no: H2395).
12. Digital camera with a hot shoe (e.g., Canon G9/Sony RX100 or similar).

13. Camera Tripod.
14. ImageJ analysis software (http://imagej.nih.gov/ij/).

3 Methods

3.1 Lateral Root Reorientation Assays

1. Sterilize Arabidopsis seeds in a 1.5 mL tube by immersion in 100% ethanol for 2 min, followed by 10% bleach for 20 min and washing five times with sterile distilled water. After the last wash, keep the seeds immersed in sterile water. Wrap the tube with foil and cold treat the seeds at 4 °C for 2–3 days to promote the rate and uniformity of germination.
2. Prepare the media plates by pouring approximately 25–30 mL of melted ATS media into sterile petri dishes (*see* **Note 4**).
3. Once the media plates have hardened, carefully drain the water from the seeds and transfer them to sterile filter paper. Use a sterile toothpick to place the seeds in a row towards the top of the plate (surface tension should hold the seed to the toothpick sufficiently well to make this possible. A 20–200μL pipettor and sterile pipette tip can also be used). Make sure that the seeds are at least 1.0–1.5 cm apart so that the lateral root system for each plant can be easily observed.
4. Incubate the plates in standard tissue culture conditions at 20 ± 2 °C, long day (16 h light/8 h dark cycle), and 400–500μmol m^{-2} s^{-2} light for 10–14 days.
5. Once lateral roots are at Stage III, reorientation assays may be performed. Obtain high-resolution images of the root systems of the seedlings using either a digital camera or flatbed scanner. If a scanner is used, take care not to leave the plates horizontal for longer than absolutely essential (*see* **Note 5**).
6. Reorientate the plates by 30° (use a spirit level and protractor to establish and then check the displacement). Scan again after at least 12 h of incubation in plant standard tissue conditions.
7. To determine if lateral roots are maintaining GSAs, quantify the growth angles of lateral roots before and after gravistimulation using the ImageJ software analysis package. Evaluate these by measuring the growth angle made by the 0.5 mm segment of the root tip before and after reorientation (*see* **Note 6**).

3.2 Lateral Shoot Branch Reorientation Assays

1. Sow Arabidopsis seeds in individual cells of soil trays (or small pots) with suitable amounts of soil in each cell. Cover the tray with foil and cold treat at 4 °C for 2–3 days.
2. Transfer the tray to a plant growth chamber at 20 ± 2 °C, long day (16 h light/8 h dark cycle), and 400–500μmol m^{-2} s^{-2} light.

3. Keep the plants suitably watered for 3–4 weeks until they have started to produce lateral inflorescence branches that are approximately 5 cm in length. In practice, lateral branches emerge from the primary axis and elongate at different times so the youngest branches at the top of the plant may still be <5 cm.
4. To perform reorientation assays, photograph individual lateral branches with a digital camera against a black background. Take care to ensure that the camera, mounted on a tripod, is absolutely parallel with the vertical plane encompassing main shoot and the lateral branch being photographed (also ensure that the ruler is positioned in this same plane). Use a hot shoe-mounted 2-axis spirit level to ensure that the camera is being held in the vertical plane (i.e., parallel with the gravity vector).
5. Reorientate the plants by 30° (use a spirit level and protractor to establish and then check the displacement). After a further 12 h of growth, photograph individual branches as described above.
6. To determine if shoot branches are maintaining GSAs, quantify the growth angles of lateral shoot branches before and after gravistimulation using the ImageJ software analysis package. Evaluate these by measuring the growth angle made by the 0.5 cm segment of the shoot tip before and after reorientation (*see* **Note 6**).

3.3 Quantification of Lateral Root Growth Angle Profiles

1. Prepare Arabidopsis seeds and media plates as described in Subheading 3.1.
2. Incubate the plates in standard tissue culture conditions at 20 ± 2 °C, long day (16 h light/8 h dark cycle), and 400–500μmol m^{-2} s^{-2} light conditions.
3. Five days post-germination, transfer the seedlings carefully to 12 cm square sterile petri dishes containing approximately 50–60 mL of ATS medium. Place the seedlings ~2 cm apart to allow for optimum visualization of the root system.
4. Once lateral roots are approximately 5 mm long (approximately 7 days after transfer), the evaluation of growth angle profiles can begin. Scan the plates using a high-resolution scanner as described in Subheading 3.1.
5. Using the ImageJ analysis software, measure ten 0.5 mm segments along the length of each lateral root to be analyzed. Analyze at least 15–20 lateral roots for each genotype/mutant/treatment (*see* Fig. 1a).
6. Measure the angle that each of these segments makes with the vertical, i.e., the direction of gravity (*see* Fig. 1a, b).

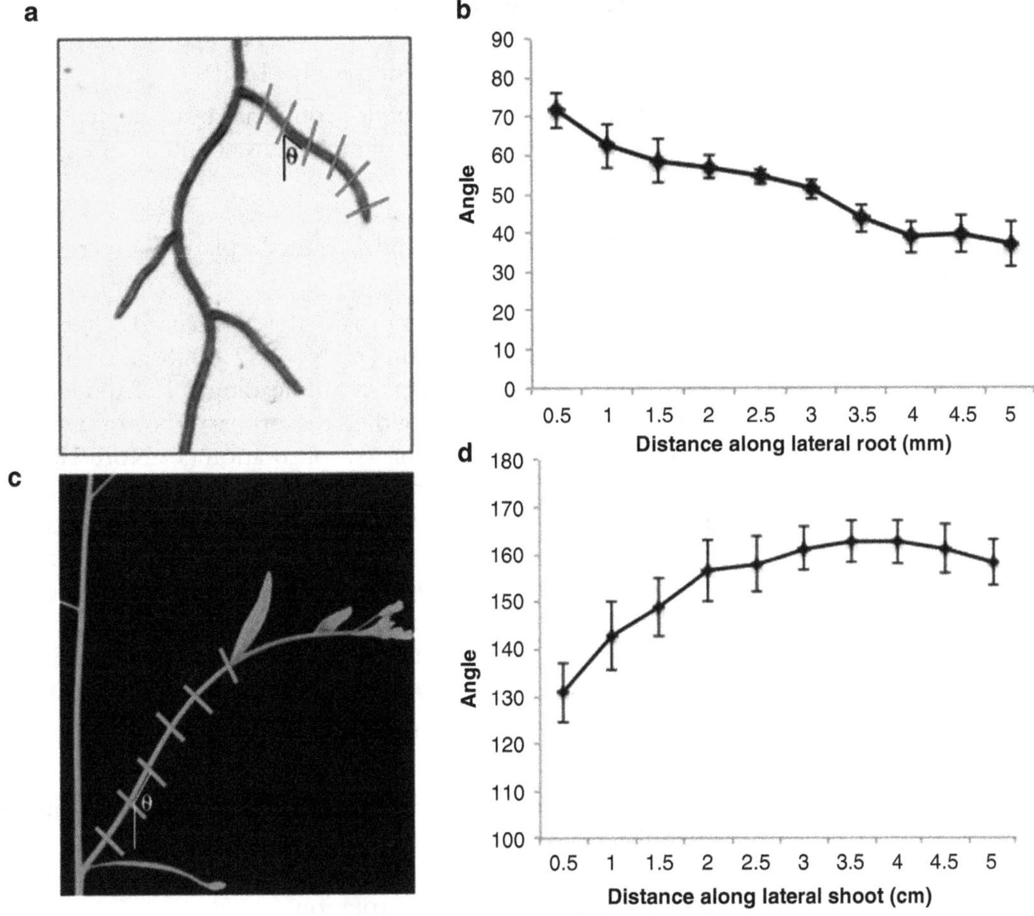

Fig. 1 Evaluation of GSA profiles of lateral root and shoot branches. (**a**, **b**) Division of a lateral root of Arabidopsis into 0.5 mm segments (**a**) and typical GSA profile of an average of 15–20 WT Col-0 lateral roots on ATS medium (**b**). Error bars represent standard error of the means. (**c**, **d**) Division of a lateral shoot of Arabidopsis into 0.5 cm segments (**c**) and typical GSA profile of an average of 15–20 WT Col-0 lateral shoots (**d**). Error bars represent standard error of the means

7. The lateral root growth angle profile of a particular mutant or ecotype may thus be calculated as the average of the growth angle of each equivalent 0.5 mm segment (*see* Fig. 1a, b).

3.4 Quantification of Lateral Shoot Branch Growth Angle Profiles

1. Germinate Arabidopsis seeds in individual cells as described in Subheading 3.2.
2. Once lateral shoots have emerged and elongated to approximately 5 cm in length, they may be photographed using a digital camera as described in Subheading 3.2 (*see* **Note 4**).
3. Using the ImageJ analysis software, divide each lateral branch into ten 0.5 cm segments (*see* Fig. 1c).

4. Measure the angle that each of these segments makes with the vertical. Analyze at least 15–20 lateral branches for each mutant/genotype/treatment (*see* Fig. 1c, d).
5. The lateral branch growth angle profile may be calculated as the average of the growth angle of each equivalent 0.5 cm segment (*see* Fig. 1c, d).

3.5 Clinorotation of Lateral Root Branches

1. Prepare Arabidopsis seeds and ATS media plates as described in Subheading 3.1.
2. Once the lateral roots are approximately at Stage III clinorotation, experiments may begin (*see* **Notes 7–9**). Scan or photograph the plates as described in Subheading 3.1. Experiments may be performed where either the primary root or a single selected lateral root is on the axis of rotation (*see* **Note 10** and Fig. 2a, b). Wrap the plates in aluminium foil to ensure that any growth pattern changes seen are not due to negative phototropism. Unclinorotated control plates should be wrapped in foil to confirm that the transition to darkness does not induce upward growth in lateral roots. Mount the petri dish securely in a horizontal orientation on a 2D clinostat and begin clinorotation at 1 r.p.m. (*see* **Notes 8, 9, and 11**, and Fig. 2a, b).
3. The plants should be clinorotated for 6 h and then re-scanned/re-photographed.
4. Changes in growth angle following clinorotation may again be quantified using ImageJ. Measure the angle that the 0.5 mm tip segment of each lateral root is oriented at with respect to gravity before and after clinorotation.

3.6 Clinorotation of Lateral Shoot Branches

1. Germinate Arabidopsis seeds as described in Subheading 3.1.
2. Once lateral branches are approximately 5 cm in length, clinorotation experiments may begin (*see* **Notes 7, 8 and 12**). Photograph individual branches as described in Subheading 3.1. Experiments may be performed where either the primary shoot or a single selected shoot branch is on the axis of rotation (*see* **Note 10** and Fig. 2c, d). Carefully mount each plant securely on a horizontal 2D clinostat and begin clinorotation at 4 r.p.h. (see **Notes 8–10**, and Fig. 2c).
3. After 6 h of clinorotation, photograph lateral branches once again as previously described.
4. Changes in growth angle following clinorotation may again be quantified using ImageJ as described in Subheading 3.3. Measure the angle that the 0.5 cm tip segment of each lateral shoot is oriented at with respect to gravity before and after clinorotation.

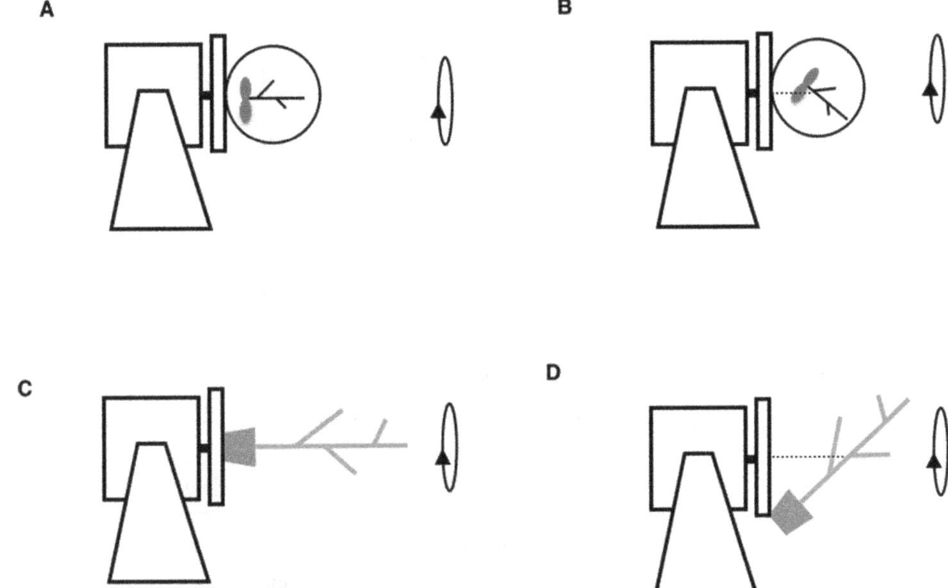

Fig. 2 Set up of clinorotation experiments. (**a**, **b**) Lateral root clinorotation experiment with primary root in the axis of rotation (**a**) or a single lateral root in the axis of rotation (**b**). The axis of rotation is represented by the dotted line. (**c**, **d**) Lateral shoot clinorotation experiment with primary shoot in the axis of rotation (**c**) or a single shoot branch in the axis of rotation (**d**)

3.7 Reorientation Kinetics of Lateral Roots

1. Perform **steps 1–4** in Subheading 3.1.
2. Set up the imaging system with the backlighting as described in Fig. 3. The box used should be kept in standard temperature-controlled conditions, in a shaded area, on a flat surface and away from sources of vibration. Secure the LED panel to one side of the box with tape or some other sufficiently secure method, checking verticality with a spirit level. Place the camera at suitable distance and height so that the imaging field is centered on the panel and that an entire plate can be accommodated within the field of view. Use a hot-shoe-mounted spirit level to check that the camera sensor is parallel with the LED panel in the vertical plane.
3. When the Arabidopsis seedlings have a well-developed root system, carefully lift the square petri dish from the holder and place it in a vertical position inside the black box. Place the lid on the box and allow the plate to remain in darkness for 1 h prior to beginning the assay.
4. Find camera exposure settings that permit light power is kept to a minimum (5–20% of maximum power), with ISO is less than 400, and shutter speed ~1 s. (*see* **Notes 13–15**).
5. Choose an appropriate angle of reorientation (*see* **Note 16**). Carefully rotate the petri dish to the chosen angle and use

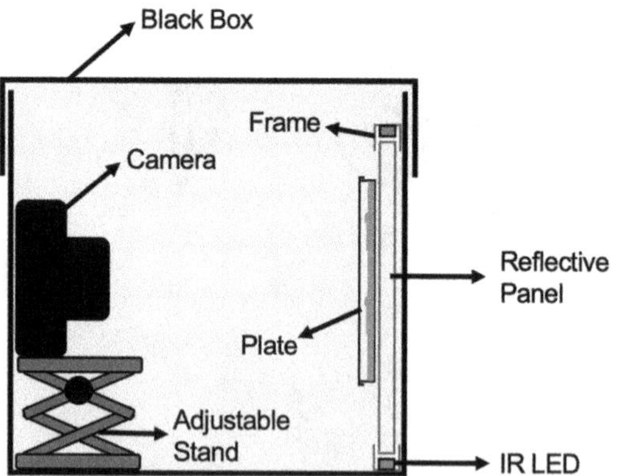

Fig. 3 Set up for imaging of root reorientation kinetics. Schematic representation of the cross section of the imaging set up for seedlings grown on agar plates. Back panels were constituted by a solid, thick reflective panel with an inward-facing strip of infrared (IR) LED lights on all four sides. The LED strip was masked by a metal frame to minimize lateral illumination. The infrared converted camera had a macro lens and was placed on an adjustable stand. Everything was contained in a black box to block any external light

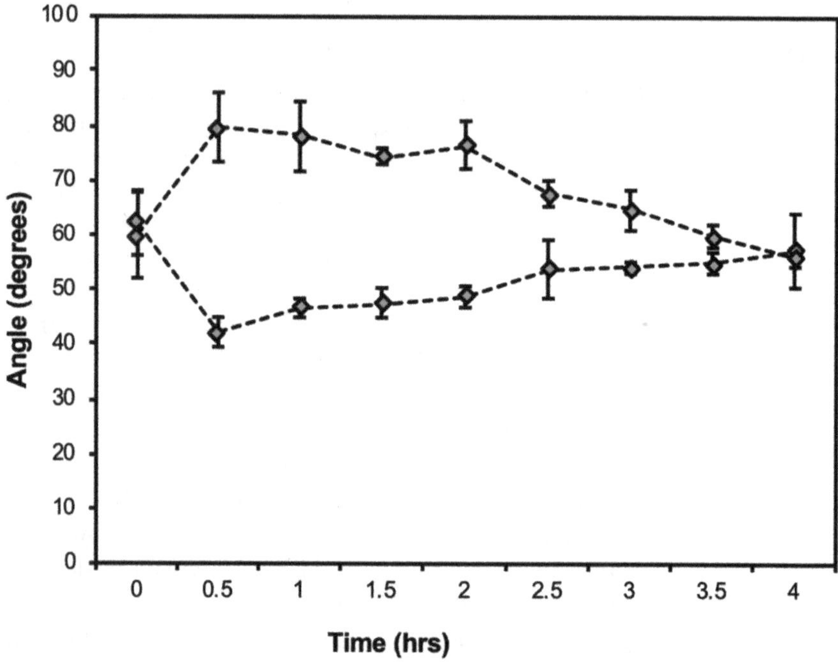

Fig. 4 Reorientation kinetics of Stage III WT Col-0 lateral roots. Kinetics of upward and downward reorientation in WT Col-0 stage III lateral roots reorientated from their original GSAs by 30°. The data are an average of kinetics of bending from three independent experiments, with 10 roots quantified for each direction of bending per experiment. Bars represent s.e.m

masking tape to secure the plate to the back panel in this position.

6. Optimize the focus of the camera to ensure that high quality clear images can be obtained (*see* **Note 17**).
7. Using the remote imaging software, set up the time lapse system to capture an image of the plate every 10 min for at least 6 h. (*see* **Note 18**).
8. Once the time lapse imaging is completed, analysis can be performed. Lateral roots will be divided into those that were reoriented above their GSA (and therefore would be effecting a gravitropic response or "bending" downward), and those that are rotated below their GSA (and hence bending upward). Using the ImageJ analysis package, measure the angle made by the last 0.5 mm segment of the root tip to the vertical a time interval of your choosing. We recommend measuring root angles every 30 min.
9. The kinetics of root bending can be plotted on a graph such as in Fig. 4, where each time point denotes the average GSA of several upward or downward bending roots, respectively. The time point of 0 min denotes the lateral root GSA before reorientation. Measure at least 10 roots, from multiple plates if needed (*see* **Note 19**), at each time point for both upward and downward kinetics. Roots can be determined to return to their original GSA when they come to within 5° of their starting angle before reorientation (*see* **Note 3**).

3.8 Lateral Root Growth Angle and Reorientation Assays in Larger Seedlings

1. Prepare and autoclave suitable growth media for the species concerned. For example, for wheat, prepare and autoclave 5 L solution of Hoagland's medium.
2. Using a pair of forceps, place a filter paper disk into the bottom of a 9 cm petri dish. Using the same pair of forceps, carefully place 4–5 seeds spaced well apart from each other on the filter paper in the petri dish. Using a dropper, carefully add a few drops of sterile water to moisten the filter paper. Place the lid on the petri dish and place the dish in standard tissue culture conditions for 3–4 days.
3. Check the petri dish regularly for signs of contamination and keep the filter paper moistened with sterile water.
4. After 3–4 days, when the seeds have germinated, transfer each one to individual seed pouches. To do this, using a pair of scissors, first cut about 2 cm off the plastic covering at the bottom of each pouch to expose the wick.
5. Next, using sterile forceps, carefully enlarge the central perforation of the wick trough (see Fig. 5). Carefully place the

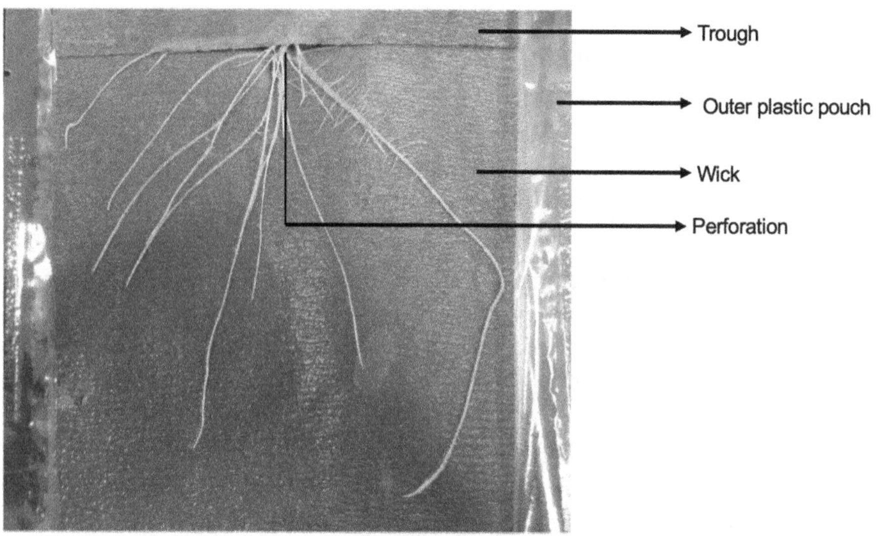

Fig. 5 Set up of larger seedling species in CYG™ seed germination pouches. Representative image of a 7-day-old bean seedling grown in a CYG ™ seed germination pouch. Note that the outer plastic pouch has been peeled back for imaging the root system

germinated seed onto the trough with the root just coming through the perforation.

6. Place each pouch upright within the holder.
7. Place the holder vertically within the large container.
8. Flood the bottom of the container with an appropriate volume of Hoagland's medium.
9. Place the container in standard tissue culture conditions (*see* **Note 20**).
10. Depending on the species of interest, allow the root systems to develop for 3–7 days until seminal or adventitious roots are at least 5–7 cm long. Check the levels of Hoagland's medium regularly and keep the base of the container flooded.
11. At an appropriate time, carefully lift each pouch out of the holder and obtain high-resolution images of each seedling root system using either a digital camera or flatbed scanner. If a flatbed scanner is used, take care not to leave the pouches horizontal for longer than absolutely necessary. The outer plastic covering of the pouch may be cut open carefully and peeled back to avoid any glare during imaging.
12. Analyze seminal or adventitious root GSA as described in points 5–7 from Subheading 3.3 using a 1 cm interval along the length of the root.
13. Analyze at least 15 roots from at least 5–6 different plants per mutant/genotype/treatment.

14. To confirm if these roots are maintaining GSAs, quantify the angles made by the final 1 cm of the roots. Rotate the pouches by an appropriate angle (*see* **Note 14**), taking care to see that one end of the pouch remains in contact with the medium. Leave the pouches in this orientation for 3 days (*see* **Note 18**) and acquire high-resolution images again. Using the ImageJ software analysis package, quantify the angles made by the final 1 cm segment of the root before and after reorientation (*see* **Note 21**).

4 Notes

1. Any camera model can be selected, but it is important to ensure it is one that your local IR conversion company are able to convert. With respect to camera resolution, we have used cameras from ~12 megapixels (Canon 450D) up to ~42 megapixels (Sony A7RIII). For downstream image analysis, more pixels are better because it means that more roots can be imaged at sufficient resolution within the field of view. Using the camera's maximum resolution settings, we imaged an area of about 14 cm × 9 cm (1 pixel = 0.176 mm); this was enough to image most of a 12 cm × 12 cm square petri dish surface and resulted in ~10 pixel wide roots. With respect to the choice of lens, we recommend a macro lens because, although imaging will typically be done above the minimum focusing distance of many non-macro lenses, the macro gives the flexibility to get closer to the subject and increase the number of pixels captured across each root, should this be required. For reference, we typically use a Sony 50 mm F/2.8 Macro lens with our Sony A7RIII cameras.
2. Most commercially available digital consumer cameras are unsuitable for photography under infrared (IR) light because they have an IR filter in front of the camera sensor. While cameras without such filters can be purchased from scientific suppliers, it is typically much more economical to buy digital cameras with an interchangeable lens (allowing easy access to the camera sensor) and to have the IR filter removed by a specialist in IR camera conversions. Since IR filters are often removed for astrophotography, it is usually easy to identify companies offering these services by searching the internet with the phrase "IR camera conversion". You can elect to have the IR filter replaced with either a filter excluding light below ~880 nm or simply a clear glass filter (full spectrum). We have typically used the latter because the box or room within which the imaging is performed should not allow any visible light in. Any light "leak" is undesirable because it introduces

the possibility of phototropic and other photomorphogenic responses and a camera lacking a visible light cut-off filter should reveal such leaks in the captured images.

3. The chosen back panels were constituted by a solid, thick reflective panel with an inward-facing strip of LED lights on all four sides. The LED strip was masked by a metal frame to minimize lateral illumination. This set up insures even lighting, while minimizing heating. Since the plates are taped to the back panel, it is important that this is secure and vertical. If the back panel has to be attached, suspended, to the inside of the box, the tape would have to sustain its weight for a long period of time and verticality would depend on the inclination of the side of the box. If the side of the box is found to not be vertical, the back panel should be lightly taped to the inner wall of the box and only securely fixed in place once inclination has been assessed using a spirit level and confirmed to be vertical.
4. Other growth media may be used in place of ATS (e.g., Murashige and Skoog medium, 8).
5. For root GSA analyses, we found that scanning plates gave us the highest resolution images. However, images could also be taken using a high quality digital camera. In order to avoid having to move plants to the horizontal for scanning we have found it is possible to fix the Hewlett Packard Scanjet G4050 Scanner on its side so that plates can remain vertical during scanning. We expect that the same will be true of many other makes and models of scanner although we cannot guarantee that malfunction will not result!
6. We noticed that Arabidopsis root and shoot branches, apparently under GSA control, consistently return to a growth angle that is slightly more vertical than the original angle of growth [4]. This is consistent with the idea, discussed in the introduction, that GSA control in Arabidopsis is dynamic, raising the possibility that this more vertical final angle post-reorientation merely reflects a biologically relevant shift in GSA for these organs. In addition, since it has been shown that auxin regulates GSA [4], it should also be borne in mind that the act of reorientating the plant might impact on the GSA of the organ under investigation. As discussed in the introduction, we consider the capacity for branches to reorientate their growth both upward and downward to be a more robust hallmark of existence of a GSA than the absolute growth angle achieved following reorientation.
7. It is important to make sure that lateral branches used for GSA analyses are all approximately the same length and therefore at the same stage of development. This is because lateral branches that elongate change their GSA rapidly, and differences in

developmental stages may lead to artificial differences in GSA between different plant lines. Also, in contrast to the root system, it is not possible to detect changes in GSA in lateral branches from their final growth profile.

8. The direction of clinorotation may be either clockwise or counter clockwise. The growth patterns of lateral roots and shoot branches we have observed are unaffected by the direction of clinorotation [4].
9. It is important to ensure that the primary root or shoot is not gravistimulated prior to clinorotation to prevent the development of gravitropic curvature in these organs; it has been shown that even brief periods of gravistimulation (as short as 10 s) are sufficient to cause bending during subsequent clinorotation [7–10]. Thus, plants or seedlings should be mounted quickly on the horizontal clinostat. Many clinostats can be adjusted between the horizontal and vertical. Ideally plants should be mounted while the clinostat is in vertical orientation and then smoothly moved to the horizontal while the stage is rotating at the desired speed.
10. Clinostats are, of course, a highly artificial and abstract system for plant growth. For this reason, it is imperative that sufficient controls are in place to avoid misinterpreting the patterns of growth that are observed during clinorotation. Clinorotation speeds need to be chosen carefully such that statoliths in gravity-sensing cells are being rotated fast enough to remain in suspension, but not so fast to cause centrifugal effects [11, 14]. There is also the possibility that vibration or other physical disturbance from the clinostat could affect growth during clinorotation. These factors can be largely ruled out by simply rotating plants at the test rotation speed in a vertical orientation. In this configuration, there should be no change in the growth patterns of lateral branches. In studying the effects of clinorotation on lateral branch growth, it is also necessary for the gravity-sensing cells of at least some branches to be orientated at an angle to the axis of rotation, a factor that has the potential to influence the experiment. Therefore, it is important to perform clinorotation experiments in two formats, one in which the primary root or shoot is in the axis of rotation and another where a single selected lateral branch is in the axis of rotation. For the latter, all other branches and the primary root or shoot will either be at an angle to the axis of rotation or at some distance from it. It is only safe to assume that patterns of growth under clinorotation reflect something of the underlying GSA biology of the branch if those patterns are observed when performed both on and away from the axis of rotation. The experimental format in which the primary root or shoot is placed at an angle to the axis of rotation provides

another useful check on the suitability of the experimental setup. Under no circumstances should the primary root or shoot bend. If it does then something is amiss and that clinorotation setup cannot be used to discern anything about the GSA biology of branches. As noted above, if bending of the primary axis is observed then it is first important to rule out gravistimulation prior to clinorotation as the cause. The clinorotation protocols described here were developed for Arabidopsis but, keeping in mind the considerations discussed above, can in principle be adapted for any species.

11. In order to exclude concerns about artifacts arising from the interaction between the effects of clinorotation and the growth of root systems on a flat "two-dimensional" surface, root clinorotation experiments may also be performed using a "three-dimensional" system. Sterile Arabidopsis seeds may be germinated on the surface of ATS medium containing 0.2% Phytagel in a glass beaker. The low density of the phytagel allows the root system to grow into the medium. Once lateral roots are produced, after 12–14 days of growth, the beaker may be placed on the clinostat in a similar orientation as the petri dishes.
12. During clinorotation, the primary shoot may be carefully supported using a thin stake in order to prevent excessive movement over the course of clinorotation.
13. Before starting experiments, test the light-tightness of the box. Prepare four plates as described in **steps 1–3** in Subheading 3.1. Place the plates in vertical position in the conditions described in **step 4** in Subheading 3.1 for a day, to stimulate germination. Transfer the plates to the box, after wrapping two of them in aluminium foil. Incubate the plates in vertical position in the box for 2–3 days and compare the etiolated seedlings in the two sets of plates.
14. Depending on fluence rate, plants have been shown to respond to up to 1000 nm light. We chose to use 940 nm LED since the response threshold is ten times lower compared to 880 nm light [12]. Run a test similar to that described in **Note 13** with the light on to confirm that plants do not respond to the IR light set up.
15. Since LED lights will tend to warm up if kept on at maximum power for prolonged length of time, the right camera settings will be a compromise between LED power, sharpness, and exposure. Choose low ISO to minimize image noise. We try to use the lowest ISO settings possible within the range 50–400. Note that sensor noise at higher ISO settings is a property of the sensor used and continual improvements in sensor technology mean that higher and higher ISO settings

are becoming usable for analysis of growth angle. The maximum usable ISO will need to be determined empirically for each camera type used. This can be done by capturing images at a range of ISOs and attempting angle measurement to determine the ISO setting at which measurement of angles is hampered by poor resolution. Set the camera to Av (Aperture priority) and set aperture to medium values [4–10, 13], according to your requirement for depth of field (*see* **Note 17**). Then, lower the light intensity until time exposure is close to 1 s, even if black bands start appearing on the screen. Capture an image to test sharpness and exposure: if the dark bands have not disappeared, overexpose of 1 f-stop at the time until the image is uniform. Modify the aperture value if exposure time gets to 2 s to avoid losing sharpness.

16. When performing reorientation assays, there is a limit to the angular displacement that should be imposed. This is because it is essential that the lateral organs under investigation are not shifted past the vertical, where the gravity sensing cells would be stimulated to repolarize such that faces of the cells that were previously on the lower side (closer to the center of the Earth) were now the faces on the upper side. In such a situation, the potential repolarization of gravity response would confound the purpose of the experiment. So for example, if a lateral root had a GSA of 40°, a downward displacement of 30° would be acceptable whereas a displacement of 45° would not. A further consideration is that one would typically hope to capture the behavior of many roots on a given plate, and so the displacement angle should be chosen to maximize the number of measureable roots (or shoots) across the entire experiment, paying attention to potential differences in starting GSA between ecotypes, genotypes, or treatments being studied.
17. It is important to ensure that roots are in focus across the image. This will be achieved by focusing the lens carefully, but also by ensuring that the aperture selected gives sufficient depth of field (DoF). The smaller the aperture (the larger the f-stop value), the greater the depth of field. In practice apertures of f/4 and above should give sufficient DoF, assuming that the seedling plate and camera sensor have been arranged so that they are as close to parallel as possible.
18. Some species or genotypes may have extremely slow kinetics of reorientation. In order to confirm if lateral roots in these lines are actively maintaining GSAs, it may be appropriate to continue imaging for up to 24 h in some cases. In all cases, it is essential to confirm that the roots are continuing to elongate. If they are not, they should be excluded from the analysis.

19. If the kinetics of lateral root orientation are to be compared between two different lines, such as for example between WT and a mutant of interest, plate both lines on one plate, and perform the experiment with multiple plates.
20. We refer to the system we have described above as "semi-sterile," because, unlike in standard tissue culture systems, the system is not self-contained, particularly after Hoagland's medium has been added. Although the seed pouches can be sterilized according to the manufacturer's instructions, we do not deem it necessary. The aim is to eliminate the growth of mold on the germinating seedlings in the petri dishes and the pouches. We did not find it necessary to set up the system in a sterile environment such as a flow hood, but we recommend following sterile technique as far as possible.
21. In general, we found that the kinetics of root gravitropism for monocot roots were significantly slower than Arabidopsis roots, at least in our experimental conditions. We therefore recommend leaving the pouches for 3–4 days prior to quantifying root tip angles again.
22. If a flatbed scanner is used for imaging, commercially available software such as RootNav can be used to analyze root tip angles to increase throughput. The advantage of using this software is that additional parameters, such as root length and degree of branching may be quantified from the same image.

References

1. Digby J, Firn R (1995) The gravitropic set-point angle: identification of an important developmentally controlled variable governing plant architecture. Plant Cell Environ 18:1434–1440
2. Roychoudhry S, Kepinski S (2015) Branch growth angle control—the wonderfulness of lateralness. Curr Opin Plant Biol 23:124–131
3. Mullen JL, Hangarter RP (2003) Genetic analysis of gravitropic setpoint angle. Adv Space Res 10:2229–2236
4. Roychoudhry S, Kieffer M, Del Bianco M, Kepinski S (2013) Auxin controls gravitropic-setpoint angle in higher plant lateral branches. Curr Biol 5:1497–1504
5. Guyomarc'h S, Leran S, Auzon-Cape M, Perrine-Walker F, Lucas M, Laplaze L (2012) Early development and gravitropic response of lateral roots in *Arabidopsis thaliana*. Philos Trans R Soc B 367:1509–1516
6. Ruiz-Rosquete M, von Wangenheim D, Marhavy P, Barbez E, Stelzer EHK, Benkova E, Maizel A, Kleine-Vehn J (2013) An auxin transport mechanism restricts positive orthogravitropism in lateral roots. Curr Biol 9:817–822
7. Murashige T, Skoog F (1962) A revised medium for rapid growth and bioassays with tobacco tissue cultures. Physiol Plant 15:473–497
8. Kiss JZ, Hertel R, Sack FD (1989) Amyloplasts are necessary for full gravitropic sensitivity in roots of Arabidopsis thaliana. Planta 177:198–206
9. Blancaflor EB, Hou GC, Mohamalawari DR (2003) The promotive effect of latrunculin B on maize root gravitropism is concentration dependent. Adv Space Res 31(10):2215–2220
10. Hou G, Mohamalawari DR, Blancaflor EB (2003) Enhanced gravitropism of roots with an enhanced cap root actin cytoskeleton. Plant Physiol 131(3):1360–1373
11. Hou G, Kramer VL, Wang YS, Chen R, Perbal G, Gilroy S, Blancaflor EB (2004) The promotion of gravitropism in *Arabidopsis* roots upon actin disruption is coupled with the

extended alkalinization of the columella cytoplasm and a persistent lateral auxin gradient. Plant J 39:113–125

12. Schafer E, Lassig T-U, Schopfer P (1982) Phytochrome-controlled extension growth of Avena sativa L. seedlings. 11. Fluence rate response relationships and action spectra of mesocotyl and coleoptile responses. Planta 154:231–240

13. Waidmann S, Ruiz Rosquete M, Schöller M et al (2019) Cytokinin functions as an asymmetric and anti-gravitropic signal in lateral roots. Nat Commun 10:3540

14. Dedolph RR, Dipert MH (1971) The physical basis of gravity stimulus nullification by clinostat rotation. Plant Physiol 47:756–764

Chapter 11

A Flat Embedding Method to Orient Gravistimulated Root Samples for Sectioning

Utku Avci and Jin Nakashima

Abstract

Microscopy is an important tool used for biological research and has played a crucial role toward understanding of cellular mechanisms and protein function. However, specific steps in processing of biological samples for microscopy warrant improvements to consistently generate data that can more reliably help in explaining mechanisms underlying complex biological phenomenon. Due to their small and fragile nature, some biological specimens such as *Arabidopsis thaliana* roots are vulnerable to damage during long sample preparation steps. Moreover, when specimens with a small diameter (typically less than 100 μm) are embedded in conventional silicone mold or capsule embedding, it is not only difficult to locate their orientation inside the capsule, but also a challenge to obtain good median longitudinal sections. Specimen orientation in particular is crucial because understanding certain plant biological processes such as gravitropism rely on precisely knowing spatial information of cells and tissues of the plant organ being studied. Here, we present a simple embedding technique to properly orient small plant organs such as roots so that the desired sectioning plane is achieved. This method is inexpensive and can be accomplished with minimal equipment and supplies.

Key words *Arabidopsis thaliana*, Fixation, Flat embedding, Microscopy, Sectioning

1 Introduction

To better understand how plants sense and respond to gravity, it is important to obtain accurate spatial information of specific cells and tissues. However, retaining precise orientation of biological specimens as they are being processed for microscopy can be a daunting task. Thus, there is a need for simple and reliable methods to properly orient samples prior to sectioning for light and electron microscopy. This is especially true in the case of electron microscopy where very thin sections are necessary because small deviations from the desired sample orientation may result in misinterpretation of structural information and spatial relationships among cell organelles and types. There are published methods that address the need for correctly orienting delicate biological samples [1–3]. The flat

Elison B. Blancaflor (ed.), *Plant Gravitropism: Methods and Protocols*, Methods in Molecular Biology, vol. 2368, https://doi.org/10.1007/978-1-0716-1677-2_11,

embedding method described here is another convenient method that we have successfully used [4–6]. In this chapter, we provide updates to a procedure that we introduced previously [7]. This method is inexpensive and can be adapted easily using minimum equipment.

2 Materials

1. Glass microscope slides (*see* **Note 1**).
2. Cover glasses (*see* **Note 2**).
3. Diamond scriber pen to score the cover glass for breaking.
4. Dip Miser for slide coating (Electron Microscopy Sciences, Hatfield, PA). Alternatively, a 50 mL centrifuge tube can also be used.
5. Liquid release agent. 70% ethanol or a glass cleaning product.
6. Loctite super glue (*see* **Note 3**).
7. Aluminum weighing dishes.
8. Slide tray or rack to dry glass slides.
9. LR White resin (*see* **Note 4**).
10. A lab oven with temperature setting.
11. Binder clips.
12. Kimwipes, gloves, and plastic pipettes.
13. Forceps and dissecting teasing needles.
14. Hot plate or slide warmer.
15. Resin blocks. These can be made in the lab using any embedding molds and leftover resin.
16. Ultraviolet cryo chamber.
17. Stereomicroscope.

3 Methods

Safety precautions and appropriate laboratory attire (*i.e.*, lab coats), gloves, and goggles should be used while performing these methods.

3.1 Preparation of Glass Slides

1. Clean microscope slides with 70% ethanol using Kimwipes (Fig. 1a, b). Set the hot plate to low (35–40 °C) and place slides on the surface of the plate (Fig. 1c).
2. Using a diamond scriber pen, score cover glasses in desired widths (Fig. 1d) and separate into small pieces (Fig. 1e). These pieces are used as a spacer (Fig. 1f).

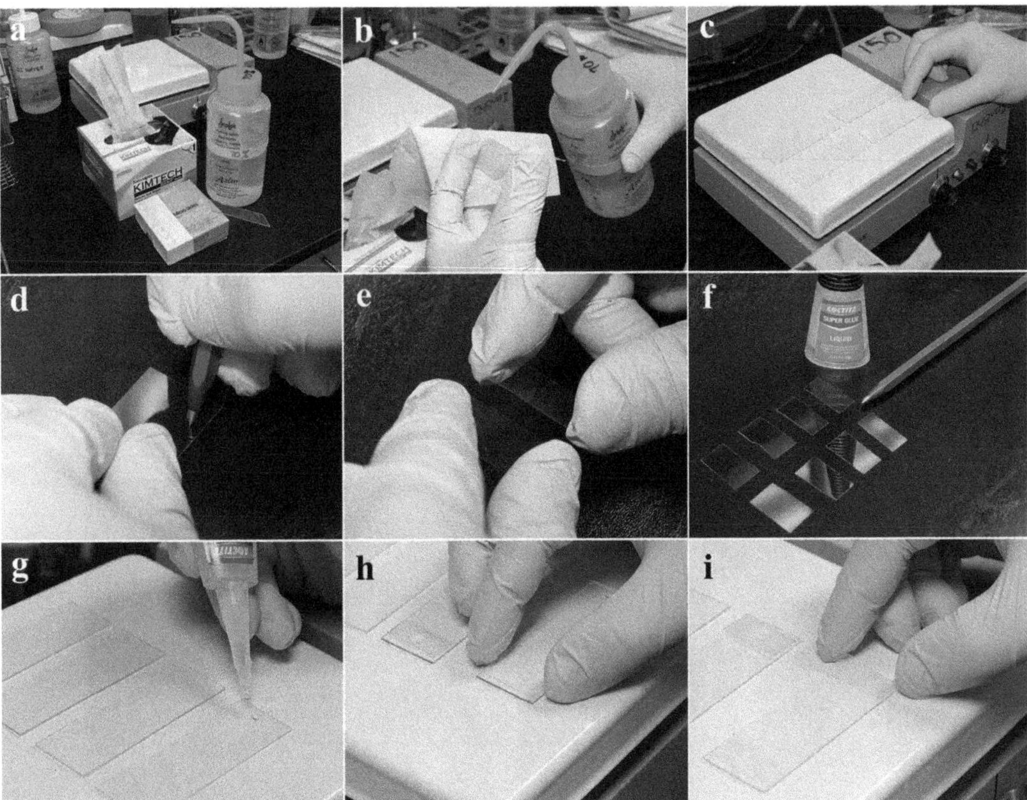

Fig. 1 Preparation of slides. Glass microscope slides, cover glasses, a diamond scriber pen, 70% ethanol, Kimwipes, and a hot plate are required for these steps (**a**). Cleaning glass slides with 70% ethanol (**b**). Slides on the surface of a hot plate set to low temperatures (35–40 °C) (**c**). Scoring cover glasses using a diamond scriber pen (**d**). Cover glasses being separated into equal halves after scoring (**e**). Pieces of cover glasses shown in (**e**) are used as spacers (**f**). For half of the glass slides, a tiny drop of Loctite super glue is applied on opposite edges of the cleaned glass slide (**g**). A piece of the scored cover glass spacer is glued onto each edge of the glass slide (**h** and **i**)

3. Apply a tiny drop of Loctite super glue on each corner of the cleaned slide for half of the glass slides since a spacer is required only on one side (Fig. 1g). Glue this spacer on each corner of the slide (Fig. 1h, i). The space needed for samples between glass slides is dependent on the diameter/thickness of the specimen. More than one layer of cover glass can be glued on top of each other if necessary (*see* **Notes 5** and **6**).
4. When the **steps 1–3** are completed and the slides are dry, remove them from the hot plate.

3.2 Coating Glass Slides with Liquid Release Agent

1. Select a dust-free work area preferably a clean bench (Fig. 2a).
2. Pour liquid release agent into a Dip Miser or a 50 mL centrifuge tube (Fig. 2b). Hold the slides with forceps at one end and immerse them into the liquid release agent (Fig. 2c). Remove

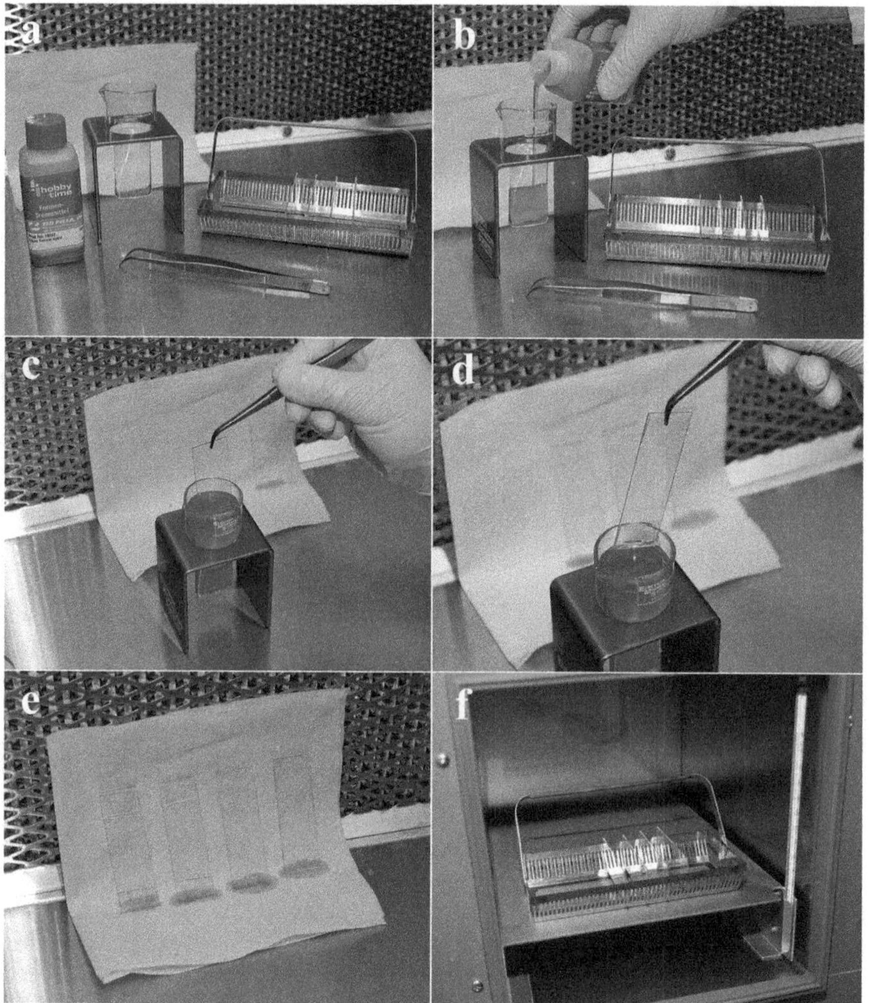

Fig. 2 Coating slide glasses with liquid release agent. A Dip Miser for slide coating, liquid release agent, glass microscope slides, forceps, slide tray, Kimwipes, and dust-free clean bench are required for these steps (**a**). Pouring liquid release agent into a Dip Miser (**b**). Immersing slides into the liquid release agent using forceps (**c, d**) Placing slides upright on Kimwipes for drying at room temperature after coating (**e**). Slides on a slide tray ready for drying in an oven (**f**)

the slides after immersion (Fig. 2d). Repeat this procedure about three times.

3. After coating, dry the slides by placing them upright on Kimwipes at room temperature for 2 h (Fig. 2e; *see* **Note 7**).
4. After slides have completely dried, transfer them into a slide tray or rack. A general purpose slide box or a slide staining dish can also be used. Dry slides overnight at 55 °C. Temperatures between 40 and 70 °C would usually work for drying (Fig. 2f).

3.3 Flat Embedding Procedure

1. Embed specimens between two slides. This step can be better performed under a stereomicroscope. Hold the slide using an aluminum weighing dish, which is also useful for catching any excess resin emerging from the slides (Fig. 3a).
2. Apply fresh resin onto a slide using a disposable plastic pipette. Using forceps or a plastic pipette position infiltrated specimens onto a slide (Fig. 3b; *see* **Note 8**). Gently separate specimens from each other using forceps or dissecting teasing needles (Fig. 3c; *see* **Note 9**).
3. When dealing with multiple samples in one glass slide, make sure that individual samples (*e.g.*, individual roots) do not touch each other. This will ensure that individual samples can be separated before mounting.
4. After samples have been laid out on the bottom glass slide, the top slide will have to be placed on top of the samples. To do this, the edge of the second slide should gently touch the edge of the bottom slide making sure that it maintains an angle of 30–45° (Fig. 3d). This approach should prevent air bubbles from forming. It is important to remove air bubbles because this may inhibit polymerization of the resin. Slowly add more resin using a plastic pipette around the edges of the slides so that air spaces are completely filled with resin (Fig. 3e; *see* **Note 10**).
5. Hold two slides together with binder clips so that specimens are flattened between the two slides (Fig. 3f). Depending on the amount of resin applied to the slides, resin may leak out between two slides due to the compression created by clips or may retract when there is not enough resin to cover a whole slide. If the resin retracts, additional resin is slowly added by a plastic pipette around the edges to fill empty spaces (Fig. 3g; *see* **Note 11**).
6. Repeat **steps 1–4** for all specimens. Slides are then ready to be transferred to an ultraviolet cryo chamber in a cold room at 4 °C (Fig. 3h) or an oven (Fig. 3i) for polymerization (*see* **Note 12**).

3.4 Extracting Specimens from Flat Embedded Resin

1. When polymerization is complete, remove binder clips and separate the two slides by inserting a razor blade in between the two slides. It is best to insert the razor blade at the corner of the two slides. Apply a twisting motion with the razor blade. This will create a force that will cause the top slide to separate from the bottom slide (Fig. 4a). If the slides do not separate completely, repeat the above procedure but this time use the opposite corner (*see* **Note 13**).
2. Only samples with the desired orientation and those that are well preserved after fixation are extracted from the slide. To

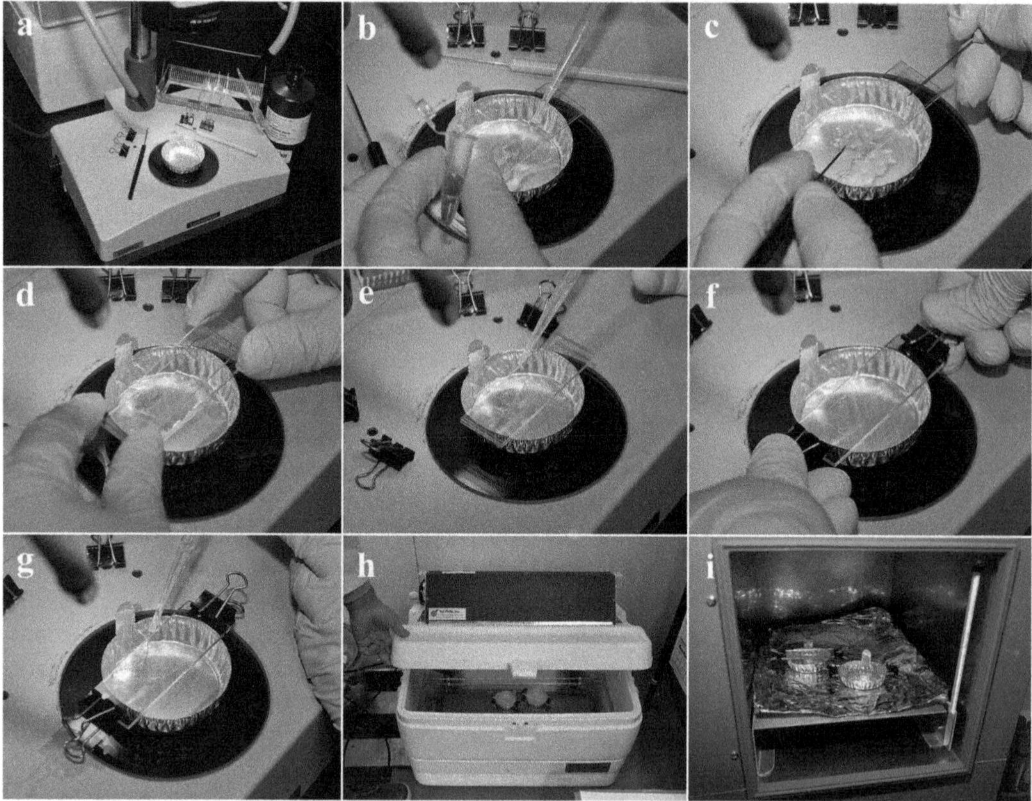

Fig. 3 Flat embedding of samples on glass slides. Aluminum weighing dishes, dissecting teasing needles, a plastic pipette, binder clips, slide tray, and stereomicroscope are required for these steps (**a**). We use an aluminum weighing dish to hold the slide. Using a disposable plastic pipette, fresh resin is applied onto a slide prior to positioning the infiltrated samples (**b**). Gently separate specimens from each other using forceps or dissecting teasing needles (**c**). After samples have been laid out on the bottom glass slide the top slide will have to be placed on top of the samples. To do this, the edge of the second slide should gently touch the edge of the bottom slide making sure that it maintains an angle of 30–45° (**d**). Slowly add more resin using a plastic pipette around the edges of the slides so that air spaces are completely filled with resin (**e**). Two slides held together with binder clips with specimens between the two slides (**f**). Additional resin is slowly added around the edges using a plastic pipette (**g**). Slides are kept in an ultraviolet cryo chamber in a cold room at 4 °C (**h**) or an oven (**i**) for polymerization

extract the specimen, carefully score the resin with a fresh razor blade. The first and inner rectangle score is made by applying gentle pressure (Figs. 4b, c). Then, the second and outer rectangle score is applied with firm pressure to separate the sample from the rest of the resin (Fig. 4d–f; *see* **Note 14**).

3. Based on the direction in which the section is cut (*i.e.*, longitudinally as shown in Fig. 4g or as a cross section as shown in Fig. 4h), the specimen is glued onto a resin block accordingly. Scraping the surface of the resin block with the razor blade may help the glue adhere tighter.

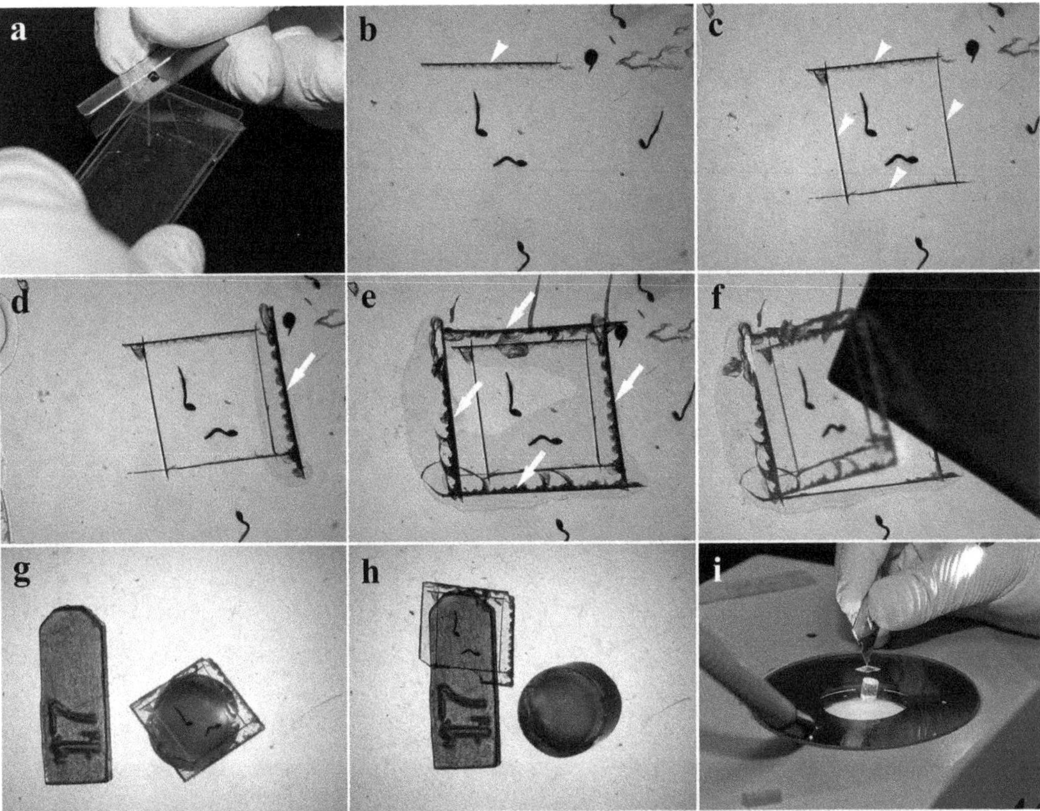

Fig. 4 Extracting specimens from flat embedded resin. Separating the top and bottom slide using a razor blade (**a**). Embedded specimens after removal of the glass slides (**b**). Extraction of the specimen and scoring the resin with a new razor blade. Shown are the first and inner rectangle score (arrowheads, **b** and **c**). The second and outer rectangle score (arrows) created with firm pressure to separate the sample from the rest of the resin (**d–f**). Mounting longitudinal (**g**) or cross sections (**h**). Placement of the thin specimen onto a resin block (**i**)

4. After the specimen is firmly in place, it can now be processed for trimming and sectioning (Fig. 4i; *see* **Note 15**). Application and representative results of the flat embedding procedure as applied to Arabidopsis roots are shown in Fig. 5.

4 Notes

1. We typically use unfrosted slides. Although frosted slides can be used, specimens that move into the frosted area of the slide while the resin polymerizes will be difficult to observe under a light microscope. This will prevent the researcher from selecting specimens that are optimally preserved.
2. Thickness of the cover glass is not critical for the procedure to succeed. Cover glasses that are 150 μm and 200 μm thick are recommended.

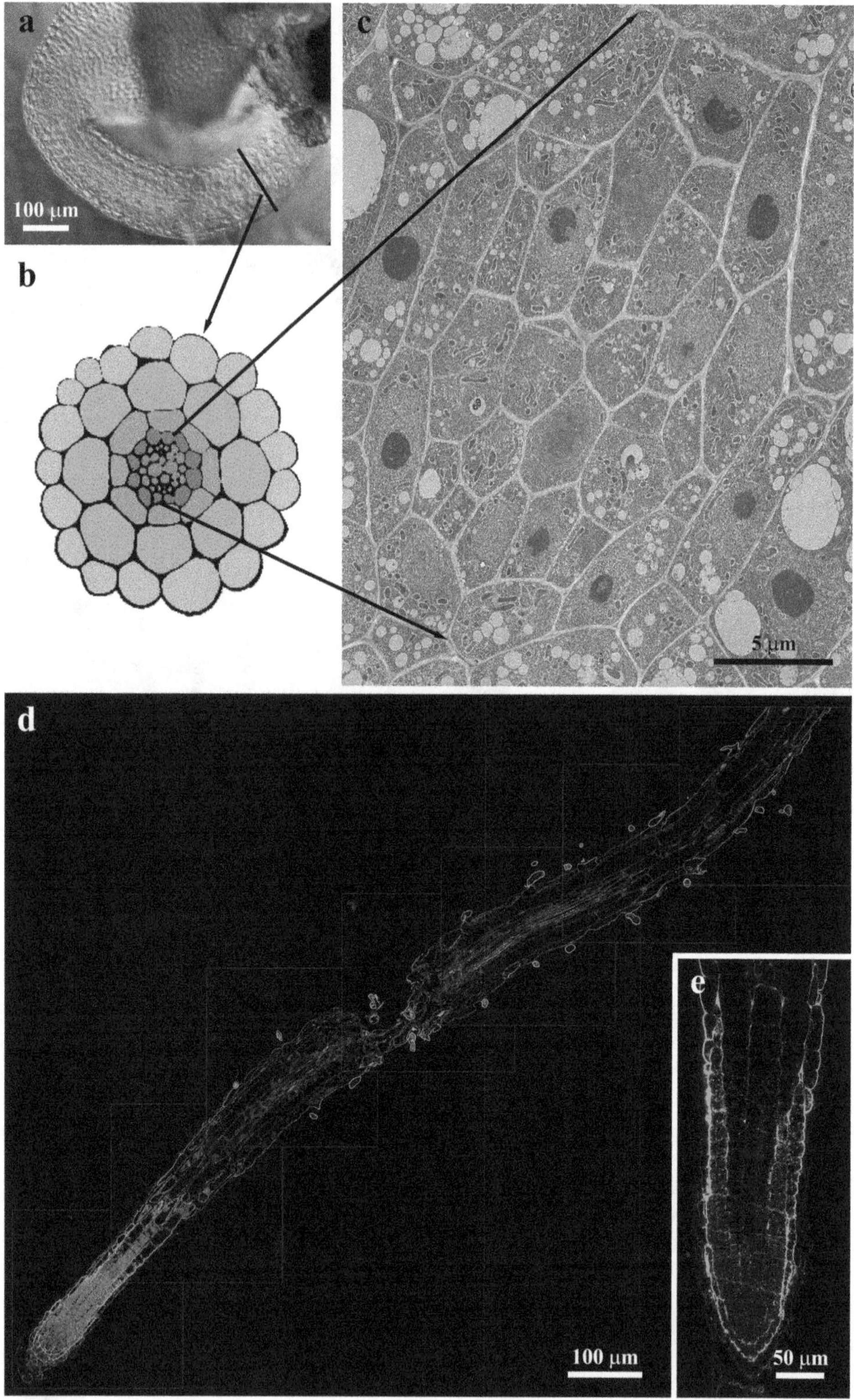

Fig. 5 Application and representative results of the flat embedding procedure as applied to Arabidopsis roots. Image of a 3-day-old Arabidopsis seedling on a flat embedded slide (**a**). An 80 nm cross section from the

3. Having tested superglue from different vendors, only Loctite superglue has been proven to work with our protocol. Other brands of superglue appear to cause severe resin retraction after polymerization.
4. LR White comes in different forms including soft, medium, and hard grades. This is important because the resin grade has to be compatible with the sample being embedded. For example, for Arabidopsis roots and other fragile tissues, we routinely use medium grade resin with catalyst. Another advantage of LR White is that this resin can be used for immunolocalization.
5. It is essential that only a tiny drop of Loctite super glue is applied so that excess glue does not leak from the edges when the cover glass is placed on top of the slide. If excess superglue leaks from the cover glass edges, it might be difficult to separate the slides from each other at the final step of the method.
6. The specimen thickness will guide you on how many cover glasses are placed between the slides. This information is crucial so that the specimen is not crushed when the pressure is applied by binder clips.
7. Liquid release agent usually comes with an instruction manual. Use the recommended times and temperatures for the given product.
8. Some investigators reuse LR White or aliquot LR White into separate vials. In our experience, however, it is better to use fresh resin from the original bottle.
9. If at all possible, try not to put too many specimens on a single slide. The number of specimens that can be accommodated in one slide will depend on the size of each sample.
10. While performing this operation, be as gentle as possible so as not too apply too much pressure on the slides. Excess pressure could easily displace the specimens.

Fig. 5 (continued) indicated region depicted in the root cartoon (**b**). Cells in stele region of the root visualized under a transmission electron microscope (**c**). Semi-thin sections (0.25 μm) taken from a median longitudinal section of a 7-day-old *Arabidopsis* primary root from seedlings fixed with 4% paraformaldehyde and 2.5% glutaraldehyde. Sections were stained with a monoclonal antibody against fucosylated xyloglucan (CCRC-M1) using indirect immunofluorescence (**d**). A different seedling of a 7-day-old *Arabidopsis* primary root grown in microgravity analog, the Space Bio-Laboratories, Inc. GRAVITE 3D-clinostat is shown. Semi-thin sections were stained with a monoclonal antibody against fucosylated xyloglucan (CCRC-M1) using indirect immunofluorescence. Magnified primary root tip image shows punctate cytoplasmic labeling reminiscent of Golgi-derived vesicles (**e**)

11. Be careful not to displace specimens by using too much pressure. When small air bubbles appear during this step, slides should be tilted to slowly remove bubbles.
12. The flat embedding procedure described here is suitable for a variety of resins. In case LR White is the preferred resin, it is important to place samples at the center of the slide because LR White cannot polymerize when it comes in contact with oxygen. LR White resin located at the extreme edge of the slide will not polymerize.
13. Embedded specimens should now be visible under a light microscope. This makes the technique particularly convenient for identifying the specific sample that needs to be sectioned (Fig. 4b). Possible defects in the specimen resulting from extended fixation times as well as poor fixation can be identified prior to sectioning. This allows the researcher to select samples that are well preserved and therefore not waste time processing samples that are poorly fixed.
14. Extracting the specimen from the resin is one of the most crucial steps of the entire procedure. When scoring around the sample, excess pressure could produce a stress crack that could destroy the specimen. Thus, it is very important to be slow and gentle while scoring. Upon completing the first square/rectangle, the inner square/rectangle protects the specimen from getting damaged while making the second outer score (refer to Fig. 4e to see how the stress crack stops at the inner square/rectangle).
15. The specimen mounted on the resin block is very fragile. Because the sample is embedded in resin while in between two glass slides, the original resin support is generally very thin. Therefore, caution should be used during trimming. A thick razor slice could lead to the loss of some or the entire sample of interest.

Acknowledgments

This work was supported by the National Aeronautics and Space Administration (NASA) grant 80NSSC18K1462 to J.N.

References

1. Reymond OL, Pickett-Heaps JD (1983) A routine flat embedding method for electron microscopy of microorganisms allowing selection and precisely orientated sectioning of single cells by light microscopy. J Microsc 130:79–84
2. Oorschot V, de Wit H, Annaert WG, Klumperman J (2012) A novel flat-embedding method to prepare ultrathin cryosections from cultured cells in their *in situ* orientation. J Histochem Cytochem 50:1067–1080

3. Wu S, Baskin TI, Gallagher KL (2012) Mechanical fixation techniques for processing and orienting delicate samples, such as the root of *Arabidopsis thaliana*, for light or electron microscopy. Nat Protoc 7:1113–1124
4. Avci U, Petzold HE, Ismail IO, Beers EP, Haigler CH (2008) Cysteine proteases XCP1 and XCP2 aid micro-autolysis within the intact central vacuole during tracheary element differentiation in *Arabidopsis* roots. Plant J 56:303–315
5. Avci U, Pattathil S, Singh B, Brown VL, Hahn MG, Haigler CH (2013) Comparison of cotton fiber cell wall structure and remodeling in two commercial cotton Gossypium species with different fiber quality characteristics. PLoS One 8: e56315
6. Nakashima J, Sparks JA, Carver JA Jr, Stephens SD, Kwon T, Blancaflor EB (2014) Delaying seed germination and improving seedling fixation: lessons learned during Science and Payload Verification Tests for Advanced Plant EXperiments (APEX) 02-1 in space. Gravit Space Res 2(1):54–67
7. Avci U, Nakashima J (2015) Plant gravitropism: methods and protocols. Humana Press, New York, NY, p 13

Chapter 12

Conducting Plant Experiments in Space and on the Moon

Tatsiana Shymanovich and John Z. Kiss

Abstract

The growth and development of plants during spaceflight have important implications for both basic and applied research supported by NASA and other international space agencies. While there have been many reviews of plant space biology, this chapter attempts to fill a gap in the literature on the actual process and methods of performing plant research in the spaceflight environment. One of the authors (JZK) has been a principal investigator on eight spaceflight projects. These experiences include using the U.S. Space Shuttle, the former Russian Space Station Mir, and the International Space Station, utilizing the Space Shuttle and Space X as launch vehicles. While there are several ways to fly an experiment into space and to obtain a spaceflight opportunity, this review focuses on using the NASA peer-reviewed sciences approach to get an experiment manifested for flight. Three narratives for the implementation of plant space biology experiments are considered from rapid turn around of a few months to a project with new hardware development that lasted 6 years. The many challenges of spaceflight research include logistical and resource constraints such as crew time, power, cold stowage, data downlinks, among others. Additional issues considered are working at NASA centers, hardware development, safety concerns, and the engineering versus science culture in space agencies. The difficulties of publishing the results from spaceflight research based on such factors as the lack of controls, limited sample size, and the indirect effects of the spaceflight environment also are summarized. Lessons learned from these spaceflight experiences are discussed in the context of improvements for future space-based research projects with plants. We also will consider new opportunities for Moon-based research via NASA's Artemis lunar exploration program.

Key words Gravitational biology, International Space Station (ISS), Microgravity, Space biology, Spaceflight

1 Importance of Plant Space Biology

The spaceflight era began with the launch in October 1957 of Sputnik 1. Shortly thereafter in November 1957, the dog Laika was the first animal sent into orbit around the Earth although fruit flies were the first animals sent to space by a V-2 rocket in 1947 from White Sands, New Mexico. The earliest questions in space biology centered on whether living organisms can simply survive in weightlessness or the microgravity environment of low Earth orbit [1]. While there are some significant physiological changes in living

Elison B. Blancaflor (ed.), *Plant Gravitropism: Methods and Protocols*, Methods in Molecular Biology, vol. 2368, https://doi.org/10.1007/978-1-0716-1677-2_12,

organisms in microgravity, animals and humans adapt remarkably well to the conditions of weightlessness.

The study of plants in space was an important part of space biology from the dawn of the spaceflight era. The first plant materials sent into space in May 1960 on Sputnik 4 were seeds, including those of maize, peas, wheat, and spring onion [2]. Plant space biology can be divided into two broad and interrelated themes: (1) understanding plants for use in bioregenerative life support systems, and (2) using microgravity as a novel research tool in order to study the fundamental aspects of plant biology. Of course, the latter field has implications for the former, and several researchers have taken a more synergistic approach since these two areas clearly are interrelated (e.g. [3]).

While plant space biology in many ways can be considered in its infancy, we have made interesting discoveries, and there are several excellent reviews on this topic [2–8]. As an example of a key discovery, in the field of bioregenerative life support, several studies have shown that it is possible to grow multiple generations of plants seed-to-seed and that microgravity does not deleteriously affect the plant life cycle [9–11].

We also have made progress in fundamental plant biology by using microgravity as a research tool. For instance, circumnutation, an oscillatory or helical growth pattern around an axis, was hypothesized by Charles Darwin to be an endogenous feature of plants while others have suggested that it is a gravity-dependent phenomenon [12]. Spaceflight experiments provided a way to resolve these controversies by providing an environment in which circumnutation could be studied without the "complications" of an effective gravity vector [13]. Thus, Johnsson et al. [14] in an experiment on the International Space Station (ISS) demonstrated that while endogenous circumnutations occur in stems, as Darwin predicted, gravitational accelerations can amplify these nutations. Another example of space research used to answer basic questions comes from our own work on the study of tropisms. We showed that red-light-based phototropism in stems occurs in seedlings grown in microgravity (Fig. 1) and that this phototropism is masked by the normal 1-g on Earth [15]. This discovery has implications for understanding the evolution of land plants since directional red-light responses occur in 1-g in older plant lineages such as mosses and ferns [16].

Recent molecular biology experiments via RNA-seq performed at the International Space Station showed that plants in microgravity and reduced gravity have altered expression for multiple genes and biosynthetic pathways when compared to 1-g controls [17, 18]. In microgravity, light-associated pathways such as photosynthesis-antenna proteins, photosynthesis, porphyrin, and chlorophyll metabolism are significantly downregulated [18]. At low-g, genes effected belonged to general, chemical, and hormone

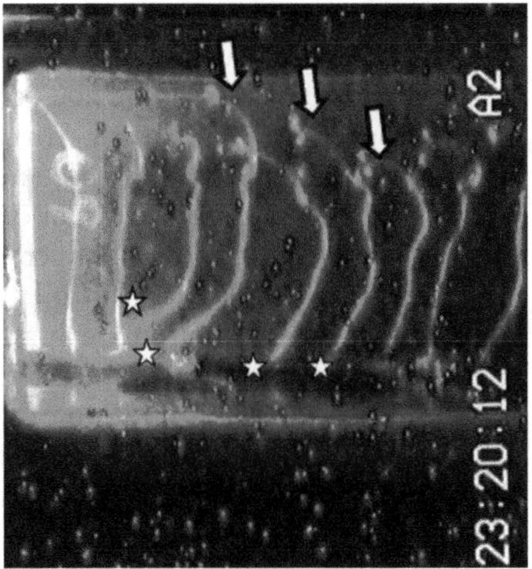

Fig. 1 Seedlings that developed in microgravity exhibited a positive phototropic curvature toward red light in hypocotyls (indicated by arrows) and roots (indicated by stars). Images were from the TROPI-2 spaceflight experiment on the ISS, and unidirectional red light was from the left side of the image. The text on the image indicates Greenwich Mean Time and Experimental Container number (A2)

stress responses, while at the Moon *g*-level, genes related to cell wall and membrane structure and functioning were shown to be alternately expressed [17].

While there is an extensive scientific literature of plant biology in space, to date, very little has been written about the challenges and unique opportunities of spaceflight research. A few reports in the peer-reviewed scientific literature consider the prospects of performing spaceflight experiments with plants [2] as well as the limits and constraints of such space experiments [7, 19]. There also are some non-reviewed technical publications of the National Aeronautics and Space Administration (NASA) and the European Space Agency (ESA) that consider the topic of conducting biological experiments in space [20, 21]. Given this dearth of literature and the dated nature of the existing publications, the aim of this chapter is to consider the technical, organizational, and logistical aspects of performing experiments in plant biology using orbiting spacecraft in platforms such as the ISS. We also will look ahead to emerging opportunities for lunar plant biology experiments in the upcoming Artemis program at NASA.

2 Principal Methods of Spaceflight Research

One of the major goals of spaceflight experiments in plant biology is to obtain conditions of near weightlessness by free fall, in which the gravity effects on objects are effectively reduced [21, 22] so that biological phenomena can be studied without a constant gravity vector. Some methods to obtain free fall conditions or microgravity in plant research include parabolic flights of airplanes, rockets, and space vehicles in low Earth orbit such as the former American Space Shuttle, Russian Soyuz, and the ISS [23, 24]. By definition, the gravitational acceleration on the surface of the Earth is 1 g (in SI units, 9.8 m s^{-2}), and microgravity conditions can be defined in the range of 10^{-6} to 10^{-4} g [19].

While methods such as parabolic flights and sounding rockets have been useful in studies of plant biology [24, 25], this review will focus on the utilization of space vehicles in low Earth orbit. In the past, these experiments were on the Space Shuttle, but now the opportunities will be in laboratories on the ISS. Space stations, including the former American Skylab and Russian Mir, provide the possibility to study plant biology in a microgravity environment for a longer duration. Other related topics beyond the scope of this chapter include the use of microgravity "analogs" such as clinostats and the use of biosatellites [4, 26].

3 Our Experience with Spaceflight Research with Plants

To date, our group (with J.Z. Kiss as Principal Investigator) has had eight spaceflight projects flown on vehicles in low Earth orbit (Table 1). These projects have been performed on the Space Shuttle as well as on the ISS, and we have used both the Space Shuttle and Space X as launch vehicles. Interestingly, seven of the eight projects (the exception being BRIC-16), were collaborative with the European Space Agency (ESA) and involved European hardware and laboratory facilities developed for spaceflight research with plants (Table 1). BRIC is an acronym for Biological Research in Canisters.

Our first two projects, PREPLASTID on STS-81 (Space Transportation System, the formal name of the Space Shuttle) and PLASTID (for the role of plastids in gravity perception) on STS-84 in 1997, were on the Biorack facility, which flew in the Spacehab module (a pressurized unit in the cargo bay of the Space Shuttle). These Biorack projects were concerned with mechanisms of gravity perception in plants [27, 28]. We used seeds of Arabidopsis wild type (WT) and a series of starch-deficient mutants, and the experiments were performed in ESA's Biorack, which was a multiuser facility for biological research [29]. In these projects, seedlings that developed in microgravity were given gravitational pulses from an on-board centrifuge. Hypocotyls of WT seedlings responded to these pulses while those of starch-deficient mutants had reduced or no responses, thereby adding support for the starch-statolith model of gravity perception. In addition, seedlings were chemically fixed in space, and subsequent electron

Table 1
Summary of the spaceflight projects flown on vehicles in low Earth orbit with J.Z.Kiss serving as the principal investigator

Project title	Year	Topic	Methods	Facility	Launched	Performed	Returned
PREPLASTID	1997	Gravity perception	Video tapes; chemical fixation	Biorack-Spacehab	STS-81	STS-81	STS-81
PLASTID	1997	Gravity perception	Video tapes; chemical fixation	Biorack-Spacehab	STS-84	STS-84	STS-84
TROPI-1	2006–2007	Tropisms; microgravity	Video tapes; freezing	EMCS-ISS	STS-121, 115	ISS Incr. 14	STS-116, 117, 120
TROPI-2	2010	Tropisms; reduced gravity	Video downlink; freezing	EMCS-ISS	STS-130	ISS Incr. 22	STS-131
BRIC-16	2010	Plant morphology; gene profiling	Chemical fixation	BRIC-Space Shuttle Middeck	STS-130	STS-130	STS-130
Seedling Growth-1	2013–2014	Phototropism; gravity perception; gene profiling	Video downlink; freezing	EMCS-ISS	Space X2	ISS Incr. 35	Space X-3
Seedling Growth-2	2014–2015	Phototropism; gravity perception; cell cycle; gene profiling	Video downlink; freezing	EMCS-ISS	Space X3	ISS Incr. 41	Space X-4
Seedling Growth-3	2017	Phototropism; gravitropism; cell cycle physiology & morphology; gene profiling	Video downlink; freezing; chemical fixation	EMCS-ISS	Space X11	ISS Incr. 52	Space X-11

microscopic studies did not show deleterious effects of microgravity on the structure of the gravity-perceiving columella cells in the root cap [30].

The interaction between phototropism and gravitropism [31, 32] was the focus of our next two space projects which flew in 2006 (TROPI-1) and 2010 (TROPI-2) on the ISS (Table 1). TROPI was a code name for the study of tropisms. We utilized the European Modular Cultivation System (EMCS) and were the first research group to use this facility on the ISS. The EMCS is an incubator with atmospheric control and high-resolution video camera system [33, 34]. In addition, the EMCS contains two variable centrifuge rotors so that a 1-g control can be performed as well as reduced gravity levels.

While TROPI-1 had a number of technical difficulties as is typical the first time space hardware is flown [35], we still were able to report a fascinating new finding. Specifically, a novel positive phototropic response to red light was observed in hypocotyls of seedlings that developed in microgravity. This type of red-light-based phototropism cannot be detected in normal 1-g conditions on Earth nor was it observed in the in-flight 1-g control [15]. Thus, these studies showed that some flowering plants may have retained a red-light sensory system for phototropism, and this discovery may have implications for understanding the evolution of light sensory systems in plants.

In TROPI-2, we had solved many of the technical and organizational difficulties encountered in TROPI-1 [36]. In addition to studying phototropism in microgravity, we also considered reduced gravity levels that were generated by using the EMCS centrifuge. The red-light-based phototropism was confirmed, and the reduced or fractional gravity studies showed an attenuation of red light due to gravitational accelerations ranging from 0.1 g to 0.3 g [16]. The transcriptome of seedlings grown in space in TROPI-2 was analyzed, and we found that 280 genes were differentially regulated (at least two-fold) when comparing spaceflight samples with ground controls [37].

The BRIC-16 project in 2010 was our only spaceflight project that was exclusive to NASA and took place in the middeck region of the Space Shuttle (Table 1). We found that an endogenous response in seedlings causes the roots to skew toward one direction in microgravity and that this default growth response is masked by the normal 1-g conditions on Earth [38].

Our most recent series spaceflight experiments (2013–2017) were termed Seedling Growth (SG). These experiments were a joint project again using the EMCS with ESA [39]. While the NASA PI (J. Z. Kiss) continued to explore phototropism in the absence of gravity, the ESA PI (F.J. Medina) was interested in the effects of microgravity on the cell cycle. In SG-1, we studied the interaction of gravitropism and phototropism by use of microgravity and reduced gravity [16]. In SG-2, the effects of Moon and Mars levels of gravity on plant development were examined as well as the effect of these gravity levels on the cell cycle [40]. We also studied gene expression of plants at various gravity levels in both SG-1 and SG-2 [17, 18]. We are currently evaluating the cell biology of Arabidopsis grown in microgravity by microscopic observations based on in-flight fixation using a new instrument available in SG-3, the FixBox.

4 How Do You Get an Experiment to Fly in Space?

There are several pathways to get biological experiments to fly into space on the ISS. Currently, four principal approaches for sponsorship of spaceflight experiments are: (a) NASA peer-reviewed research, (b) National Laboratory research, (c) educational activities, and (d) ISS international partner research. This list is not exclusive as there have been other methods to gain access to space such as through NASA's Small Business Innovation Research (SBIR) and Small Business Technology Transfer (STTR) Programs, among others [41]. In 2005, the US segment of the ISS was designated as a National Laboratory available for use by other federal agencies, non-profit organizations and the private sector [42, 43], and this effort is managed by the Center for the Advancement of Science in Space (CASIS).

This chapter will focus only on NASA-funded spaceflight research since the authors have direct experience with this approach. A key characteristic of this method of securing space biology flight opportunities is that the science is peer reviewed by experts in the field [44]. The concept of peer-reviewed science is very important since some of the other ways to get experiments into space do not always involve the strict process of peer review, which, despite its limitations is considered the hallmark of mainstream scientific investigations [45]. In the past, some spaceflight projects that did not undergo peer review had received unfavorable commentaries from some groups of scientists [46] as well as in the news media.

In the NASA-funded spaceflight approach, following scientific peer review, proposals then receive a technical feasibility review [44]. However, there have been some highly ranked scientific projects, but they were not rated feasible due to the lack of instrumentation and/or other resources (e.g., crew time). Thus, it is important for investigators to pay attention to descriptions of existing spaceflight hardware and capabilities when this information is provided in the call for proposals. Generally following peer and technical reviews, there also is a programmatic review to see how the project fits into NASA's broader goals at the time. An interesting twist to the peer review and the technical review is that one or both of these panels may have a group of international scientists and engineers.

The ISS has modules from the US, Japan, Europe, and Russia as well as other international partners, and facilities and crew time are shared, so it can be important to have scientists and engineers from these partners represented on the panels. In general, most of the scientific cooperation involves the United States Orbital Segment (USOS) of the ISS [47], which includes NASA, ESA, the Canadian Space Agency (CSA) and Japan Aerospace Exploration Agency (JAXA). When scientific peer review was first conducted by international panels, some adjustments had to be made due to the differing approaches used by peer reviewers from different countries and cultures.

5 Three Narratives for Implementation of Spaceflight Experiments

5.1 Biorack

Our first experience with spaceflight research was with the joint NASA/ESA Biorack project in 1997 (Table 1). The Biorack module, developed by ESA was a multiuser facility which serves as a small laboratory for the study of cell and developmental biology in unicellular organisms, plant seedlings, and small invertebrates [29], and Biorack flew on six Space Shuttle missions. This facility had two incubators with variable speed centrifuges (for use as a 1-g control), a glovebox, and external cameras.

The call for proposals was made in November 1994 jointly by NASA and ESA, and the proposals were due in January 1995 [48]. Peer review was conducted in March 1995, which was followed by a technical review, and proposals were selected for a "definition" phase starting in June 1995. Following this definition stage in which the feasibility of the space experiment and the resources required were determined, projects were formally accepted by NASA for spaceflight.

Biorack flew in the Spacehab module of the cargo bay of the Space Shuttle and was part of the Shuttle-to-Mir (the former Russian space station) missions [29, 48]. Our first experiment, PR EPLASTID, flew and was completed on mission STS-81 in January 1997, and the second experiment PLASTID was completed on STS-84 in May 1997 [49]. PREPLASTID was considered a small scale, preliminary experiment to the larger main experiment PLASTID. Unfortunately, the two projects were too close together to make significant changes between the two experiments. For instance, we discovered an ethylene effect on seedlings in the first experiment [50] and did not have time to potentially mitigate this effect in the second experiment [49].

Since the turn around time was considered short by NASA standards, principal investigators were to use already developed hardware from a list of spaceflight hardware that had flown on

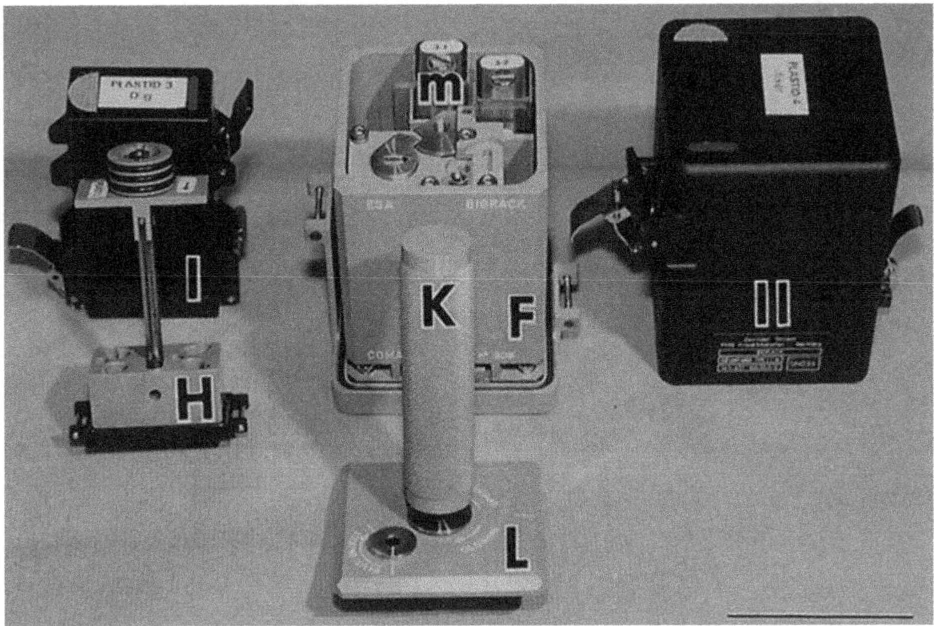

Fig. 2 Spaceflight hardware utilized in the Biorack PLASTID experiments on the Space Shuttle. Type I containers (I) house two mini-containers (m) in which seeds developed into seedlings, and two mini-containers were placed into a metal holder (H). Each type II container (II) houses a fixation unit (F) which can accommodate three of the mini-containers. The fixation unit was used to inject a glutaraldehyde solution into the mini-containers (two are shown which are ready for fixation) by turning a key (K). L = lid which covers the fixation unit. Scale bar = 5 cm

three previous Biorack missions. In our case, we used the lentil-roots hardware (Fig. 2) developed by CNES (Centre National d'Etudes Spatiales), the French Space Agency, for use by Perbal and coworkers [51]. We had to modify the non-permanent internal configuration of the hardware to accommodate the Arabidopsis seeds, which were much smaller than lentil seeds that were used by Perbal [52].

Our Biorack hardware set consisted of two parts (Fig. 2). Standard Biorack type I containers housed the units for plant growth, and the type II containers had the device for chemical fixation. The type I container, which had a drilled hole that allowed for gas exchange, accommodated two mini-containers for seed germination and seedling growth. The growth medium was injected by an astronaut with a syringe adapted for fitting to one of the mini-containers, which had a transparent plastic cover to allow for observations. During the flight experiments, the astronaut mission specialists transferred mini-containers from the type I container to the fixation device in the type II container.

The concept of a standard external container with specialized internal components has been used by several ESA spaceflight facilities including Biorack, BioPack, Biopan, and Kubik [29, 53–56], and all of these facilities used the same Type I (internal dimensions 81 × 40 × 20 mm) and Type 2 (87 × 63 × 63 mm) containers. This idea was extended to the newer facilities on the ISS, including Biolab and EMCS [57], which used standard containers that, however, were different from the earlier Types I and II containers. These standardized containers are appealing to engineers who then can design experimental unique equipment into containers with known parameters that can provide an interface to the larger facility.

We also conducted full scale "dress rehearsals" of the experiments on both flights at about six months prior to the flight. ESA termed these activities ESTs, or Experimental Sequence Tests, which performed the full time line of the experiments in order to determine if there were any technical or logistic problems. In fact, in the EST for the STS-81 mission, we did discover that seeds did not receive enough red illumination from the resident Biorack fluorescent lamps. Seeds of many plant species require red-light pretreatment to stimulate germination [31]. Thus, working with NASA, we had to develop a light box with red LEDs to promote seed germination to robust levels [52]. This light box worked well, and seed germination on the Biorack missions ranged between 92–100% in the spaceflight experiments [49].

5.2 EMCS

In contrast to our experience with Biorack, when the time between selection and flight was about 1.5 years, the EMCS project termed TROPI-1, which was selected in 2000 and flew in 2006 (Table 1), had a development period of about 6.5 years. There were three major factors in this protracted period. The first was the loss of the Space Shuttle *Columbia*, which occurred in February 2003 causing system-wide delays, and the second was that TROPI-1 was the first experiment to utilize the EMCS on the ISS. Finally, while EMCS had standardized hardware termed Experiential Containers (ECs), we had to develop the internal EUE (experimental unique equipment) for the growth of Arabidopsis seedlings (Fig. 3).

Similar to Biorack, EMCS was an ESA facility, so we had to work closely with both NASA and ESA engineers on hardware development [55]. ESA provided the ECs (Fig. 3a) while NASA had to develop the internal EUE (Fig. 3b). The concept for the seedling cassettes was largely based on the Biorack Type I mini-containers in which seeds were affixed to a black membrane [33]. Both the Biorack and EMCS experiments were launched with dry seeds, and the experiment was activated by hydration of the seeds, which then developed into seedlings. In the case of Biorack, hydration was performed manually by the astronauts with pre-filled syringes [52], and with EMCS, the hydration was programed or occurred via telemetric commands from the ground [35].

Our experiment TROPI-1 was the first to utilize the EMCS, and in fact, was launched in July 2006 on STS-121, the same flight on which the EMCS facility was launched [55]. During the course

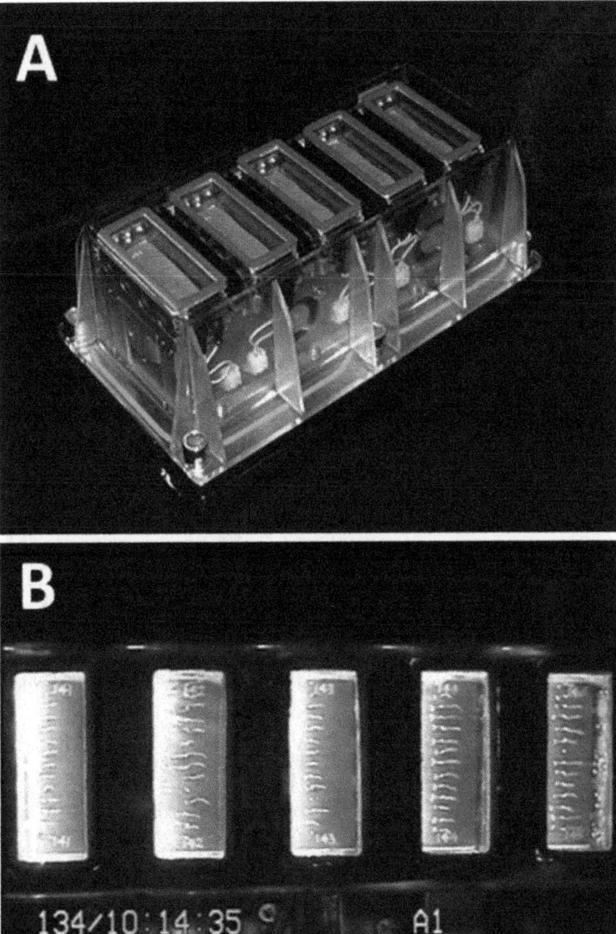

Fig. 3 TROPI hardware used to grow seedlings in EMCS spaceflight experiments. **(a)** Hardware in an EMCS Experimental Container (EC) showing five cassettes which are used for seedling growth. The dimensions of the EC are 186 mm (length) · 100 mm (width) · 90 mm (height). **(b)** View of the top of the EC showing the growth of seedlings of *Arabidopsis thaliana* in the cassettes during a spaceflight experiment. The text on the image indicates Greenwich Mean Time and Experimental Container number (A1)

of hardware development from 2001 to 2005, we traveled several times to Germany to test the interface between the EUE, the ECs, and the flight model of the EMCS.

One major difference from the Biorack experiments is that TROPI-1 was a largely automated experiment with limited crew time. We had to be proactive with obtaining high-quality imaging that was needed to study growth, development, and phototropism of the seedlings. Thus, the transparent plastic cover (Fig. 3a) on the five seedling cassettes (per EC) had a transparent anti-fogging heater membrane (similar to a rear window defogger in an automobile) attached so that condensation did not interfere with video observations [33]. In addition, the plastic cover had four slots with a gas permeable membrane covering them to allow for gas exchange with minimal moisture loss.

A comprehensive series of biocompatibility and other related tests were performed during the definition and development phase of the project [33, 58]. Additional tests were conducted because we designed new EUE and also because this was the first project scheduled on the EMCS. These tests examined numerous factors such as: types of materials used in the EUE, optimal LED illumination (quality and quantity), temperature (and temperature tolerance), humidity in the EC, placement of red illumination period in the timeline to optimize seed germination, video compression ratios during downlinks, quality of optical surfaces used in imaging, and numerous other parameters.

While TROPI-1 had some success and we were able to identify a novel red-light-based phototropism in microgravity [15], there were a number of serious issues and concerns. For instance, due to operational constraints of NASA and the ISS program, seeds were stored in sealed ECs for 6, 7, and 8 months in runs 1, 2, and 3, respectively [35]. During the spaceflight experiments, seed germination dropped dramatically between runs 1 (58%) and 2 (23%), and was even lower in run 3 (11%). A major factor resulting in this low germination was that seed storage for these long periods allowed for accumulation of gasses from materials, including the conformal coating of circuit boards that is required for safety considerations.

Other problems encountered in TROPI-1 included lack of hydration of seeds in the first attempt, issues with image capture from video tapes, difficulties with cryo-transfer procedures in flight as well as other operational issues [35]. Many of these difficulties could be attributed to the fact that this was the first mission to use the TROPI hardware and the first operational mission of the EMCS.

Since we had such low seed germination in TROPI-1, we had to modify the experiment in near real time by telemetric commands. The original plan called for studying seedlings in reduced gravity or fractional gravity (provided by the EMCS centrifuges) as well as in microgravity. However, we cancelled the reduced gravity runs so as to boost the sample size in the microgravity and 1-g control seedlings [15].

Another problem in TROPI-1 was that we did not conduct a full EST, Experimental Sequence Test, or "dress rehearsal" for the spaceflight experiments. The main reasons for the lack of an EST were: the lack of hardware availability, expense of conducting this test in Europe at the Norwegian Users Operations Control Center (Trondheim, Norway), and organizational constraints at NASA. We believe that some of the issues encountered would have been mitigated by performing an EST.

Despite these problems, the experiment had enough successes that NASA allowed us to develop TROPI-2, which was conducted on the ISS in 2010 (Table 1). Given that TROPI-2 was considered a reflight, we had significant technical improvements in this project that allowed for a vastly improved experiment [36]. Highlights of changes in procedures included decreased storage time in hardware, which increased seed germination and growth of the seedlings [16]. We also eliminated the use of video tapes (that were used in TROPI-1) and utilized direct downlinking of digital images from the ISS to obtain better quality images. In addition, the cryo-storage and cryo-transfer procedures during TROPI-2 maintained the low temperatures needed for good-quality RNA for use in microarray analyses [37].

While we did not perform an EST for TROPI-2, we performed a more abbreviated OVT, Operations Verification Test, which did help us optimize a number of parameters including focusing and imaging of the specimens [36]. The OVT had a limited number of experimental containers compared to a full EST. Nevertheless, having this OVT did help us to achieve a better operational success, which improved the scientific yield of TROPI-2 [16, 37].

It is interesting to note that an upgrade to the TROPI hardware system (during the Seedling Growth Project) was made by adding the FixBox (Fig. 4), which allowed for chemical fixation. This allowed us to study cell structures via microscopic methods and opened a new avenue of research with the EMCS.

5.3 BRIC-16

While our Biorack project proceeded in a relatively quick manner and our EMCS project took much longer, the BRIC-16 experience occurred in a greatly accelerated time frame. The general approach used by NASA in BRIC-16 was for a "Rapid Turn Around" spaceflight projects, and BRIC-16 was the first test of this method [59]. The solicitation was released in September 2009, and proposals were due in November 2009. Both the scientific peer review and the technical feasibility review were conducted in December 2009 with final proposal selection in January 2010 in which three projects were selected. The spaceflight experiments were scheduled to launch on mission STS-131 in March 2010. Fortunately (at least in this case), there was a flight delay, and the BRIC-16 payload launched in April 2010, allowing us more time to prepare for the project.

Similar to the scenario with the Biorack projects, we were able to conduct a "dress rehearsal" for our BRIC-16 experiments. Unfortunately NASA and ESA seemingly use different terms for these events, and the ground experimental test for the time line of BRIC-16 was called a PVT, or Payload Verification Test. In some NASA projects an additional ground test is performed, which is termed an SVT or Science Verification Test. The SVT considers the

Fig. 4 FixBox (FB) used in the Seedling Growth-3 Experiment using the EMCS. **(a)** FB shown in a closed configuration in a ground study. The carrousel (Cr) assembly holds seedling cassettes and is placed inside the FB by the astronauts during a space experiment. Dimensions of the FB are 95 mm diameter × 74 mm height. **(b)** FB shown in an open configuration with the top (T) and the bottom (B). The carrousel (Cr) assembly is shown with a single seedling cassette (SC). Diameter of the FB is 95 mm. **(c)** FB shown in space in the Columbus module of the ISS. The FB (with seedlings) was attached to Closing Tool (CT) platform by the astronaut. A wrench (W) will be attached to the FB by a Closure Adapter (CA). The wrench will be turned by the astronaut in order to release the fixative into the FB, which provides the required three levels of containment. Thus, seedlings will be fixed in space. Diameter of the FB is 95 mm

science specifically and is a precursor to the PVT, which is a full dress rehearsal of the time line and the post-flight analysis.

The three PIs selected for the BRIC-16 project in January 2010 were summoned to Kennedy Space Center in February 2010 to plan for the PVT. There was no time for an SVT, which was unfortunate as the chances of success typically improve with further ground testing. While we had some results [38], there were difficulties with post-flight processing of our samples for microscopy since we lacked the time to fully analyze the results from the PVT. Nevertheless, we were able to document changes in cell wall morphology [60] and gene expression in microgravity [61] in our BRIC-16 studies. Thus, while a rapid turnaround from flight selection to implementation is desirable, this experience was in fact "too fast." Fortunately, the BRIC-17 and BRIC-18 projects had a longer time to prepare for the actual spaceflight experiments [59], and these missions allowed for both SVTs and PVTs to be performed.

In contrast to our other experiences, BRIC-16 was a NASA-only project that was performed on the middeck of the Space Shuttle (although STS-131 docked to the ISS). The hardware system, term BRIC-PDFU (Petri Dish Fixation Unit) was relatively simple compared to our Biorack and EMCS projects (Fig. 5). The

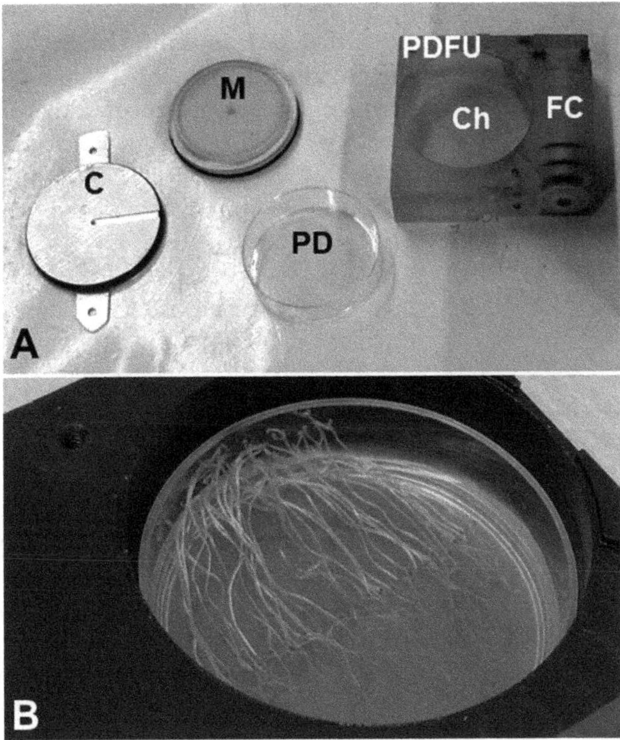

Fig. 5 Petri Dish Fixation Unit (PDFU) hardware used in the BRIC-16 project. **(a)** The bottom of a standard polystyrene Petri dish (PD) was placed into a PDFU, which has a chamber (Ch) for the dish and a fluid chamber (FC) filled with fixative. The Petri dish was directly covered with a manifold (M), and an additional cover (C) was attached to the PDFU polycarbonate body. Note the dimensions of the Petri dish are 60 · 15 mm. **(b)** Seedlings of *Arabidopsis thaliana* were grown on nutrient agar in the dark in a PDFU during a ground control, and the covers were removed to show that seedlings exhibited vigorous growth

BRIC system lacked an imaging system, atmospheric control, and video downlink capabilities. In addition, another key difference in BRIC compared to the Biorack and EMCS experiments, is that there was no on-board centrifuge. This instrument has been critical in distinguishing between spaceflight effects and true microgravity effects on living systems [48, 49, 62, 63]. In general in the recent past, many NASA laboratory facilities lack centrifuges as a control while ESA has typically included this instrument in their spaceflight laboratories [64].

Despite its simple nature, the BRIC-PDFU was robust and at that time, had already successfully flown on STS-87 in 1997 [65] as well as on the ill-fated STS-107 mission in 2003 [66]. The hardware consisted of a chamber with a single Petri dish (in which seedlings were grown), and fixation occurred with fluids from another chamber (Fig. 5). In our experiments on STS-131, we used three different fixation fluids: paraformaldehyde, glutaraldehyde, and RNAlater® in order to study growth, development, and gene expression [38].

Other colleagues in the BRIC-16 project had successes with this hardware and the "Rapid Turn Around" approach. For instance, one group reported that endogenous growth patterns of seedling roots in microgravity are suppressed by the actin cytoskeleton [67]. The other investigators obtained data to support to the hypothesis that undifferentiated plant cells could detect changes in gravity in the absence of more specialized tissues that are found in whole plants [68]. In addition, there have been many successful investigations using BRIC hardware [6, 59] following its reintroduction on BRIC-16.

6 New Approaches to Genomics Research

6.1 NASA GeneLab Project

Space conditions have significant effects on molecular building blocks of life, such as DNA, RNA, proteins and other metabolites. Since omics data from these space experiments are very rare and valuable, NASA developed an interactive and open-access resource called GeneLab where scientists can collaborate and find new uses for this data [69]. Epigenomics, genomics, transcriptomics, proteomics, and metabolomics data from spaceflights and corresponding analog experiments can be uploaded, downloaded, stored, shared and analyzed with the GeneLab research platform. The newest analytical and visualization tools as well as tools for collaboration also are available through GeneLab. This platform aims to provide scientific knowledge essential not only for human space exploration, but also for understanding biological processes on the Earth as well.

6.2 TOAST AstroBotany Database

For plant space biology omics research, the Test Of Arabidopsis Space Transcriptome (TOAST) database is another great tool [70]. It was developed by the University of Wisconsin-Madison researchers as a free, public, and educational resource [71]. With TOAST, researchers can analyze curated meta-data from the GeneLab and incorporate data from the following: SUBcellular location database for Arabidopsis (SUBA4) for genes, The Arabidopsis Information Resource (TAIR10) for Arabidopsis genome annotations, the Kyoto Encyclopedia of Genes and Genomes (KEGG) for known cellular pathways, Ensembl Orthologous Matrix for comparisons between species, and the National Center for Biotechnology Information (NCBI) data-repositories. They can use these databases to examine different spaceflight factors that may have effects on gene expression. The interactive data visualization environment allows researchers to make quick gene-specific comparisons across spaceflight experiments, but also to explore genetic networks of interest by using several filtering systems such as functional gene ontology, KEGG pathway analysis, or others. In the future, the TOAST database is expected to be extended for similar data exploration across biological systems (other than plants) that are being studied in spaceflight experiments [72].

6.3 Moving from DNA Microarrays to RNA-seq Methods

In order to effectively grow plants during spaceflight, scientists must understand how DNA expression for specific biosynthetic pathways is altered in these conditions. For many years, gene profiling through DNA microarrays have been used in transcriptomic studies of plants grown in space (e.g. [37, 61]). However, more recently, improved methods include RNA-seq techniques, which allow scientists to find changes in gene altered activity in conditions of reduced gravity or microgravity (e.g., [17]). If several genes within the same pathway have altered expression, these results indicate that this metabolic pathway can be either downregulated or upregulated in conditions of spaceflight. For example, using this approach, we found that the photosynthesis-antenna protein pathway has reduced expression in microgravity (Fig. 6), and understanding this pathway can be important for insights into plant adaptation to space [18].

7 Challenges of Space Research

To study changes in DNA transcription and other omics data, scientists need to fix their samples right after the experiment termination. The BRIC- Petri Dish Fixation Unit (PDFU) chemically fixes tissues with glutaraldehyde, RNALater, formaldehyde or other preservatives [38]. Several studies already used this system to their advantage [61, 74,].

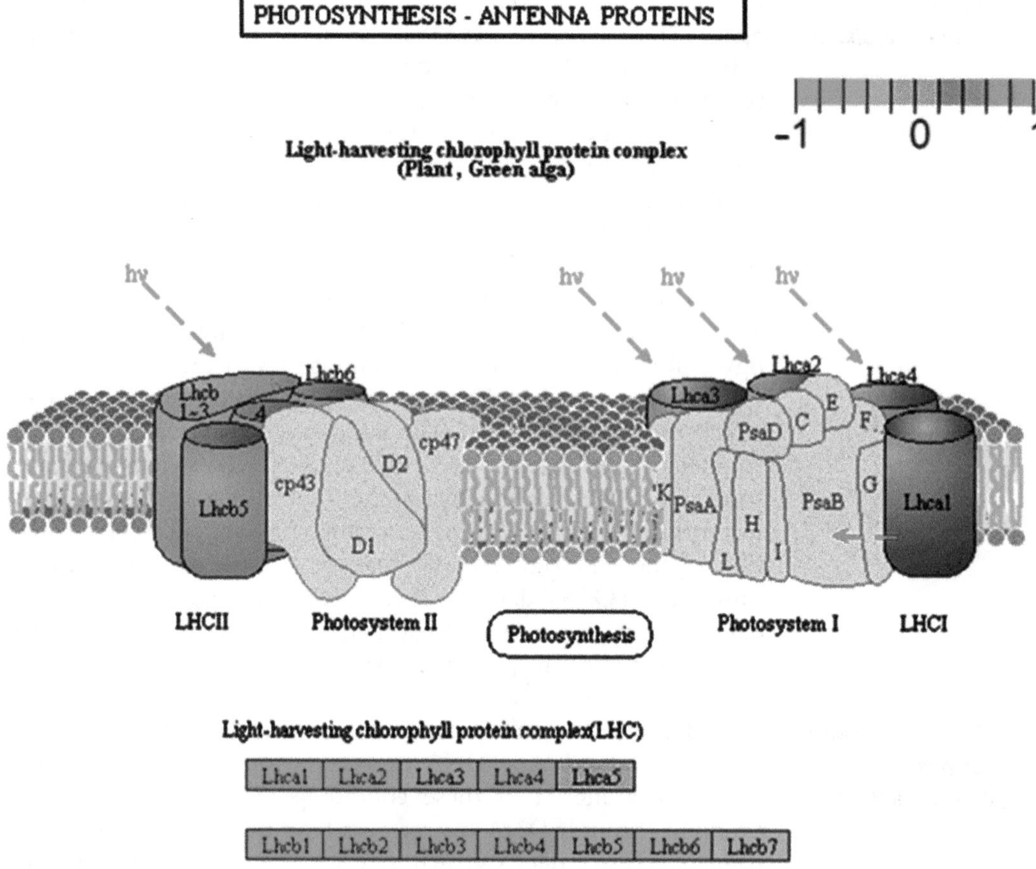

Fig. 6 Photosynthesis-antenna proteins pathway of differentially expressed genes identified in conditions of microgravity when compared to 1-g control. Genes were identified using the HISAT2-Stringtie-DESeq analysis pathway ($p = 1.61E-03$). Genes highlighted with green indicate reduced expression when compared to 1-g control

A new modification is the BRIC-Light Emitting Diode (LED) hardware has been developed for use with plant experiments [38, 73]. This new design based on BRIC-PDFU model, but has capacities to illuminate each Petri dish within a canister with blue, red, far-red, or white LEDs. It also has capacities to control temperature within ±3 °C from the surrounding conditions and even provide up to 1.5 °C difference across six canisters. Two fluids can be used for an experiment start or chemical tissue fixation at termination. Experiment start and stop times, real time telemetry, tray and canister temperatures, and LED status can be controlled from the KSC Ground Station [73]. This improved designed will be welcome by the community of plant space biologists.

7.1 Logistical Issues

The fundamental challenge of spaceflight research is that your work is being performed outside of your laboratory. In fact, the project is a long way from your lab, and once it is launched you have very little control over its fate. One of the most respected staff scientists at ESA (who help to implement spaceflight experiments) came from a scientific career of having performed ecological research in remote desert regions, and this background certainly was part of his success [63].

As stated above, once a project passes peer review, technical feasibility review, and meets programmatic relevance criteria, there then is a period of definition and development. Of the three spaceflight examples given, Biorack (1 year) and BRIC-16 (2 months) had accelerated times of definition and development while the EMCS had a protracted period of 6.5 years. Biorack and BRIC-16 had less formal stages of definition and development while the EMCS project had several stages including pre-Phase A, Phase A, Phase B, and Phase C/D [74]. Thus, while there appears to be a formal process for life science spaceflight experiments as we experienced with the EMCS project, the requirements and exact procedures differ from project to project.

The processes of the implementation of a spaceflight project have been delineated in a recent NASA document [74]. However, this section outlines the general approaches and issues considered in the execution of such projects. The definition portion of the space project occurs when the preliminary plan that was in the grant proposal becomes a more detailed time lined experiment. Thus, experimental requirements are defined, schedules are determined, and the specifications of the experiment become more precisely delineated [7, 55]. In the development phase, the exact hardware configuration is selected, and the precise time during the flight is finalized. At the end of the development phase, the dress rehearsal activities (or ground simulations of the experiment) such as SVTs, PVTs, and ESTs are performed.

A major part of definition and development is to create a mountain of paperwork. One of the key documents is termed an ERD, or Experiment Requirements Document [48]. In ESA's terminology, this document is called an ESR, Experiment Scientific Requirements. The goal of these documents is to capture all of the experimental and resource requirements for the successful completion of a spaceflight experiment. Thus, the Principal Investigator (PI) and his team will need to work closely with NASA scientists and engineers to provide very detailed inputs on all aspects of their space experiment. These details may include what they consider routine aspects of their research that may typically be taken for granted. This type of detailed consideration makes already precise scientists think more precisely.

Related to the space experiments is the need for a laboratory at a NASA facility in which the experiment has to be prepared. In the case of Biorack and BRIC-16, our laboratory was at Kennedy Space Center (KSC) while for the EMCS project, the lab was at Ames Research Center (ARC). Even in the latter case when the main lab was at ARC, we needed additional space for the processing of frozen samples retrieved from the return of the spaceflight experiments at KSC [39].

One of the many challenges is having your group work in one of these NASA facilities both prior to the experiment and upon the return of your experiment [19, 75]. Firstly, all items that you routinely need have to be clearly defined since it is difficult to order laboratory items expeditiously. In addition, all safety standards are strictly enforced, and documentation of potential risks is very strict in the NASA setting compared to the situation in most university labs. As an example, it was necessary for us to file the Material Safety Data Sheet (MSDS) for water when we worked at a NASA laboratory. Access also is more strictly controlled at NASA sites compared to the freewheeling nature of typical academic settings and requires some degree of security clearance and background check. If a member of your team is a foreign national, then the security clearance takes much longer than it does for US citizens.

Other logistical challenges include travel and housing your research team. The uncertainties of launch dates in the Space Shuttle era seem to be extending to the post-Shuttle era with commercial rocket launches, such as Space X, carrying experimental payloads into orbit. Thus, the research team may have to travel to the launch site several times and may have to travel home if there are extended delays. The costs of airline tickets, rescheduling, and housing on site can become expensive.

7.2 Development of Spaceflight Hardware and Procedures

We have had the entire continuum of experience in terms of hardware development (Table 1). In the BRIC-16 project, we used the hardware in the exact configuration it was provided as it had flown several times prior to our mission [38, 60]. During the Biorack experience, we used the lentil-roots hardware in Type I and Type II containers that was developed by Perbal and his colleagues [76], but we modified the internal configuration to accommodate small seeds of *Arabidopsis thaliana* [52].

In contrast, in the EMCS project, we were on the new frontier in that NASA ARC, along with our close cooperation, developed hardware specific for our project [33, 58]. In addition, in the last EMCS project (SG-3), ESA developed a FixBox (Fig. 4) to allow for in-flight fixation of seedlings. We also note that our experiment was the first one to use the EMCS facility which was launched to the ISS on the same Space Shuttle mission (i.e., STS-121) that launched the facility [36, 55].

Certainly, the simplest method was to use already established, proven, and flown hardware [19] as we did in our BRIC-16 project, and the most difficult and time-consuming method is to develop completely new hardware as in the EMCS. In most cases, the investigators have little choice in this matter as the call for proposals for space experiments will list available hardware for the proposed missions.

Since our TROPI-1 experiment was the first to utilize EMCS, we had an extended opportunity to develop spaceflight hardware for our project and to test the hardware in the flight model of the EMCS, which was located at the manufacturer in Germany, prior to its launch and delivery to the ISS. However, this aspect of hardware development added to the logistical complexity of the project [33]. As noted above, in our case, the general concept for the new hardware was largely based on our Biorack experience with the PLASTID project. In recent years, as a cost saving and practical measure, NASA and ESA have been moving away from hardware specific to single spaceflight experiment, and these agencies have been developing both modular and reusable hardware [55, 56, 75].

An important caveat regarding hardware development is that it is common for payloads to have problems that develop during the first spaceflight. Risk is reduced when there is a reflight of successfully flown hardware. Thus, in the Biorack project, we used the lentil-roots hardware with relatively minor adaptations for the smaller seeds [52]. Our TROPI-1 experiment had potentially large risks due to the fact that it was the first time that both the TROPI hardware flew, and this mission (launched on STS-121) was the first for the EMCS facility as well.

In addition to hardware risks, there also are risks inherent in developing new operations in spaceflight experiments. As an example, the TROPI-1 experiment was activated by an automated sequence of programed events in the EMCS beginning with the hydration of seeds. During the mission's first hydration attempt, telemetry from the EMCS indicated that the hydration command was successfully performed, but visual information from the downlinked images indicated that the seeds were, in fact, not hydrated [35]. A programming issue appeared to cause the problems, which were resolved by telemetric commands to the EMCS, and a second hydration attempt was successful. Fortunately, due to the robust nature of our procedures (i.e., an additional video downlink), the problems of hydration were discovered, and the experiment proceeded nominally.

7.3 Resource Constraints in Spaceflight Experiments

Both during hardware development and the execution of a plant spaceflight experiment, the PI is faced with a large number of resource constraints. One of the first issues confronting the investigator is the relative lack of crew time available to perform experiments. In our Biorack projects on the Space Shuttle, there seemed to be a generous amount of astronaut time, and we did not face major hurdles in terms of this parameter. For instance, a relatively large amount of time of three astronauts was dedicated to the Biorack payloads on STS-84 [49]. However, ironically, in the ISS era, crew time constraints have become very apparent.

EMCS was an early utilization facility that is largely automated and can be controlled by telemetry from the ground [18, 33, 57]. Even with this automation, EMCS projects were limited by lack of crew time to initiate and terminate the experiment. However, from the author's perspective, while NASA indicated serious issues with crew time, an ample amount of time somehow becomes available during the performance of space experiments. Due to the relatively simple nature of the hardware, crew time was not a major issue in the BRIC-16 project [38, 61].

The availability of video downlinks from the ISS is another example of a resource issue common to space experiments. As stated in the previous section, having the images from video downlinks helped resolve problems encountered during the hydration of seeds in TROPI-1 and saved the experiment [35]. However, NASA personnel initially objected to the inclusion of video downlinks early in the time of the space experiment, arguing that these downlinks were not needed because of the telemetry coming from the EMCS. Fortunately, we were able to successfully defend our position on the importance of downlinks to monitor the progress of the space experiments.

Crew training also changed dramatically between the Shuttle era and the advent of the ISS. In our Biorack projects, the PIs met with the astronauts involved in the space experiments in order to give them a theoretical framework for the importance of the research while hands-on training was provided by NASA and ESA staff. In the ISS era, the model was to avoid interaction between PIs and the astronauts, and crew training was done by NASA staff primarily on DVDs and web-based methods. In the opinion of the authors, there is something lost by lack of direct interaction between the scientists and the crew members performing the experiments.

Crew training also involves the preparation of very detailed procedures taking into account aspects of the experiment that may be taken for granted by PIs and the unique aspects of microgravity (i.e., tools floating away). While NASA personnel typically prepare the crew procedures based on input from the scientists, it is important for the PI team to carefully evaluate and review these procedures. As an example, in TROPI-1, we lost part of the samples that were placed into the −80 °C freezers on the ISS during the transfer from the ISS freezer to a cold bag for the transfer to the ground due to lack of instituting an explicit limit of keeping the frozen specimens at ambient temperature [35].

Related to this issue, another typical resource constraint has been cold temperature storage that is required by many plant biologists to preserve the samples resulting from a spaceflight experiment. In the Space Shuttle era, this problem appeared to be more acute due to the lack of space on the shuttle. Fortunately in the ISS era, the international partners have installed three freezers termed MELFI (Minus Eighty-Degree Laboratory Freezer for ISS), which is designed to keep samples at −80 °C [77]. However, several independent compartments in MELFI allow for higher temperatures such as −20 °C, which may be more suitable for some experiments.

The problem since the decommissioning of the Space Shuttle is keeping samples cold in the downloading of experiments from the ISS. The only current option is the unmanned Space X Dragon capsule, which is capable of both up mass and down mass from the ISS [78] although other vehicles will be added through NASA's COTS (Commercial Orbital Transportation Services) program. Space X has been fitted with freezers including MERLIN (Microgravity Experiment Research Locker/Incubator) with temperatures down to −15 °C and GLACIER (General Laboratory Active Cryogenic ISS Experiment Refrigerator) with capabilities from −80 °C to −180 °C [79].

Another issue, more general than cold stowage, for spaceflight experiments is the availability of power, which has been a problem for plant biology experiments since the early days of the space program [19]. We faced the issue of power constraints when we were working with NASA to design the TROPI hardware. While the final version of the hardware had 5 seedlings cassettes per EC, and earlier version had 6 cassettes, but the power requirements could not support the higher number of cassettes [33]. We also faced power issues regarding the quantity of light output from the LEDs in the TROPI cassettes.

Safety issues also fall into the general categories of a constraint of performing a spaceflight experiment [35, 80]. A common requirement of biological experiments is to have some type of fixative such as an aldehyde or RNAlater®. Hardware must have three levels of containment for fixatives as was possible with the BRIC-PDFU hardware we used in one of our space experiments [38, 60]. The quantity of fixatives and the restraints on corrosive or radioactive materials that are commonly used in ground-based experiments may pose additional limitations on spaceflight investigators.

7.4 Organizational Issues with NASA and Other Space Agencies

One issue that becomes readily apparent to an investigator of a spaceflight experiment is that there are a large number of scientists and engineers involved with this effort. As an example, in our TROPI-2 experiment on EMCS [36], we had the following parts of NASA affiliated with the experiment: Ames Research Center (ARC) as the science lead, Johnson Space Center (JSC) for ISS crew interface, Marshall Spaceflight Center (MSFC) for ISS operations, Kennedy Space Center for launch operations, and NASA Headquarters for overall supervision. In addition, since EMCS was an ESA facility, we also were involved with ESA Headquarters (Paris), ESTEC (European Space Research and Technology Center in Noordwijk, the Netherlands), N-USOC (Norwegian User Support and Operations Center in Trondheim, Norway), and Astrium Space Transportation in Friedrichshafen, Germany.

Whenever there is such a large, multinational group working on a project, conflicts and communication issues inevitably will arise. Even within NASA, the individual centers have a rivalry and competition for resources. In the course of these spaceflight projects, one of the authors (JZK) has been caught in the middle of infighting between two NASA centers. In addition, there have been conflicts about approaches and resources between NASA and ESA, and he has had to navigate in these waters as well. These potential disagreements can be managed, and the PI needs to keep his/her concerns about the integrity of the science as paramount.

Another issue that became evident in our spaceflight management experiences is the difference between the engineering and the scientific perspectives. As PI, the driving force is to maximize the success and scientific output from an experiment. However, while engineers certainly want to have a successful experiment, they may be more concerned with operational and safety aspects that will make the experiment feasible. At times, the different approaches will have the scientists wanting to increase complexity and the engineers to favor simplicity. One comment was that the engineers are always concerned about biologists ruining their elegant hardware–so it was better to fly it up and down empty! However, it is clear that there are a number of good engineers and scientists who want to break the barriers between these two cultures by trying to understand other perspectives and working together for a successful spaceflight project.

7.5 Interpretation and Publishing Data from Spaceflight Experiments

Compared to typical laboratory research, spaceflight research is difficult and typically requires a significant effort of time on the part of investigators. PIs can wait many years before their experiment flies in space. In addition, because many aspects of the environment are difficult to characterize, it also has been difficult to distinguish between "spaceflight effects" and true microgravity effects on biological systems [5, 30, 37]. Some of these space effects not directly related to microgravity include the vibrations, the gaseous atmosphere, elevated radiation, and other environmental factors.

In terms of plant biology, several observations that were reported to be effects of microgravity [4] were shown to be a limitation of the spaceflight hardware required to grow plants. As an example, growth of plants in spaceflight experiments has been shown to be slower, faster, and the same as on the ground. It turns out that the consensus is the growth rates of plants in microgravity are greater than on the ground when the hardware is optimized [5]. The slower growth reports were largely due to problems such as poor gas exchange, accumulation of ethylene as well as other factors.

Some of the issues in distinguishing spaceflight effects from true microgravity effects can be resolved by having a 1-g centrifuge as control [51, 55, 62, 63]. Thus, investigators can compare among flight-microgravity, flight 1-g (centrifuge), and 1-g (normal ground conditions). If biological phenomena are the same in the flight-microgravity and flight 1-g (and differ from the ground), then it is likely to be an environmental effect rather than a true microgravity effect. In fact, in the Biorack missions, in which there was an on-board centrifuge, several investigators reported differences between the 1-g flight and 1-g ground (e.g., [63, 81]).

The availability of a centrifuge in our Biorack projects helped resolve whether the observed differences in plant morphology were due to environmental conditions during the spaceflight or to true microgravity effects [30, 49]. For example, we observed an anomalous, exaggerated hypocotyl hook structure (Fig. 7) of seedlings grown in microgravity [49]. However, while this structure was observed in the 1-g in-flight control, it was not observed in seedlings in the 1-g ground control (Fig. 7). Subsequent ground studies

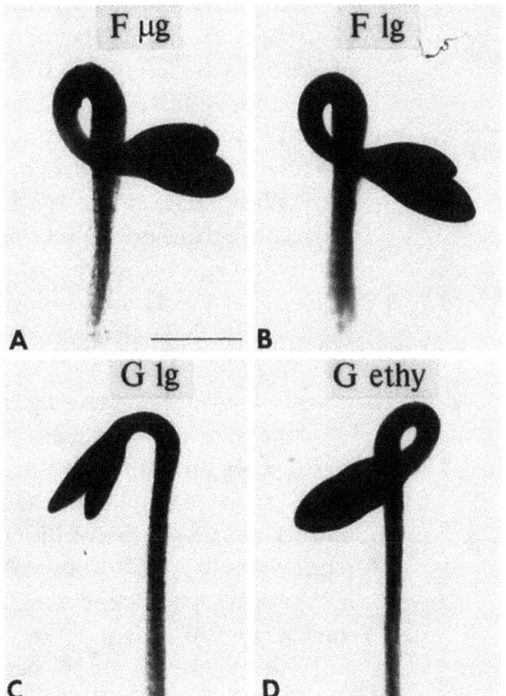

Fig. 7 Images of seedlings from the Biorack space project and the ground controls. Scale bar = 0.5 mm. **(a)** Light micrograph of an exaggerated hypocotyl hook from a seedling grown in-flight in microgravity (F-μg). **(b)** Seedling with an exaggerated hypocotyl from the in-flight, l g control (F-1 g; with the 1 g provided by the on-board centrifuge). **(c)** Seedling with a normal hypocotyl from the ground control (G-1 g). **(d)** Seedling with an exaggerated hypocotyl from the ground control with exogenous ethylene (G-ethy)

showed that the anomalous hypocotyl structures observed in the seedlings in the space experiments (Fig. 7) were due to relatively high levels of ethylene present aboard the orbiting spacecraft [49]. While there are a number of centrifuges aboard the ISS [24], many space experiments with plants have lacked a 1-g centrifuge (reviewed in [2, 5–7]), and certainly the availability of this control would benefit spaceflight research with plants and aid in the goal of publishing results in peer-reviewed publications.

Another difficulty of publishing spaceflight data in plant biology is the lack of replication. Due to the limited nature of spaceflight opportunities, it is often difficult to replicate and refine experiments as would be possible in typical ground-based research. In addition, the one-time nature of many space research projects puts enormous pressure on investigators to succeed on the first attempt. Sometimes during the time of the spaceflight, investigators have been asked to estimate the percentage of success even before they have any real data.

Fortunately, our group has had progression of experiments with improvements in design and procedures that resulted in studies with replication, improved analyses, and more refined conclusions. Specifically, we started with TROPI-1 during the first flight of the EMCS facility and had many technical problems [35], but with some interesting results on mechanisms of phototropism [15]. We were able to continue with TROPI-2 and to confirm the essential discovery of red-light-based phototropism in microgravity [36]. With the lessons learned from TROPI-1, we also were able to expand the experiment to consider plant phototropism at reduced gravity levels, which is relevant to the NASA agenda of exploration of other planets such as Moon and Mars [16, 24]. The last part of the Seedling Growth project built upon our previous technical and scientific knowledge to consider the inter-relationships among gravity sensing, light sensing, and the cell cycle [39].

Several other investigators in plant space biology from international space programs have been able to maintain a steady progression of space research. For example, on the ESA side, Perbal and his coworkers had a series of Biorack experiments on gravity sensing in lentil-roots [51, 76]. On the NASA side, Musgrave and her group had several space experiments to study the potential effects of microgravity on plant reproduction [9, 82]. More recently, Ferl and his colleagues have been using the Advanced Biological Research System (ABRS) as well as other hardware to examine gene expression in spaceflight grown plants in a progression of experiments [83–85]. In fact, one of the concepts behind the NASA "Rapid Turn Around" spaceflight projects was to potentially use the BRIC hardware in a series of several space experiments [59]. In any case, it is certainly desirable to continue to find opportunities to replicate and extend experiments in plant space biology as the field continues to mature.

8 Lessons Learned

Our group has been fortunate to have had eight spaceflight projects that have been completed (Table 1). We have worked with NASA, ESA, other international space agencies, and industry with scientists, engineers, astronauts, and managers in a complex structure in a collaborative manner in order to be successful as space biologists.

One of the first lessons learned is to keep experiments as simple as possible while maintaining the scientific objectives and integrity of the research. This concept is more important in experiments, which involve a potentially large amount of crew time. As an example, in our proposal for the Biorack project, we had approximately twice as many treatments as was performed in the final spaceflight experiments [49, 52]. Fortunately, while this was our first experiment in space, we had a mentor in Gerald Perbal who was an experienced investigator with several flights with Biorack [51]. While our initial inclination was to maximize the scientific yield by developing a complex experiment, Perbal recommended reducing the number of treatments in order to ensure the completion of the project in space.

Our project on Biorack was operationally successful and resulted in several publications in peer-reviewed journals [30, 49, 50, 86]. While a seemingly obvious concept, it is important for spaceflight investigators to publish in regular, peer-reviewed journals. A great deal of progress has been made in this area since the early period from 1960 to 1980 when many results from space experiments were non-peer reviewed and appeared in obscure outlets such as NASA technical reports. In recent years, space experiments have been published in very good journals with strong impact factors (reviewed in [5–7, 87, 88]), and this trend needs to continue.

As mentioned above, in the course of a space project, there can be an inherent tension between the investigator and NASA officials. While caution is needed in asking for "too many" resources, investigators should clearly indicate the spaceflight resources that are needed. In the case of our TROPI experiments, when we first requested video downlinks to monitor the hydration of seeds to confirm activation of the experiment, we were told that this was not needed since telemetry already was provided for this confirmation [35]. In fact, the telemetry was erroneous, and the downlinks saved the experiment. Thus, it was important that our team was firm in our pursuit of this important resource in the face of the initial opposition of our request.

In the course of such a complex endeavor such as a spaceflight experiment, inevitably some mistakes will be made. The important concept in these situations is to learn from the errors and to improve the project/procedures for future experiments. In our experience with working with NASA centers, there appears to be a tendency to want to assign "blame" to an individual, to a group of individuals, or to another NASA center. However, this approach is not helpful and needs to be avoided.

In the TROPI-1 project, we had a problem with cold stowage and cryo-transfer procedures. Seedlings frozen at −80 °C at the end of the experiment were returned on three Space Shuttle missions, and we had problems during the first sample return on STS-116 in that RNA was degraded and not usable (Table 1). A post-flight analysis following the first mission suggested that the issue was due to a problem in cold transfer and crew procedures [35]. The plant samples remained frozen in the −80 °C freezer but needed to be transferred from the freezer to the NASA cold stowage bags for return on the space shuttle. NASA officials estimated that the transfer time by the astronauts ranged from 7 to 29 min, but our analysis showed that the samples would warm to unacceptable temperatures within 3 min.

The initial reaction was to try to assign blame, but the investigators insisted that a solution needed to be found for the remaining two shuttle flights (STS-117 and STS-120), which would be returning samples. We had excellent cryopreservation once we instituted a "3-minute rule", which stated that this was the maximum time that frozen samples retrieved from the −80 °C freezer could be at ambient temperature prior to insertion into the NASA cold bag [35]. To meet this objective, two astronauts were required to perform the cold transfer procedures, these resources were provided, and the remaining samples yielded good-quality RNA that was suitable for gene profiling studies [37].

Spaceflight experiments need to be robust, which is defined by the Merriam-Webster dictionary as "successful or impressive and not likely to fail or weaken." Our TROPI-1 project was a robust experiment in two aspects: sample size and the plan we used for imaging [15, 35]. Initially, we had designed our experiments to study phototropism in microgravity as well as in reduced gravity (from the on-board EMCS centrifuge). However, we had to cancel the reduced gravity runs due to low seed germination. Nevertheless, we had a successful experiment with data since we added sample size to the microgravity runs by using the seeds originally intended for the reduced gravity runs. As described, we also had a redundant system to assay for hydration of the seeds for the activation of the space experiment. When one failed (telemetry), we still had a second system (video downlinks) to monitor the hydration events. In a sense, the concept of redundancy, which is used by NASA for rockets and other space vehicles, can also be applied on a smaller scale to the design of spaceflight experiments.

Spaceflight experiments can be difficult to replicate and refine as is possible in a typical laboratory setting on Earth primarily due to the lack of flight opportunities. This approach puts tremendous pressure on the principal investigator and his/her group to have a flawless experiment. However, many of the more successful investigators have had multiple spaceflight opportunities, and NASA and the international partners should encourage repetition as much as is possible. As mentioned above, there are several examples of repetition and refinement such as the projects on plant developmental and reproductive biology involving a series of six spaceflight experiments between 1989 and 1995 [9].

Our own experience in these matters may be instructive. Initially, in the Biorack project, we were offered one spaceflight opportunity, but during the definition and development phase, we successfully negotiated for a small preliminary experiment. Thus, PREPLASTID flew on STS-81 [50], and the larger, main experiment flew on STS-84 [49]. TROPI-1 led to a more refined TROPI-2, which greatly increased the scientific output [15, 16, 37]. As we move ahead, impetus for projects which involve multiple spaceflight opportunities should come from both the space agencies as well as the principal investigators. This approach would help to make space biology more similar to the standard practices in scientific research–execution, refinement, and repetition of experiments.

9 Outlook and Future Prospects

While plant space biology has made significant contributions to both basic and applied plant biology, there are a considerable number of important questions that still need to be addressed. How do plants integrate gravity and light sensory systems? What are the direct versus the indirect effects of microgravity on plants? How does the high radiation environment of spaceflight influence plant growth and development? How does spaceflight and microgravity affect plant–microbe interactions in terms of both pathogenic microbes and beneficial symbiotic microbes? While the answers to these questions have important implications for basic science, they also are compelling since plants will likely serve as part of bioregenerative life support systems during long-range space travel and colonization of other planets such as Mars [3, 89].

The approach to solving these issues in plant space biology will need to come from a robust program in terms of using both ground-based and spaceflight research methods. The NASA commitment to continue the ISS program recently has been extended from 2020 to 2024 and is likely to last beyond this date. However, compared to its international partners, NASA has under-utilized microgravity opportunities such as sounding rockets and biosatellites, and access to space via these methods should be increased. While the ISS certainly has facilitated international cooperation in space research, research collaboration between American investigators and emerging nations with space programs, such as China and

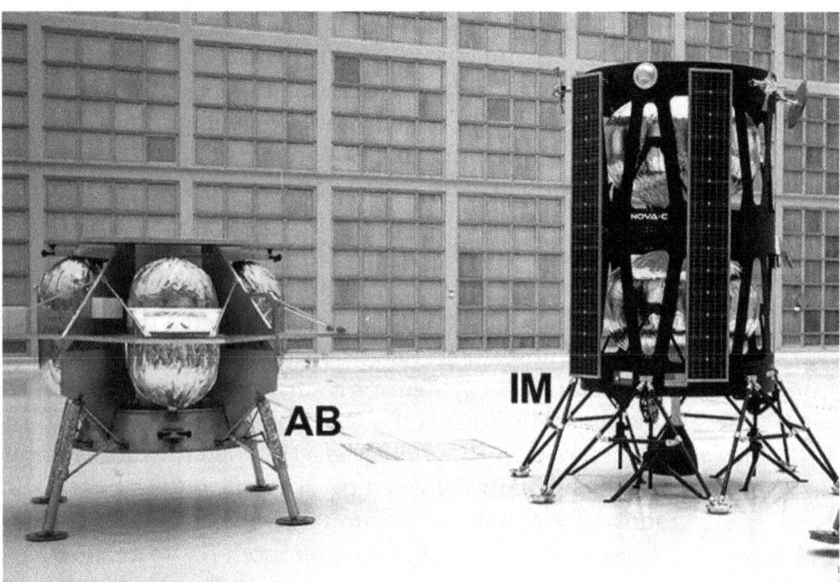

Fig. 8 Robotic moon landers designed by Astrobotic (AB) and Intuitive Machines (IM) for the Artemis lunar program. For scale, length of IM lander is 3 m

India, should also be developed and encouraged. The challenges for spaceflight researchers will be to maintain access to the microgravity environment for high-quality, peer-reviewed science so that the progress made in the last decades will continue in the future.

NASA in collaboration with several international partners has developed the Artemis program (2019–2028), which aims to accelerate Moon exploration and establish a sustainable human presence in order to develop various techniques before a Mars mission [90]. Initially, unmanned robotic landers will be launched to the Moon by Commercial Lunar Payload Services (CPLS). Astrobotic and Intuitive Machines (Fig. 8) have contracts to deliver at least 23 payloads to the Moon's surface [91]. This approach will potentially allow researchers to perform some plant experiments within small volumes such as by using CubeSats [92]. In 2024, the human Moon exploration program is scheduled to begin [93] and will provide greater opportunities for plant research projects as we move, in the long term, toward projects and human habitation on Mars.

Acknowledgments

Thanks to NASA for continued financial support of our spaceflight research and to ESA for providing excellent research laboratories for space research. Over the years, we have had fine support from NASA centers (ARC, KSC, JSC, and MSFC) and European facilities (ESTEC and N-USOC). We also wish to acknowledge our colleagues, friends, students, and the many astronauts who have contributed to the successes of our spaceflight projects.

References

1. Clément G, Slenzka K (2006) Fundamentals of space biology: research on cells, animals, and plants in space. Springer, New York
2. Halstead TW, Dutcher FR (1984) Status and prospects. Ann Bot 54(S3):3–18
3. Ferl, R.J., Wheeler, R., Levine, H.G, and Paul, A.-L. (2002) Plants in space. Curr Opin Plant Biol 5, 258–263
4. Halstead TW, Dutcher FR (1987) Plants in space. Ann Rev Plant Physiol 38:317–345
5. Wolverton SC, Kiss JZ (2009) An update on plant space biology. Gravit Space Biol 22:13–20
6. Paul AL, Wheeler RM, Levine HG, Ferl RJ (2013a) Fundamental plant biology enabled by The Space Shuttle. Am J Bot 100:226–234
7. Vandenbrink JP, Kiss JZ (2016) Space, the final frontier: a critical review of recent experiments performed in microgravity. Plant Sci 243:115–119
8. Kordyum EL (2014) Plant cell gravisensitivity and adaptation to microgravity. Plant Biol 16:79–90
9. Musgrave ME, Kuang A (2001) Reproduction during spaceflight by plants in the family Brassicaceae. J Gravit Physiol 8:29–32
10. De Micco V, Pascale S, Paradiso R, Aronne G (2014) Microgravity effects on different stages of higher plant life cycle and completion of the seed-to-seed cycle. Plant Biol 16:31–38
11. Musgrave ME, Kuang A (2003) Plant reproductive development during spaceflight. Adv Space Biol Med 9:1–23
12. Darwin C, Darwin F (1881) The power of movement in plants. John Murray, London
13. Kiss JZ (2009) Plants circling in outer space. New Phytol 182:555–557
14. Johnsson A, Solheim BGB, Iversen T-H (2009) Gravity amplifies and microgravity decreases circumnutations in *Arabidopsis thaliana* stems: results from a space experiment. New Phytol 182:621–629
15. Millar KDL, Kumar P, Correll MJ, Mullen JL, Hangarter RP, Edelmann RE, Kiss JZ (2010) A novel phototropic response to red light is revealed in microgravity. New Phytol 186:648–656
16. Kiss JZ, Millar KDL, Edelmann RE (2012) Phototropism of *Arabidopsis thaliana* in microgravity and fractional gravity on the International Space Station. Planta 236:635–645
17. Herranz R, Vandenbrink JP, Villacampa A, Manzano A, Poehlman W, Feltus FA, Kiss JZ, Medina FJ (2019) RNAseq analysis of the response of Arabidopsis thaliana to fractional gravity under blue-light stimulation during spaceflight. Front Plant Sci 10:1529
18. Vandenbrink JP, Herranz R, Poehlman WL, Feltus AF, Villacampa A, Ciska M, Medina JF, Kiss JZ (2019) RNA-seq analyses of Arabidopsis thaliana seedlings after exposure to blue-light phototropic stimuli in microgravity. Am J Bot 106:1466–1476
19. Krikorian AD (1996) Strategies for "minimal growth maintenance" of cell cultures: A perspective on management for extended duration experimentation in the microgravity environment of a space station. Bot Rev 62:41–108
20. Looft FJ (1986) The design of flight hardware. In: NASA Conference Publication 2401. National Aeronautics and Space Administration, Washington DC, pp 109–116
21. Briarty LG, Kaldeich B (1989) Biology in microgravity. ESA Publications, Noordwijk, The Netherlands, A guide for experimenters
22. Klaus DM (2001) Clinostats and bioreactors. Gravit Space Biol Bull 14:55–64
23. Claassen DE, Spooner BS (1994) Impact of altered gravity on aspects of cell biology. Int Rev Cytol 156:301–373
24. Kiss JZ (2014) Plant biology in reduced gravity on the Moon and Mars. Plant Biol 16 (S1):12–17
25. Limbach C, Hauslage J, Schafer C, Braun M (2005) How to activate a plant gravireceptor. Early mechanisms of gravity sensing studied in characean rhizoids during parabolic flights. *Plant Physiol* 139:1030–1040
26. Kiss JZ, Wolverton C, Wyatt SE, Hasenstein KH, van Loon J (2019) Comparison of microgravity analogs to spaceflight in studies of plant growth and development. Front Plant Sci 10:1577

27. Kiss JZ, Wright JB, Caspar T (1996) Gravitropism in roots of intermediate-starch mutants of *Arabidopsis*. Physiol Plant 97:237–244
28. Kiss JZ, Guisinger MM, Miller AJ, Stackhouse KS (1997) Reduced gravitropism in hypocotyls of starch-deficient mutants of *Arabidopsis*. Plant Cell Physiol 38:518–525
29. Manieri P, Brinckmann E, Brillouet C (1996) The Biorack facility and its performance during the IML-2 Spacelab mission. J Biotech 47:71–82
30. Guisinger MM, Kiss JZ (1999) The influence of microgravity and spaceflight on columella cell ultrastructure in starch-deficient mutants of *Arabidopsis*. Am J Bot 86:1357–1366
31. Molas ML, Kiss JZ (2009) Phototropism and gravitropism in plants. Adv Bot Res 49:1–34
32. Kiss JZ, Mullen JL, Correll MJ, Hangarter RP (2003) Phytochromes A and B mediate red-light-induced positive phototropism in roots. Plant Physiol 131:1411–1417
33. Correll MJ, Edelmann RE, Hangarter RP, Mullen JL, Kiss JZ (2005) Ground-based studies of tropisms in hardware developed for the European Modular Cultivation System (EMCS). Adv Space Res 36:1203–1210
34. Brinckmann E, Schiller P (2002) Experiments with small animals in BIOLAB and EMCS on the International Space Station. Adv Space Res 30:809–814
35. Kiss JZ, Kumar P, Millar KDL, Edelmann RE, Correll MJ (2009) Operations of a spaceflight experiment to investigate plant tropisms. Adv Space Res 44:879–886
36. Kiss JZ, Millar KDL, Kumar P, Edelmann RE, Correll MJ (2011) Improvements in the re-flight of spaceflight experiments on plant tropisms. Adv Space Res 47:545–552
37. Correll MJ, Pyle TP, Millar KDL, Sun Y, Yao J, Edelmann RE, Kiss JZ (2013) Transcriptome analyses of *Arabidopsis thaliana* seedlings grown in space: implications for gravity-responsive genes. Planta 238:519–533
38. Millar KDL, Johnson CM, Edelmann RE, Kiss JZ (2011) An endogenous growth pattern of roots is revealed in seedlings grown in microgravity. Astrobiology 11:787–797
39. Kiss JZ, Aanes G, Schiefloe M, Coelho LHF, Millar KDL, Edelmann RE (2014) Changes in operational procedures to improve spaceflight experiments in plant biology in the European Modular Cultivation System. Adv Space Res 53:818–827
40. Valbuena MA, Manzano A, Vandenbrink JP, Pereda-Loth V, Carnero-Diaz E, Edelmann RE, Kiss JZ, Herranz R, Medina FJ (2018) The combined effects of real or simulated microgravity and red-light photoactivation on plant root meristematic cells. Planta 248:691–704
41. NASA. (2014) Office of Small Business Programs. http://osbp.nasa.gov/SBIR-STTR.html. Accessed on April 1, 2020
42. Ruttley TM, Evans CA, Robinson JA (2011) The importance of the International Space Station for life sciences research: past and future. Gravit. Space Biol. 22:67–81
43. Giulianotti MA, Low LA (2020) Pharmaceutical research enabled through microgravity: perspectives on the use of the International Space Station US National Laboratory. Pharm Res 37 (1):1. https://doi.org/10.1007/s11095-019-2719-z
44. NSPIRES (2014) NASA Solicitation and Proposal Integrated Review and Evaluation System. http://nspires.nasaprs.com. Accessed on April 1, 2020
45. Raff H, Brown D (2013) Civil, sensible, and constructive peer review in APS journals. J Appl Physiol 115:295–296
46. National Research Council (1995) Peer Review in NASA Life Sciences Programs. National Academy of Sciences Press, Washington DC
47. Voels SA, Eppler DB (2004) The International Space Station as a platform for space science. Adv Space Res 34:594–599
48. Brillouet C, Brinckmann E (1999) Biorack facility performance and experiment operations on three Spacehab Shuttle to Mir missions. In: Perry M (ed) Biorack on Spacehab (SP-1222). ESA Publications, Noordwijk, The Netherlands, pp 3–21
49. Kiss JZ, Edelmann RE, Wood PC (1999) Gravitropism of hypocotyls of wild-type and starch-deficient *Arabidopsis* seedlings in spaceflight studies. Planta 209:96–103
50. Kiss JZ, Katembe WJ, Edelmann RE (1998) Gravitropism and development of wild-type and starch-deficient mutants of *Arabidopsis* during spaceflight. Physiol Plant 102:493–502
51. Perbal G (2009) From ROOTS to GRAVI-1: Twenty five years for understanding how plants sense gravity. Microgravity Sci Technol 21:3–10
52. Katembe WJ, Edelmann RE, Brinckmann E, Kiss JZ (1998) The development of spaceflight experiments with *Arabidopsis* as a model system in gravitropism studies. J Plant Res 111:463–470
53. Brinckmann E (1999) Spaceflight opportunities on the ISS for plant research- the ESA perspective. Adv Space Res 24:779–788
54. Willemsen HP, Langerak E (2007) Hardware for biological microgravity experiments in

Soyuz missions. Microgravity Sci Technol 19:75–79

55. Kittang, A.-I. ,Iversen, T.-H. , Fossum, K. R. , Mazars, C. , Carnero-Diaz, E. , Boucheron-Dubuisson, E., Le Disquet, I. , Legué,V., Herranz, R., Pereda-Loth, V., and Medina, F. J. (2014) Exploration of plant growth and development using the European Modular Cultivation System facility on the International Space Station. Plant Biol 16, 528–538
56. Astrium (2012) Space biology product catalog. Astrium, Friedrichshafen, Germany
57. Brinckmann E (2005) ESA hardware for plant research on the International Space Station. Adv Space Res 36:1162–1166
58. Kiss, J.Z., Kumar, P., Bowman, R.N., Steele, M.K., Eodice, M.T., Correll, M.J., and Edelmann, R.E. (2007) Biocompatibility studies in preparation for a spaceflight experiment on plant tropisms (TROPI). *Adv. Space Res.* **39**, 1154–1160
59. Camacho JR, Manning-Roach SP, Maresca EA, Levine HG (2012) BRIC-PDFU rapid turnaround spaceflight hardware. ASGSR Meeting, Abstract Book, p 87
60. Johnson CM, Subramanian A, Edelmann RE, Kiss JZ (2015) Morphometric analyses of petioles of seedlings grown in a spaceflight experiment. J Plant Res 128:1007–1016
61. Johnson CM, Subramanian A, Pattathil S, Correll MJ, Kiss JZ (2017) Comparative transcriptomics indicate changes in cell wall organization and stress response in seedlings during spaceflight. Am J Bot 104:1219–1231
62. Brown AH (1992) Centrifuges: evolution of their uses in plant gravitational biology and new directions for research on the ground and in spaceflight. Gravit Space Biol Bull 5:43–57
63. Brinckmann E (2012) Centrifuges and their application for biological experiments in space. Microgravity Sci Technol 24:365–372
64. Dutcher FR, Hess EL, Halstead TW (1994) Progress in plant research in space. Adv Space Res 14:159–171
65. Kern VD, Sack FD (1999) Irradiance dependent regulation of gravitropism by red light in protonemata of the moss *Ceratodon purpureus.* Planta 209:299–307
66. Kern VD, Schwuchow JM, Reed DW, Nadeau JA, Lucas J, Skripnikov A, Sack FD (2005) Gravitropic moss cells default to spiral growth on the clinostat and in microgravity during spaceflight. Planta 221:149–157
67. Nakashima J, Liao F, Sparks JA, Tang Y, Blancaflor EB (2014) The actin cytoskeleton is a suppressor of the endogenous skewing behaviour of *Arabidopsis* primary roots in microgravity. Plant Biol 16(S1):142–150
68. Paul AL, Zupanska AK, Ostrow DT, Zhang Y, Sun Y, Li J-L, Shanker S, Farmerie WG, Amalfitano CE, Ferl RJ (2012) Spaceflight transcriptomes: unique responses to a novel environment. Astrobiology 12:40–56
69. NASA GeneLab (2020). https://genelab.nasa.gov. Accessed on April 1, 2020
70. Gilroy Life Science Lab TOAST (2020). https://astrobiology.botany.wisc.edu/astrobotany-toast. Accessed on April 1, 2020
71. Astrobotany (2020). https://astrobotany.com. Accessed on April 1, 2020
72. Barker R, Lombardino J, Rasmussen K, Gilroy S (2020) Test of Arabidopsis space transcriptome: a discovery environment to explore multiple plant biology spaceflight experiments. Front Plant Sci 11:147. https://doi.org/10.3389/fpls
73. Caron AR (2016) Biological Research in Canisters-Light Emitting Diode (BRIC-LED) ISS R&D Conference. San Diego, CA
74. Jet Propulsion Laboratory (2014) Basics of space flight. http://www2.jpl.nasa.gov/basics/bsf7-1.php. Accessed on 14 January 2020
75. Porterfield DM, Neichitailo GS, Mashinski AL, Musgrave ME (2003) Spaceflight hardware for conducting plant growth experiments in space: the early years 1960–2000. Adv Space Res 31:183–193
76. Perbal G, Driss-Ecole D (1994) Sensitivity to gravistimulus of lentil seedling roots grown in space during the IML 1 mission of Spacelab. Physiol Plant 90:313–318
77. De Parolis MN, Crippa G, Chegancas J, Olivier F, Guichard J (2006) MELFI ready for science – ESA's −80 °C freezer begins work in space. ESA Bull 128:26–31
78. Stern SA (2013) The low-cost ticket to space. Sci Amer 308:68–73
79. Robinson JA, Thumm TL, Thomas DA (2007) NASA utilization of the International Space Station and the Vision for Space Exploration. Acta Astronaut 61:176–184
80. Lewis ML, Reynolds JL, Cubano LA, Hatton JP, Lawless BD, Piepmeier EH (1998) Spaceflight alters microtubules and increases apoptosis in human lymphocytes (Jurkat). FASEB J 12:1007–1018
81. Van Loon JJ, Bervoets D-J, Burger EH, Dieudonné SC, Suzanne C, Hagen J-W, Semeins CM, Doulabi BZ, Veldhuijzen PJ (1995) Decreased mineralization and increased calcium release in isolated fetal mouse long

bones under near weightlessness. J Bone Min Res 10:550–557

82. Kuang A, Popova A, McClure G, Musgrave ME (2005) Dynamics of storage reserve deposition during *Brassica rapa* L. pollen and seed development in microgravity. *Int. J.* Plant Sci 166:85–96
83. Paul A-L, Ferl RJ (2011) Using green fluorescent protein (GFP) reporter genes in RNALater fixed tissue. Gravit. Space Biol. 25:40–43
84. Paul AL, Zupanska AK, Schultz ER, Ferl RJ (2013) Organ-specific remodeling of the Arabidopsis transcriptome in response to spaceflight. BMC Plant Biol 13:1–11
85. Paul A-L, Popp MP, Gurley WB, Guy C, Norwood KL, Ferl RJ (2005) Arabidopsis gene expression patterns are altered during spaceflight. Adv Space Res 36:1175–1181
86. Kiss JZ, Brinckmann E, Brillouet C (2000) Development and growth of several strains of *Arabidopsis* seedlings in microgravity. Int J Plant Sci 161:55–62
87. Choi WG, Barker RJ, Kim SH, Swanson SJ, Gilroy S (2019) Variation in the transcriptome of different ecotypes of *Arabidopsis thaliana* reveals signatures of oxidative stress in plant responses to spaceflight. Am J Bot 106:123–136
88. Zabel P, Bamsey M, Schubert D, Tajmar M (2016) Review and analysis of over 40 years of space plant growth systems. Life Sci in Space Res 10:1–16
89. Paradiso R, De Micco V, Buonomo R, Aronne G, Barbieri G, De Pascale S (2014) Soilless cultivation of soybean for Bioregenerative Life-Support Systems: a literature review and the experience of the MELiSSA Project--Food characterisation Phase I. Plant Biol 16 (S1):69–78
90. NASA Moon to Mars (2020). https://www.nasa.gov/specials/moontomars/index.html. Accessed on April 1, 2020
91. Clark, S. (2019) https://spaceflightnow.com/2019/06/04/nasa-picks-three-companies-to-send-commercial-landers-to-the-moon. Accessed on April 1, 2020
92. NASA CubeSat Launch Initiative (2017) CubeSat101: basic concepts and processes for first-time CubeSat developers. https://www.nasa.gov/sites/default/files/atoms/files/nasa_csli_cubesat_101_508.pdf Accessed on April 1, 2020
93. NASA Artemis (2020) https://www.nasa.gov/what-is-artemis. Accessed on April 1, 2020

Chapter 13

Plant Proteomic Data Acquisition and Data Analyses: Lessons from Spaceflight

Proma Basu, Colin P. S. Kruse, Darron R. Luesse, and Sarah E. Wyatt

Abstract

Proteomics has the capacity to identify and quantify the proteins present in a sample. The technique has been used extensively across all model organisms to study various physiological processes and signaling pathways. In addition to providing a global view of regulatory processes inside a cell, proteomics can also be used to identify candidate genes and retrieve information on alternative isoforms of known proteins. Here, we provide protocols for protein extraction from *Arabidopsis thaliana* seedlings and describe analysis techniques used after data collection. This approach was originally used for the Biological Research in Canisters (BRIC) 20 spaceflight experiment but is valid for any ground-based or flight experiment. Extraction protocols for soluble and membrane proteins and basic analysis and quality metrics for MS/MS data are provided. Avenues for data analysis post-MS/MS data acquisition and details of software that can be used in gathering structural data on proteins of interest are also included. Use of differential abundance and network-based approaches for proteomics data analyses can reveal regulatory patterns not apparent through differential abundance or transcriptomic data alone.

Key words Proteomics, Protein extraction, Post-translational modifications, Membrane proteins, BRIC 20, Spaceflight, Arabidopsis, iTRAQ, Network-based proteomics

1 Introduction

Proteins are the culmination of gene expression; however, gene expression provides an indirect measure of protein status in the cells. Genes may be transcribed, but translation inhibited; the transcripts degraded; or if translated, proteins may not be functional due to post-translational modifications. To provide a clearer picture of the physiological status and responses involved in plant growth and development, proteomics is needed. Characterization of the plant proteome provides a more accurate view of the functional status of the tissue or organism and allows for identification of mechanisms inaccessible through transcriptomics alone. The proteomic status of the sample provides not only an indirect record of gene transcription, but also the integration of pre- and

Elison B. Blancaflor (ed.), *Plant Gravitropism: Methods and Protocols*, Methods in Molecular Biology, vol. 2368,
https://doi.org/10.1007/978-1-0716-1677-2_13,

post-transcriptional regulation via epigenetics, RNA polymerase regulation, small RNAs, mRNA stability, mRNA localization, alternative splicing, and translation inhibition.

Comparing the proteome of an organism between experimental and control conditions can uncover a plethora of information ranging from a broad overview of systemic changes to selecting candidate proteins by focusing on a pathway of interest. Depending on the desired information, various approaches for data analysis can be used. Statistical tests like Analysis of Variance (ANOVA) and *t*-tests (with appropriate multiple-testing correction such as Benjamini-Hochberg) can be used to determine differential abundance providing insight into the global change in the proteomic landscape for any given condition. Network-based approaches can provide another useful tool for identifying candidate genes for further study. Proteins identified, or a subset of those, can be provided as input to generate network analyses. These can examine gene regulatory pathways or physiological processes that were activated or disrupted by the experimental conditions. Alternately, *k*-means clustering can be used to cluster proteins by expression values into correlated groups, and clusters used to generate networks based on expression values. Modern proteomic platforms also allow for identification of protein isoforms, adding another dimension to our understanding of protein regulation and function at a cell-, organ-, and/or environmental condition-specific level. Identifying the isoform that is expressed in an experimental condition and studying its primary sequence can reveal more information about the function of a protein of interest. And, of course, many genes and proteins still lack functional characterization. A first step toward identification of proteins still classified as unknown or putative is protein domain analysis via SMART, InterPro [1, 2], and folded structure prediction using software and web-based interfaces like I-TASSER and Phyre2 [3–6]. The information can help develop potential hypotheses addressing protein function, which can then be tested by wet lab experiments.

Here, we describe the proteomic analysis as used in the spaceflight experiment BRIC 20 [7]. The experiment assessed gene expression (via RNAseq) and proteomic status (via protein mass spectrometry) of Arabidopsis seedlings germinated and grown for 3 days aboard the International Space Station. The protocols presented here include extraction of soluble and membrane proteins, quantification, iTRAQ labeling, quantification of peptides, identification of proteins, identification of proteins with post-translational modifications, and analyses of differentially abundant proteins and modifications, network-based strategies, and possible approaches to the study of unknown proteins (Fig. 1).

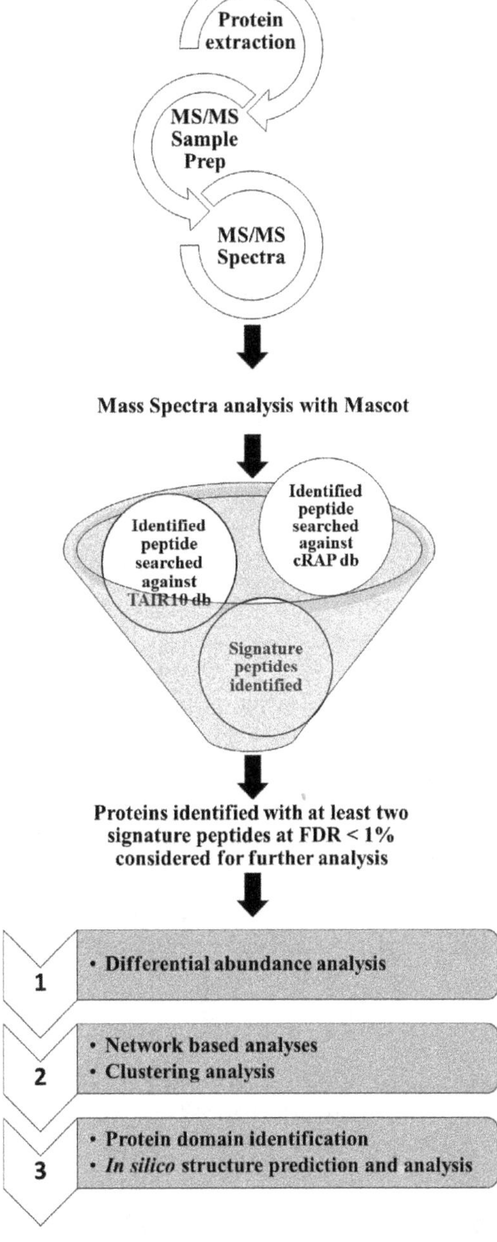

Fig. 1 Workflow of proteomics analyses discussed in this chapter. The quality of mass spectrometry data is dependent on the quality of protein extraction and the subsequent sample prep. During sample prep, samples are cleaned to remove residual salt contamination from the extraction buffers to facilitate quality data acquisition. Once the mass spectrometry is done the spectra are processed and matched against contaminant (cRAP db) and the proteome database of the target species (in our case TAIR10 was used). Proteins identified with high confidence are then used for further analysis like differential abundance, network analysis, domain analysis, etc.

2 Materials

2.1 Plant Material

1. Three- to 10-day-old Arabidopsis seedlings (*see* **Notes 1** and **2**).

2.2 Equipment

1. Mortar and pestle.
2. Vortex Genie 2 (or equivalent).
3. Eppendorf Centrifuge 5415R (or equivalent).
4. Sorvall RC5B centrifuge using SS34 rotor (or equivalent).
5. 50 mL centrifuge tube
6. Optima L-80 XP Ultracentrifuge.
7. Type 70 Ti Fixed angle Rotor.
8. Thick Wall Polycarbonate 32 mL tubes.
9. Microcentrifuge tubes 2 mL and 1.5 mL.

2.3 Extraction of Soluble Protein

1. Extraction buffer: 0.1 M Tris–HCl pH 8.0, 10 mM EDTA, 0.4% β-ME, 0.9 M sucrose. For 200 mL of buffer, add 20 mL of 1 M Tris, pH 8.0, 4.0 mL of 500 mM EDTA, 61.6 g sucrose, and bring up to 200 mL with water (*see* **Notes 3** and **4**). The buffer may be stored at RT without β-Mercaptoethanol (β-ME). Add β-ME immediately prior to use (80 μL β-ME per 20 mL of buffer) (*see* **Notes 5** and **6**).
2. Resuspension buffer: 1 M urea, 0.5 M thiourea, 0.03 M Tris–HCl, pH 7.0. To make 100 mL of buffer, add 6.06 g urea; 7.6 g thiourea and 3 mL of 1 M Tris–HCl to 45 mL of water. Once the urea dissolves completely, adjust the pH to 7.0 using 1 N HCl. Take the volume up to 100 mL with water (*see* **Notes** 7 and **8**).
3. Liquid N_2.
4. Tris-buffered phenol, pH 8.0.
5. 0.1 M ammonium acetate in 100% methanol.
6. Methanol.
7. Acetone.

2.4 Extraction of Microsomal Membrane Proteins

1. 1 M BTP-MES solution: Add 28.23 g of bis-Tris-Propane to 50 mL of water and adjust the pH to 7.8 by adding MES in powder form. Bring the volume up to 100 mL.
2. 0.5 M EGTA.
3. 0.5 M EDTA.
4. 0.5 M DTT.
5. 2 M potassium chloride.
6. 0.2 M potassium phosphate: Add 0.2 M K_2HPO_4 to 0.2 M KH_2PO_4 at a 10:1 ratio.

7. 1 M PMSF.
8. 1 M MOPS-KOH: Add 20.92 g of MOPS to 50 mL of water and adjust the pH to 7.6 by adding KOH pellets. Bring the volume up to 100 mL.
9. Protease inhibitor tablets.
10. Microsome Crushing Buffer: 250 mM sucrose, 2 mM EGTA, 10% (v/v) glycerol, 0.5% BSA, 50 mM Bis Tris Propane (BTP)-2-(N-morpholine)-ethanesulphonic acid (MES), pH 7.8, 0.25 M KI, 2 mM DTT, 1 mM phenylmethylsulfonyl fluoride (PMSF) and 5 mM β-ME (*see* **Notes 9** and **10**). To make 100 mL of fresh buffer, add 5 mL of 1 M BTP-MES buffer pH 7.8, 0.5 g of BSA, 8.56 g of sucrose, 4 mL of 0.5 M EGTA, 10 mL glycerol, 100 μL of 1 M PMSF, 4.15 g of KI, and 400 μL of 0.5 M DTT to 50 mL of water. Add 35 μL of β-ME immediately prior to use. Bring the volume up to 100 mL with water. Supplement the extraction buffer with 1 tablet of c*O*mplete protease and phosphatase inhibitor cocktail (Roche, Milwaukee, Wisconsin) per 50 mL of buffer.
11. Microsome Resuspension Buffer: 10 mM BTP-MES pH 7.8, 0.33 M sucrose, 3 mM KCl, and 5 mM potassium phosphate, pH 7.8. To make 100 mL of fresh buffer, add 1 mL of 1 M BTP-MES buffer, pH 7.8, 11.29 g of sucrose, 150 μL of 2 M KCl solution, 2.5 mL of 0.2 M potassium phosphate solution to 50 mL of water. Bring the volume to 100 mL by adding water. Add 100 μL of 1 M PMSF and 2 tablets of protease inhibitor immediately prior to use.

2.5 Proteomic Analysis

1. Protein Assay kit for protein quantification.
2. LTQ Orbitrap Velos, Thermo Scientific (West Palm Beach, FL) (*see* **Note 11**).
3. Mascot (Matrix Science, London, UK).
4. Scaffold (version Scaffold_4.3.2, Proteome Software Inc., Portland, OR).
5. Protein Prophet.

3 Methods

3.1 Plant Material

Plant material can either be fresh and processed immediately or frozen at −80 °C until ready for extractions. Treatment with RNA-later prior to freezing is acceptable, although this may alter the results [8].

3.2 Extraction of Microsomal Membrane Proteins

Chill all instruments and reagents used in this protocol at 0–4 °C (*see* **Note 12**).

1. Grind 1 g of fresh or frozen tissue in a mortar and pestle using liquid N_2.
2. Add enough chilled Microsome Crushing Buffer to wet the ground tissue (about 4–5 mL).
3. Centrifuge at 8000 × *g*, 4 °C for 15 min.

   The pellet formed after centrifugation is loose. Carefully pour off the supernatant and discard the pellet (*see* **Note 13**).
4. Centrifuge the supernatant from **step 3** at 100,000 × *g* at 4 °C for 50 min to 1 h. Save both the pellet (membrane proteins) and supernatant (soluble proteins).
5. Wash the pellet with 100 μL of 160 mM sodium carbonate solution, as described in Subheading 3.3, to remove excess soluble protein contamination.
6. Submission of this pellet formed after **step 5** is sufficient for most mass spectrometry facilities, which will then perform a sample clean-up to reduce the amount of salt in the sample (*see* **Note 14**). This is the Microsomal Preparation.

3.3 Carbonate Wash

The carbonate wash is performed to remove any soluble protein contamination from the microsomal preparation pellet.

1. Prepare 160 mM sodium carbonate solution and add 500 μL (per gram of starting material) to each fraction collected above.
2. Keep all fractions on ice for 30 min.
3. Centrifuge at 100,000 × *g* for 25 min.
4. Discard the supernatant and save the pellet.

3.4 Extraction of Soluble Proteins

Extract proteins via a phenol extraction followed by methanolic ammonium acetate precipitation (*see* **Note 15**). The protocol has been optimized to handle 0.75–1.5 g frozen tissue.

1. Using a mortar and pestle, grind 0.75–1.5 g frozen tissue to a powder in liquid nitrogen. Transfer the tissue to a 2 mL microfuge tube on ice. Keep the tissue frozen at all times.
2. Add 600 μL chilled Tris pH 8.0 buffered phenol and 600 μL chilled extraction buffer per 50–200 mg of starting tissue. If the total protein being extracted from the sample is to be fractionated between membrane and soluble proteins, then the phenol and the extraction buffer can be added in the same proportion to the supernatant formed after **step 4** in Subheading 3.2.
3. Vortex on high for 30 s.
4. Centrifuge for 10 min at 16,000 × *g*, 4 °C.
5. Transfer phenol layer (top layer) to a new 2 mL tube. Store on ice.

6. Re-extract the aqueous phase (the lower layer after **step 5**): Add 400 μL chilled Tris pH 8.0 buffered phenol to aqueous phase and vortex until thoroughly mixed (1–2 min).
7. Centrifuge for 10 min at 16,000 × *g*, 4 °C.
8. Remove the phenol layer and combine with the phenol layer collected in **step 5**.
9. Centrifuge the combined, extracted phenol layers (from **steps 5** and **8**) for 5 min at 16,000 × *g*, 4 °C then transfer the top, phenol layer (leaving behind the lower layer that previously contaminated the phenol layer) to a clean 50 mL centrifuge tube.
10. Precipitate proteins by adding 5 volumes of COLD (chilled at −20 °C) 0.1 M ammonium acetate in 100% methanol to the phenol layer.
11. Vortex until thoroughly mixed and incubate at −20 °C overnight.
12. Collect the precipitate by centrifugation (20 min, 16,000 × *g*, 4 °C). Remove the supernatant and dispose in non-chlorinated waste.
13. Wash pellet with cold 0.1 M ammonium acetate in methanol. Add 1.5 mL per 50 mg of starting tissue weight. Disrupt the pellet as much as possible by vortexing and sonication. Do not sonicate longer than 30 s at a time or longer than 1.5 min total time. If the sample begins to warm, chill before sonicating again. Place the resuspended sample at −20 °C for at least 15 min, then centrifuge in a Sorvall RC5B (or equivalent) centrifuge using SS34 rotor for 20 min, 16,000 × *g*, 4 °C. Remove the supernatant.
14. Repeat **step 13** as stated two times.
15. Repeat **step 13**, using cold 80% acetone instead of the 0.1 M ammonium acetate in methanol.
16. Repeat **step 15**, using cold 70% methanol instead of acetone.
17. After the last wash, the sample can be stored at −20 °C in 70% methanol until solubilization. Submission of this pellet is sufficient for most mass spectrometry facilities (*see* **Note 14**).

3.5 Proteomic Analysis

If sending the samples to a facility for proteomic analysis, **steps 1–4** will be performed by the facility (*see* **Notes 16** and **17**). For protein samples that are difficult to solubilize, such as microsomal proteins, the yield per sample can be increased by adding 0.1% *Rapi*gest SF solution to the protein pellet prior to digestion with Trypsin. *Rapi*gest SF solubilizes proteins in the pellet making more of the protein available for enzymatic cleavage.

1. Quantify each sample using a protein assay (e.g., Bradford or BCA assay). A modification of the Bradford Coomassie dye is mixed with the sample and absorbance is measured at 595 nm. A minimum of 50 μg of protein is required per replicate for iTRAQ-based analysis.
2. Trypsin digest and label: Reduce equal amounts of total protein from each of three replicates of each sample with Tris (2-carboxyethyl) phosphine (TCEP) and alkylate with iodoacetamide followed by overnight trypsin digestion. The digested samples are then labeled with a unique Isobaric Tag for Relative and Absolute Quantification (iTRAQ) reagent (*see* **Note 18**).
3. Pool the individually labeled peptide reactions (one pool for soluble, one for membrane, *see* **Note 19**) and subject to a series of liquid chromatography (LC) separations: Based on charge using strong cation exchange (SCX) LC; and second based on polarity using reversed-phase liquid chromatography (RP-LC). LC is necessary to reduce the complexity of the proteome which allows for more protein identifications during MS/M.
4. Inject samples directly into the LTQ Orbitrap Velos for analysis. LTQ Orbitrap Velos is a hybrid ion trap mass spectrometer. It uses two methods for the fragmentation of peptides, high energy collision dissociation (HCD) and electron transfer dissociation (ETD). ETD is optional and can be used if de-novo sequencing or information on post-translational modifications is desired. This allows for the identification and quantification of the peptides, and thus the representative proteins in the fractions.

3.6 MS/MS Data Analysis

1. Analyze the resultant MS/MS spectra (.RAW files) using Mascot Daemon v 2.5 (Matrix Science, London, UK, 17), which allows for both quantification of the peptides by analyzing the reporter ion released by the isobaric tags and identification of the protein sequence. The settings used to run Mascot will depend on which enzyme was used to digest the proteins during sample preparation and the type of isobaric tagging (iTRAQ or TMT) chosen for proteomics. For example, trypsin digests proteins by cleaving the C-terminus of lysine or arginine residues except when either is followed by proline, therefore those sites are set as cleavage sites [9]. The N-hydroxysuccinimide group in the iTRAQ label is amine reactive and reacts with the ε-amine group of lysine residues and amine groups of the N-terminus of the peptides, attaching the labels to the peptides [10] and hence the modification of lysine is considered a fixed modification while running Mascot.
2. Quantification of the peptides is achieved through the release of the characteristic iTRAQ reporter ions (114–117 m/z)

during MS/MS of the samples. The software calculates iTRAQ ratios using the abundance of the reporter ion released for each iTRAQ reagent. Evaluate statistical significance of individual peptides using the provided p-value. The intensity of the reporter ion is directly correlated to the abundance of the particular peptide cleaved in MS/MS.

3. Identify proteins: The b and y ions released by the peptide during MS/MS provide sequence information which is used to match the peptide to a protein [11]. Every protein after digestion with a specific enzyme produces signature peptides that are specific to that particular protein and Mascot uses these signature peptides to ascertain which proteins are present in the sample. This principle is used to identify common contaminants like keratin, dust and contact proteins from the proteins sample. Mass spectra are searched against the common Repository of Adventitious Proteins (cRAP), a database of common contaminants found in samples. The current version can be downloaded using this URL ftp://ftp.thegpm.org/fasta/cRAP (Copyright © 2004–2011, The Global Proteome Machine Organization). For identifying proteins from the target organism, the spectra are matched against the TAIR10 proteome database (target and decoy sequences) using Mascot [12]. After a signature peptide is identified, the decoy sequences in TAIR10 are searched by adjusting Mascot to a peptide identification false positive rate of 1%. Only proteins identified with two or more signature peptides should be considered for further analysis.
4. Use the Scaffold software for loading and analyzing the data from Mascot. Once peptide sequencing and protein identification is performed with Mascot, the same spectra are analyzed once more using Scaffold, and the proteins interpreted from the spectra are compared and combined.
5. Use the Protein Prophet algorithm to confirm the correct protein identity [13]. The algorithm assigns the amino acid sequence identified from the fragment ions to a matching protein with 90% confidence.
6. Enrich the modified peptides using affinity chromatography or antibodies specific to the modified amino acid prior to MS/MS analysis to identify peptides with post-translational modifications (*see* **Note 20**).

3.7 Differential Abundance Analysis (See Note 21)

1. After protein expression data has been quantified by MS, begin identifying the differentially abundant proteins by performing log or square root transformation using "log" or "sqrt" functions in R [14, 15].

2. Test for normality and homogeneity of variance to determine appropriate significance tests to apply.
3. If the data show normal distribution or can be mathematically transformed to normal distribution, then an ANOVA or a Student's *t*-test can be used, followed by multiple-testing corrections (e.g., Benjamini-Hochberg correction using "p. adjust" in R) to reduce the risk of abundant false positives. If data cannot be normalized, use the Welch's *t*-test or Mann-Whitney *U* test to determine statistical significance.
4. Use the abundance of proteins in each sample (as determined by isobaric tags) to determine fold change in expression between experimental and control conditions.
5. Convert the fold change expression data to $\log_2$ of the fold change such that (a) positive values equal more abundant proteins and negative values are less abundant proteins as caused by the experimental treatments, (b) a value of 1 or −1 represents a two-fold difference in expression and (c) values are more readily visualized.

3.8 Network-Based Proteomics

Adding a network-based proteomics (NBP) in addition to differential expression can help identify a protein complex or a series of proteins acting in a pathway that are co-expressed to focus on identification of functional units within a set of proteins (*see* **Note 22**) [16].

1. If the experimental plant is Arabidopsis or another reasonably well-characterized organism, begin by using VisANT to build powerful networks with edges determined by experimentally predicted or validated interactions from multiple databases [17] (*see* **Note 23**).
2. As an alternative (or in addition to) a PPI-based NBP, perform a co-expression network analysis (*see* **Note 24**). Use the web-based platform STRING (Search Tool for the Retrieval of Interacting Genes/Proteins) DB v 11 to perform straightforward co-expression NBP analyses [18].

3.9 Studying Unknown Proteins of Interest

1. If a protein annotation of "unknown protein" or "putative protein" is returned, first retrieve the amino acid sequence for the protein of interest.
2. Input the primary sequence into I-TASSER, Phyre2, SWISS-MODEL, or RaptorX [3–6, 19, 20] to predict the tertiary structure of the protein. These tools are reliant on the homology of primary sequence amino acid properties to previously characterized proteins with X-ray crystallographic, or CryoEM structures. Carefully select a template for homology-based structural modeling and evaluate the differences caused by

altering template selection to determine regions of uncertainty and confidence in the prediction.

3. Once generated, structural predictions ubiquitously use the Protein Data Bank (PDB) file format. Download and carefully label the homology-based model to insure reproducibility.
4. Use PyMol [21] or similar software tools to visualize the protein structure. The structure can help identify ligand binding sites as well as show residual electrostatic charge mapping on the protein surface revealing acidic and basic charge clustering.
5. If simple modeling and function prediction is insufficient, perform more advanced in silico analyses using Rosetta or InterevDock2 [22, 23] to identify ligand docking, structural perturbations through post-translational modification, and protein-protein interactions.

4 Notes

1. A detailed experimental design streamlines a proteomic experiment (e.g., [24]). Knowing the amount of protein required for successful identification of maximum proteins in a single run of mass spectrometry helps determine how much tissue is required per replicate. A minimum target for mass spectrometry is approximately 50 μg per replicate. The amount of tissue required for extraction should be experimentally determined as extraction efficiency can change based on tissue type, age, and preservation method. After extraction, the amount of protein can be measured using colorimetric assays such as the Bradford assay or the bicinchoninic assay (BCA) [25]. After mass spectrometry, acquiring statistically significant results requires a minimum of three replicates per sample. However, identification of smaller changes may necessitate the increased statistical power afforded by additional replicates (*see* Subheading 3.7 for details on data analysis). The design provides a necessary framework for the experiment and should be the starting point of experimental planning.
2. The protocol provided has been used to extract protein samples from etiolated Arabidopsis seedlings for spaceflight experiments where limited amounts of plant tissue are available. It can, however, be used with inflorescence stems and leaves. Use with other plant species may require modification of the extraction procedure to account for differences in cell wall composition, membrane composition, and secondary metabolites. In addition, downstream analysis of MS/MS results will require distinct protocols. The approach described in the subsequent sub-sections is an adaptation of [26].

3. If phosphoproteins are to be studied after protein identification, then a phosphatase inhibitor cocktail needs to be added to the extraction buffer. For 100 mL of buffer, 1 mL of the phosphatase inhibitor cocktail is sufficient. Phosphoproteomics requires a minimum of 200–500 mg of starting tissue, as 1 mg of protein is required before phosphoenrichment to obtain enough protein for further analysis.
4. Nanopure water is used for preparation of all the solutions and buffers to minimize the presence of contaminating ions during MS/MS.
5. β–ME is used as a reductant in protein extraction buffers. It has the same function as dithiothreitol (DTT) or dithioerythreitol (DTE).
6. β–ME has a buffering effect above pH 8. Commercial β–ME often is contaminated with keratin which can be seen on SDS-PAGE gels [27].
7. Urea is hygroscopic. While making a solution with urea, always start with half the final volume. Urea absorbs atmospheric moisture and the solution increases in volume while urea dissolves.
8. A high concentration of urea is used in protein resuspension buffers as it helps keep the protein in a single conformation, canceling out the secondary and tertiary structures. The high concentration of urea also helps keep the hydrophobic proteins in solution and stops protein–protein interactions.
9. Phenylmethyl sulfonyl fluoride (PMSF) is toxic. It must be added to the buffers before adding β–ME, DTT or DTE.
10. 4-benzenesulfonyl fluoride hydrochloride (AEBSF) can be used instead of PMSF. Although less toxic than PMSF, it causes modifications in some proteins.
11. During mass spectrometry each peptide is fragmented further. For this fragmentation different mass spectrometers use different mechanisms, for example, Collision Induced Dissociation (CID), Electron Transfer Dissociation (ETD). High Energy Collision Dissociation (HCD) is a form of CID. HCD is the fragmentation method used in Orbitrap instruments to fragment isobaric tagged peptides. Since the fragmentation affects the peaks obtained in mass spectrometry results and all the experiments in which the above proteomic protocols have been used have utilized the LTQ Orbitrap Velos instrument we are including the name of the manufacturer in the text.
12. If possible, the extraction should be carried out in a cold room maintained at 4 °C.
13. If some of the pellet gets dispersed while pouring off the supernatant, strain the supernatant through 4–5 layers of

miracloth to remove the dispersed debris before proceeding to the next step.

14. Most facilities prefer the pellet formed after **step 17** (in case of soluble proteins), but if a resuspended sample is required, the pellet can be dissolved in Resuspension buffer using approximately 15 μL per 50 mg of starting tissue.

15. Protein precipitation methods are preferred for extracting proteins from plant cells. Plant tissue has a higher abundance of polysaccharides, nucleic acids, and lipids compared to proteins. Precipitation methods help in extracting proteins from such tissue. The method also helps inhibit protease activity as it uses phenol for extracting the proteins from the tissue.

16. To ensure that enough protein can be extracted from the available tissue, preliminary protein extractions can be quantified using the Bradford assay to determine the appropriate amount of starting tissue needed per replicate.

17. The quality of the protein extracted can be evaluated based on proteins identified and quantified. This can vary based on the tissue and species being analyzed. For Arabidopsis seedlings, identification of >1000 after trypsin digest and a single LC–MS/MS run would be considered excellent with 500–999 acceptable. Less than 500 proteins identified through this initial analysis would be marginal at best, and new extractions should be considered. The acceptable values would equate to >1500 and 1000–1499, respectively, proteins identified and quantified after fractionation into 12 LC-MS/MS with iTRAQ.

18. Sample treatment before iTRAQ labeling: Proteins are digested with Trypsin overnight. The cysteine residues are methylated by methylmethanethiosulfonate (MMTS). After blocking the cysteine, iTRAQ peptide labeling is done.

19. iTRAQ reagents allow for differential labeling of multiple experimental and control samples for simultaneous quantification and identification of proteins in a single step: the LC–MS/MS (liquid chromatography–tandem mass spectroscopy). If the number of samples exceed the multiplexing capabilities and have to be analyzed in multiple runs, then an internal control can be formulated by mixing all samples across all runs in equimolar proportion and injecting the internal control with each run so that the expression of proteins can be compared across multiple runs. However, soluble and membrane proteins should be analyzed separately. Because the protein patterns for soluble and membrane proteins are likely to be different, normalization won't be possible if they are combined.

20. When the MS/MS spectrum is analyzed, the presence of post-translational modifications causes a shift in the retention time of the fragment ions. This shift can be detected by the software (e.g., Scaffold) analyzing the MS/MS spectra to identify the PTM.
21. While protein quantification is now readily accessible to labs without specialized technology and techniques, it remains a challenging, yet rewarding, pursuit. There is no "one size fits all" experimental design or analytical pipeline to characterize the proteomic landscape of a sample. However, the tools (and their increasing usability) enable researchers to more readily characterize the "functional" profile of a sample. Careful attention to the sample preparation must always be the highest priority to ensure the data gathered are as useful and representative as possible. The greatest challenge when first attempting to analyze proteomic data is mindset. There is an inherently steep learning curve to begin performing "big data" computational analyses, and perseverance is essential. Take the necessary time to learn the basic elements of working in terminal (bash) and to "play" with the data. Remember that as long as you have the raw data backed up, computational analyses are infinitely forgiving and require no reagents to test or alter protocols. In a living organism, molecules work together in synchrony to adapt and respond to environmental cues, and network-based proteomics (NBP) can help identify these interactions. In NBP approaches, the proteins form the nodes of a network and the evidence connecting any two proteins form the edges. The most common NBP approaches use one or multiple pre-calculated databases of protein–protein interactions (PPI) determined by co-expression or experimental assays.
22. VisANT is a feature-rich tool that enables researchers to build functional networks that put a set of differentially expressed proteins in the context of previously established interactions. This approach can be highly effective in determining regulatory consequences of and causative relationships between differentially expressed proteins; however, PPI-based networks are inherently incapable of characterizing previously unstudied proteins which lack PPI information.
23. Co-expression approaches to NBP entail forming a network out of all proteins identified in a proteomics experiment by comparing it to a reference network generated from co-expression in previous experiments. The approach calls for a probability-based model for analyzing and visualizing proteomics data. The probability-based approach helps improve feature selection reproducibility and class prediction accuracy. Class prediction accuracy is comparing protein complexes or a group of proteins acting in a pathway, identified from a

network analysis, to a random selection of proteins (number of proteins equal to the initial query set) from a molecular signature database [28].

24. The main aim of a large-scale experiment is to identify novel proteins of interest. Unknown proteins, those not previously characterized for structural conformation or biological function, are often overlooked because little is known about them. With the improvement of bioinformatic resources many web-based tools are available to gather preliminary information on unknown proteins of interest to determine protein homologs and infer hypothetical protein activities.

Acknowledgments

This work was partially funded by NASA # NNX13AM48G awarded to SEW and DRL. The authors would like to thank Dr. Frans Maathuis, University of York, for his valuable input and Dr. Sophie Alvarez and Mike Naldrett, Donald Danforth Plant Science Center, St. Louis, MO, for their inputs with the proteomics protocols and protein quality metrics.

References

1. Schultz J, Copley RR, Doerks T et al (2000) SMART: a web-based tool for the study of genetically mobile domains. Nucleic Acids Res 28:231–234
2. Mitchell AL, Attwood TK, Babbitt PC et al (2019) InterPro in 2019: improving coverage, classification and access to protein sequence annotations. Nucleic Acids Res 47: D351–D360
3. Zhang Y (2008) I-TASSER server for protein 3D structure prediction. BMC Bioinform. https://doi.org/10.1186/1471-2105-9-40
4. Roy A, Kucukural A, Zhang Y (2010) I-TASSER: a unified platform for automated protein structure and function prediction. Nat Protoc 5(4):725–738
5. Yang J, Yan R, Roy A, Xu D et al (2015) The I-TASSER Suite: protein structure and function prediction. Nat Methods 12:7–8
6. Kelley LA, Mezulis S, Yates CM et al (2015) The Phyre2 web portal for protein modeling, prediction and analysis. Nat Protoc 10:845–858
7. Kruse CPS, Meyers AD, Basu P, Hutchinson S, Luesse DR, Wyatt SE (2020) Spaceflight induces novel regulatory responses in Arabidopsis as revealed by combined proteomics and transcriptomic analyses. BMC Plant Biology 20:237. https://doi.org/10.1186/s12870-020-02392-6
8. Kruse CPS, Basu P, Luesse DR et al (2017) Transcriptome and proteome responses in RNAlater preserved tissue of *Arabidopsis thaliana*. PLoS One. https://doi.org/10.1371/journal.pone.0175943
9. Somiari RI, Renganathan K, Russell S et al (2014) A colorimetric method for monitoring tryptic digestion prior to shotgun proteomics. Int J Proteomics. https://doi.org/10.1155/2014/125482
10. Rauniyar N, Yates JR (2014) Isobaric labeling-based relative quantification in shotgun proteomics. J Proteome Res 13:5293–5309
11. Cox J, Hubner NC, Mann M (2008) How much peptide sequence information is contained in ion trap tandem mass spectra? J Am Soc Mass Spectrom 19:1813–1820
12. The Arabidopsis Information Resource (TAIR). https://www.arabidopsis.org/download/indexauto.jsp?dir=%2Fdownload_files%2FSequences%2FTAIR10_blastsets, *on* www.arabidopsis.org. Accessed 23 Apr 2020
13. Nesvizhskii AI, Keller A, Kolker E et al (2003) A statistical model for identifying proteins by tandem mass spectrometry. Anal Chem 75:4646–4658

14. Mertens BJA (2017) Transformation, normalization, and batch effect in the analysis of mass spectrometry data for Omics studies. In: Datta S, Mertens BJA (eds) Statistical analysis of proteomics, metabolomics, and lipidomics data using mass spectrometry. Springer International Publishing, Switzerland, pp 1–21
15. Trautwein-Schult A, Maaß S, Plate K et al (2018) A metabolic labeling strategy for relative protein quantification in *Clostridioides difficile*. Front Microbiol. https://doi.org/10.3389/fmicb.2018.02371
16. Goh WW, Wong L (2016) Integrating networks and proteomics: moving forward. Trends Biotechnol 34:951–959
17. Hu Z, Snitkin ES, DeLisi C (2008) VisANT: an integrative framework for networks in systems biology. Brief Bioinform 9:317–325
18. Szklarczyk D, Gable AL, Lyon D et al (2019) STRING v11: protein-protein association networks with increased coverage, supporting functional discovery in genome-wide experimental datasets. Nucleic Acids Res 47(D1): D607–D613
19. Schwede T, Kopp J, Guex N et al (2003) SWISS-MODEL: an automated protein homology-modeling server. Nucleic Acids Res 31:3381–3385
20. Peng J, Xu J (2011) RaptorX: exploiting structure information for protein alignment by statistical inference. Proteins 79(Suppl. 10):161–171
21. Schrödinger LLC (2015) The {PyMOL} Molecular graphics system, version~1.8
22. Conchuir SO, Barlow KA, Pache RA et al (2015) A web resource for standardized benchmark datasets, metrics, and rosetta protocols for macromolecular modeling and design. PLoS One. https://doi.org/10.1371/journal.pone.0130433
23. Quignot C, Rey J, Yu J et al (2018) InterEvDock2: an expanded server for protein docking using evolutionary and biological information from homology models and multimeric inputs. Nucleic Acids Res 46:W408–W416
24. Hutchinson S, Basu P, Wyatt SE et al (2016) Methods for on-orbit germination of *Arabidopsis thaliana* for proteomic analysis. Gravit Space Res 4:20–27
25. Olson BJSC (2016) Assays for determination of protein concentration. Curr Protoc Pharmacol 73:A.3A.1–A.3A.32
26. Basu P, Luesse DR, Wyatt SE (2015) Proteomic approaches and their application to plant gravitropism. Methods Mol Biol 1309:119–132
27. Westermeier R, Naven T (2002) Expression proteomics. Proteomics in practice. Wiley-VCH Verlag GmbH & Co. KGaA, Weinheim, Germany
28. Beck D, Thoms JAI, Perera D et al (2013) Genome-wide analysis of transcriptional regulators in human HSPCs reveals a densely interconnected network of coding and non-coding genes. Blood 122:e12 LP–e12e22

Chapter 14

Blueprints for Constructing Microgravity Analogs

Karl H. Hasenstein

Abstract

The desire to understand gravitational effects on living things requires the removal of the very factor that determines life on Earth. Unfortunately, the required free-fall conditions that provide such conditions are limited to a few seconds unless earth-orbiting platforms are available. Therefore, attempts have been made to create conditions that simulate reduced gravity or gravity-free conditions ever since the gravity effects have been studied. Such conditions depend mostly on rotating devices (aka clinostats) that alter the gravity vector faster than the biological response time or create conditions that compensate sedimentation by fluid dynamics. Although several sophisticated, commercial instruments are available, they are unaffordable to most individual investigators. This article describes important considerations for the design and construction of low cost but versatile instruments that are sturdy, fully programmable, and affordable. The chapter focuses on detailed construction, programming of microcontrollers, versatility, and reliability of the instrument.

Key words Clinostat, Random positioning machine, 3D printing, STL files, Microcontroller programming, LED illumination, Accelerometer

1 Introduction

Starting with the employment of water wheels to investigate the effect of gravity on plants [1], the development of devices designed to eliminate or minimize the effects of unidirectional gravity became increasingly complex. The first functional clinostat was developed by Ciesielski [2] and von Sachs [3] and the goal was the desire to understand gravitational effects. However, already von Sachs realized that clinorotation imposes mechanical stress to plants and later the effect was traced to elevated ethylene that caused curvature of petioles [4, 5]. Despite clearly different effects of clinorotation from "true" microgravity (i.e., free-fall conditions), this methodology is a valuable substitute for space flight experiments as long as proper boundary conditions are considered.

Electronic Supplementary Material: The online version of this chapter (https://doi.org/10.1007/978-1-0716-1677-2_14) contains supplementary material, which is available to authorized users.

Elison B. Blancaflor (ed.), *Plant Gravitropism: Methods and Protocols*, Methods in Molecular Biology, vol. 2368,
https://doi.org/10.1007/978-1-0716-1677-2_14,

Remarkably, the gravisensing mechanism can be fairly well studied and that aspect led to the discovery and analysis of presentation time [6], threshold of gravisensitivity [7, 8], onset of gravisensitivity [9], and the persistence of the gravistimulus [10]. In addition, clinorotation provided insights in transcriptional effects [11–13], and hook formation [14]. Although the majority of clinorotation studies are performed on plants, it is important to point out that the effect of clinorotation on other organisms are being studied in fungi [15], unicells [16–19], xenopus [20, 21], and a host of mammalian and human cells [22–24].

The proliferation of rather uncontrolled experiments prompted a quality assessment (Bonn criteria) for clinorotation experiments [25]. Considering the relative convenience of clinostat experiments and the desire to not only minimize gravity effects but to study fractional gravities (e.g., 0.166 g and 0.38 g for Moon and Mars, respectively), have led to increased use of ever more sophisticated instruments that include one and two axial systems (often referred to as one- and two-dimensional clinostats, see [26, 27] and Random Positioning Machines (RPMs). Unfortunately, this increase in sophistication results in considerable expenses to investigators that cannot be easily met. Commercial RPM systems result in an investment of many thousands of dollars. The purpose of this chapter is to provide a blueprint that allows anyone with a minimal technical inclination to build a laboratory-grade, low-cost RPM that is suitable for most seedling or tissue culture experiments and includes fully programmable controls.

2 Materials

2.1 Support Structure

The hardware consists of a range of 3D-printed pieces that are assembled into the system depicted in Fig. 1. Although it is possible to print the items on any Fusion Deposition Modeling (3D)-printer, a good, and possibly less-expensive alternative is to order the prints from a commercial supplier.

The main support structures are two trapeze-shaped frames (20 mm thick, 254 mm high, 212 mm wide) that are connected by two braces (264 × 40 × 8 mm) to provide proper stability and distance; they also serve as conduits for wires. Both braces and frames can be made from two pieces because standard printers may not be able to produce single pieces of the required size. The assembly supports two rotating frames. The outer frame (two pieces of 220 × 45 mm and 215 × 45 mm) support an inner frame (two sides each, 190 × 45 and 176 × 45 mm). The ring-like surface of the lateral inner sides secures the experimental frame at two possible orientations, perpendicular (as shown in Fig. 1) or parallel to the inner frame. Attached to one of the trapeze-shaped frames is a box that contains all electronic parts.

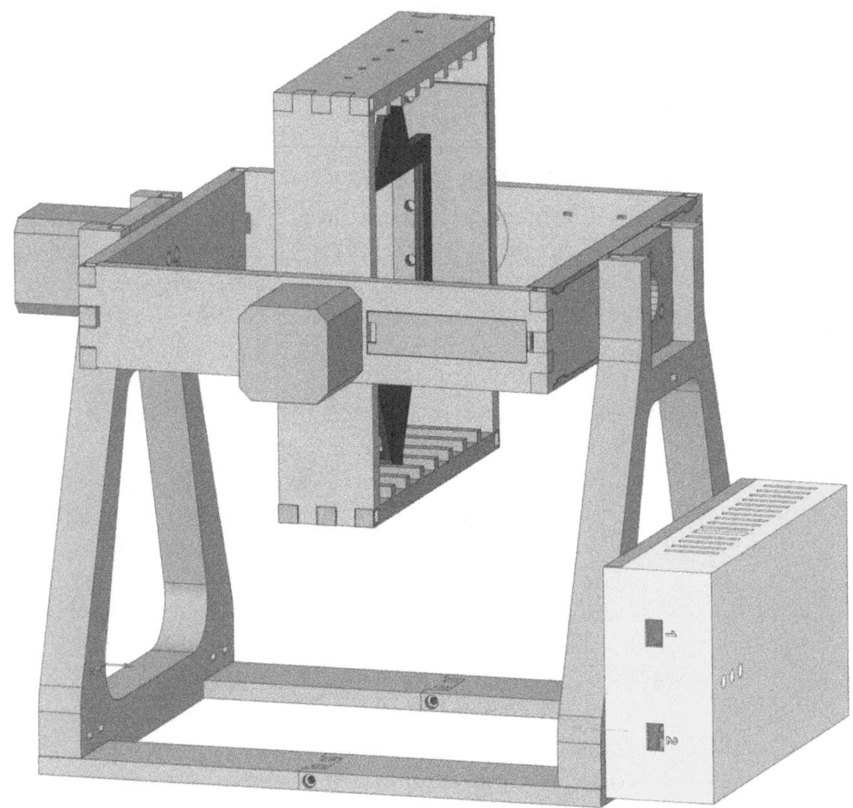

Fig. 1 Design of a Random Positioning system based on two independently rotating axes, each driven by a separate motor. The inner frame houses exchangeable LED lights and a payload frame that is suited for (stacks of) standard square Petri dishes or other (custom-made) containers. An accelerometer can be attached and record the experimental acceleration profile (not shown)

Because 3D printing supplies (motors, connectors, etc.) are typically in metric units, the assembly for all parts is based on M3 Stainless Steel Hex Socket of various lengths (see Bill of Materials). Nuts are typically of the hex type but, in some cases, square nuts are necessary.

2.2 Drives

The motors (shown as dark extensions in Fig. 1) are bipolar (i.e., four leads) stepper motors of the NEMA 17 type. Motors should be based on the expected load and based on currents of 1, 1.5, or 2 A, which typically corresponds to a torque of 13, 45, and 59 Ncm, respectively. The motors are mounted to a base plate (stationary motor) that is attached to the main support structure or the outer, rotatable frame. This design allows for easy modifications or changes to non-NEMA or geared motors. However, such modifications are not provided in the provided STL (stereolithography) files.

The power distribution necessitates slip rings. At least one such device is needed if no illumination is required. The slip ring (located opposite of the outer motor) serves as rotary support and conduit. If illumination is desired, a second slip ring accommodates the power distribution to the inner frame where the LED lights are located. Although the power requirements for LEDs are significantly lower than for the motor, it is economical to use the same type of slip ring for both pivots. The connection between motor shaft and frame requires "Rigid Motor Flange Couplings." The flanges match pre-designed holes of the inner and outer frame (Fig. 1).

2.3 Power System and Electronics

The electrical system is based on a 12 V, 50 Watt power supply (LRS-50-12) that is providing power to two microcontrollers (Arduino Nano), two motor drivers (A4988), and LED lights. The microcontrollers and drivers run on 5 V and therefore a 12–5 Volt converter (Smakn or equivalent) is needed.

The Arduino Nano controllers were chosen because they provide the smallest form factor with full programmability and direct, mini USB-based accessibility. Although it is possible to run two motors from one controller, one controller for each axis provides independent programmability. To facilitate connectivity of the microcontrollers, Screw IO Shields (adaptor) are highly recommended. The power box contains provisions to attach the shields directly to the base plate.

3 Methods

3.1 Frame Assembly

Detailed instructions are provided with the STL files. Once the printed material has been obtained, the assembly needs to start with the outer support structure (including power and electronics box), followed by inner and outer frame including slip rings, flanges and, if desired, LED lights.

The support frame can be produced as single piece or, alternatively, consists of an upper and lower part that will have to be mounted together. The motor (opposite the power box) must be mounted to its baseplate and the leads threaded through the conduit of the support frame followed by threading through the spacers. Each frame is secured by two 16 mm bolts to each of the braces. Their attachment relies on inserted (square) nuts that accommodate two 16 mm bolts on each corner. It is recommended to attach the power box to the support base (with spacers) before mounting the frame to the braces.

The assembly of the frames includes proper mounting of slip rings; each is secured by three 12 mm bolts. It is highly recommended to use connectors (Dupont or similar detachable types), not soldering, to allow for corrections and easy re-assembly. The frames are joined together and are secured by four 45 mm bolts on each corner.

3.2 Attachment of the Motors

The stationary motor and the slip ring on the opposite side must be attached to their respective baseplate before attaching the combination to the support frame. Frame and baseplate are secured by two 45 mm bolts each. Flanges should be attached to the outer (larger) frame before connecting the assembly to the motor shaft. The second motor needs to be mounted to the larger (outer) frame. The inner frame should have the flange attached before connecting to the motor.

The motor on the outer frame must be balanced to minimize load changes for the stationary motor. This balancing can be done by various means. Appropriate weights can be attached to the length of the frame and two indentations are provided that accommodate balancing weights. Mounting holes for standard cable ties hold suitable weights, e.g., metal rods, (sections of large nails), lead shots, or anything that can be secured such that no load shifts occur during rotation. It helps to weigh the assembly (motor and frame) with a spring or top-loading balance. All mass, including the cover that is part of the frame must be included in the balancing weight. Upon final assembly, the frame must remain horizontal once rotated to that position.

3.3 Power Considerations

Because the stationary motor moves the entire inner frame including the secondary motor, the moveable, inner frame, motor could be smaller; however, the added mass of a heavier motor smoothens the movements of the mobile parts. The power consumption of the motor determines the gauge for the leads, including those on the slip rings, and a capacity of 5 A per lead is recommended but 2 A is sufficient.

3.4 Assembly of Electrical Components

The “power box” contains the AC/DC power supply, a 12 V to 5 V DC converter, and one microcontroller (Arduino Nano) per axis. Two motor drivers are each connected to the 12 VDC power source and one microcontroller each. The cover is equipped with three toggle switches that control the two motors and LED bank. With the box attached to the support frame, it is best to install the power supply followed by the DC converter and the drivers. The last elements to be inserted are the Arduino microcontrollers. Connections to the microcontrollers are easier to achieve if screw IO shields (adaptors) are used. The connection between Arduino, Driver, and power units are illustrated in Fig. 2.

3.5 Adjustments of the Stepper Drivers

The ability of the drivers (A4988) to provide power to the motor coils is determined by the reference voltage. To provide a maximum current of 1.5 A, the reference voltage (between potentiometer and ground (see Fig. 2) should be set to 0.75 V when the A4988 is

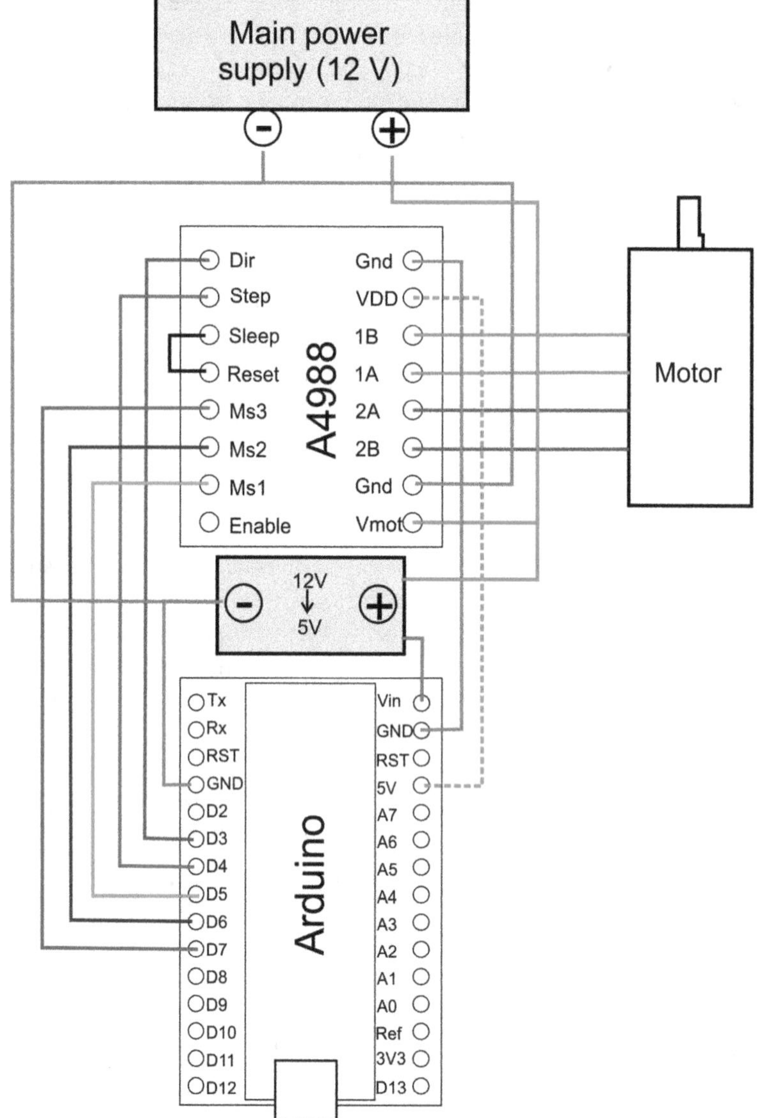

Fig. 2 Wiring diagram for an Arduino-controlled motor driver. Gray symbols (power supply and converter) are single items and provide power to all other components. Arduinos, A4988 and motors are required in duplicate, one per axis

supplied with 5 V. Since several clones of the driver exist, it is recommended to follow the detailed explanation online [https://forum.arduino.cc/index.php?topic=415724.0].

3.6 Illumination

Plant metabolism and photosynthesis depends on light quality and fluence (emitted radiation energy). Clinorotation can be accomplished in darkness or under the influence of external illumination. If clinostat or RPM-independent light is used, the resulting development will be the result of constantly changing gravity and light vectors. While these conditions may not matter for some experiments, the presence of a light gradient that is not affected by rotational movements also allows for studies of light effects during clinorotation. Such questions can only be answered if the relative position between plant material and light source does not change. The design of the RPM system uses light sources attached to the inner frame, which assures a constant direction between light source and biological material.

3.6.1 Light Emitting Diodes (LEDs)

To accommodate a light system with constant direction relative to the plant material is illustrated below. Studying clinorotation and light interactions depends on light energy and spectral composition. Because a generic "white light" setting may not be adequate for all purposes, the system described below contains two light bars; one with six and one with twelve 5 mm LEDs. Either bank is attached to the inner frame. The modular setup allows for easy and quick exchange of the entire panel such that different light modules can be utilized (or turned off on the power box). The LEDs can all be of the same type or provide different emission maxima. The 12 V power supply feeds two or four sets of three LEDs each with a forward voltage of ca. 3 V, but this value varies with LED type. The recommended resistors (see Subheading 3.6.2) limit the current to 20 mA but the values depend on the specific LEDs and their combinations. The proposed high-efficiency LEDs have a standard form factor, are small, and cover emission spectra ranging from UV to infra-red and "white" light. The illustrated system consists of 6 or 12 LEDs that can be of the same color type or mixed among desired wavelengths. Various combinations result in complex light configurations, including lateral gradients, that go beyond the scope of this description.

3.6.2 LED Connection and Power Considerations

The connection must be part of the slip-ring module that supports the inner frame and provides 12 V power. It is important to carefully consider the "Forward Voltage" (FV) of the LEDs and care should be taken that the three LEDs that form a section (two in a single row, or four in two rows) have the same value as they are connected to the same current-limiting resistor. FV ranges from 1.7 to 3.3 V for 5 mm diameter LEDs, with the higher FV typical for high-efficiency white or blue LEDs. The value of the resistor must be chosen based on the power supply voltage (12 V), the FV of each LED and the desired current, e.g., 20 mA. Based on an FV of 3 V a resistor for a set of three LEDs needs to be:

$$\text{Resistor value} = 12\ \text{V} - (3 \times 3\text{V})/0.02\ \text{A} = 150\ \text{Ohm}.$$

If a higher brightness is desired (and the LEDs support the current!) 25 mA can be used, which requires 120 Ohm resistors (Fig. 3).

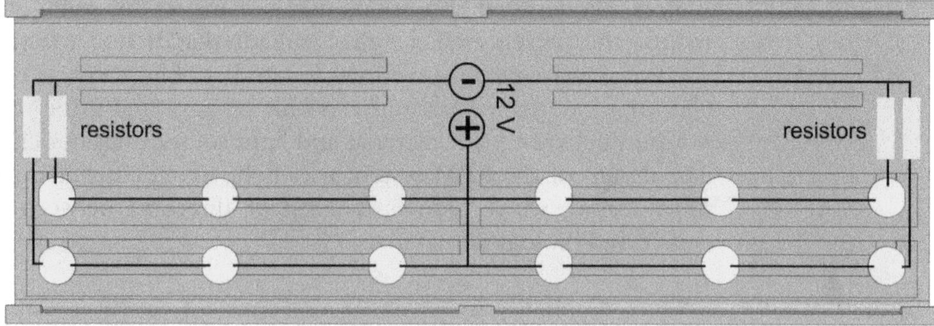

Fig. 3 Suggested light panel for the inner frame of the described RPM. For each set of three LEDs (5 mm diameter), a resistor is needed that limits the current and thus controls the light output. The polarity of the LEDs is important but easily identified as the negative electrode is indicated by the flattened side of the mounting ring

3.7 Bill of Materials

STL files for all parts are available as Electronic Supplemental Data 1:

Proceed only after all specified parts are available; the Bill of Material describes the individual parts.

Tables 1–3 list all parts required for the assembly, powering, and balancing of the motor and frames. They are provided to facilitate acquisition, organization, assembly and decisions regarding motors, hardware, and balancing weights.

3.8 Assembly Instructions

The assembly of the various parts should follow the steps outlined below and require the printed parts to be completed (*see* **Note 2**). Before assembly of individual parts, make sure that all connections, joining surfaces and mounting holes are free of printing artifacts or obstructions.

3.8.1 Frame

1. Attach stepper motor to the motor panel (4 M3 × 12).
2. Attach slip ring to slip-ring panel (3 M3 × 12, washer may be necessary if slip-ring mounting holes are too big).
3. Insert the assembly to the top part of the support frame such that the panel faces each other.
4. Feed wires (4 leads) through opening of the support frame, start with a small thread to facilitate this process.
5. Feed wires through the braces, then combine the braces (2 M3 × 12 per brace).
6. Attach the braces to the bottom of the support frames.
7. Attach the base of the power box to the frame that holds the slip ring; insert spacers between base and support frame (4 M3 × 16).

Table 1
3D-printed parts

Description	Pieces	STL file name (*.stl)
Motor panel	1	Motor panel
Slip-ring panel	1	SR-panel
Support frame 1 (*see* **Note 1**)	1	SFrame1
Support frame 1B	2	SFrame1B
Support frame 2 (*see* **Note 1**)	1	SFrame2
Support frame 2B	2	SFrame2B
Braces (4)	2	Brace
Spacers between power box and frame	4	Spacer
Power Box cover	1	PWR-lid
Power Box base	1	PWR-base
Outer rotating frame drive	1	OF-1
Outer rotating frame weight mount	1	OF-3
Outer rotating frame drive	1	OF-2
Outer rotating frame slip ring	1	OF-4
Inner rotating frame drive	1	IF-1
Inner rotating frame and LED panel	1	IF-3
Inner rotating frame sides	2	IF-2
Payload support	2	PL

Table 2
Electrical components

Stepper motor (Nema 17)	2	1.8°, 1.4A, > 40 Ncm, bipolar
Power supply (12 V, 5A) LRS 50–12	1	Mean Well or equivalent
12 V to 5 V converter	1	Smakn or equivalent
Arduino Nano	2	One per axis
Terminal Adaptor Board for Arduino Nano	2	Arduino Nano Screw Terminal Shield
Slip ring, 22 mm diam.	2	6 wires, max 240 V @ 5A,
Stepper driver A4988	2	
Wire, 20–28 AWG		Silicone coating recommended
Capacitor, 47 to 100 μF	1	
Toggle switch	3	2 Position Switch 6A 125 V or equivalent.
Dupont connectors, 0.1"	26	The purchase of a kit is advised

Table 3
Other parts

Counter weights (weight of motor)	Lead shots, iron rods, etc.
45 mm × 3 mm Hex socket screws #	12
16 mm × 3 mm Hex socket screws #	10
12 mm × 3 mm Hex socket screws #	10
3 mm nuts, square	30

Hex Socket Head Cap Screws kit is advised

3.8.2 Power Box Assembly

1. Attach power converter (120 AC to 12 DC with one M12 screw or equiv.)
2. Feed through bottom opening the AC power leads (120 or 240 V).

   Consider soldering together the pos. leads for the two A4988s, and the voltage converter to a common spade connector; repeat for neg. leads.
3. Connect the positive leads from the 12/5 V converter, the pos. wires to the two A4988s, the positive lead of the LED wires and the pos. end of the capacitor (47–100 μF) to the pos. terminal (+V) of the power supply.
4. Connect the negative leads from the 12/5 V converter, the neg. wires to the two A4988's, the neg. lead(s) of the LED supply and the neg. lead of the capacitor (47–100 μF) to the neg. terminal (−V) of the power supply.
5. Feed the AC power cord through the top slot of the power box base; connect to the L and N terminal of the power supply.
6. Attach voltage converter (Smakn 12 to 5 V DC) with two screws (M3 × 12, or 0.5″ × 8/32) to base.
7. Feed through top opening the four wires to the slip ring for the inner frame motor.
8. Attach the power box base to the clinostat frame (M2 × 12, top screws first).
9. Connect the Arduino base plates to the 5 V DC supply and control leads to the A4988s.
10. Connect the motor leads to the A4988s (see Fig. 2).

3.9 Programming

The flexibility of programming of the behavior of the RPM is its greatest asset. The following section provides one possible solution that will cover numerous possibilities of controlling motors and thus the rotation of the entire system. Importantly, the programming requires handling the Arduino IDE (Integrated Development Environment). The free software is available from https://www.

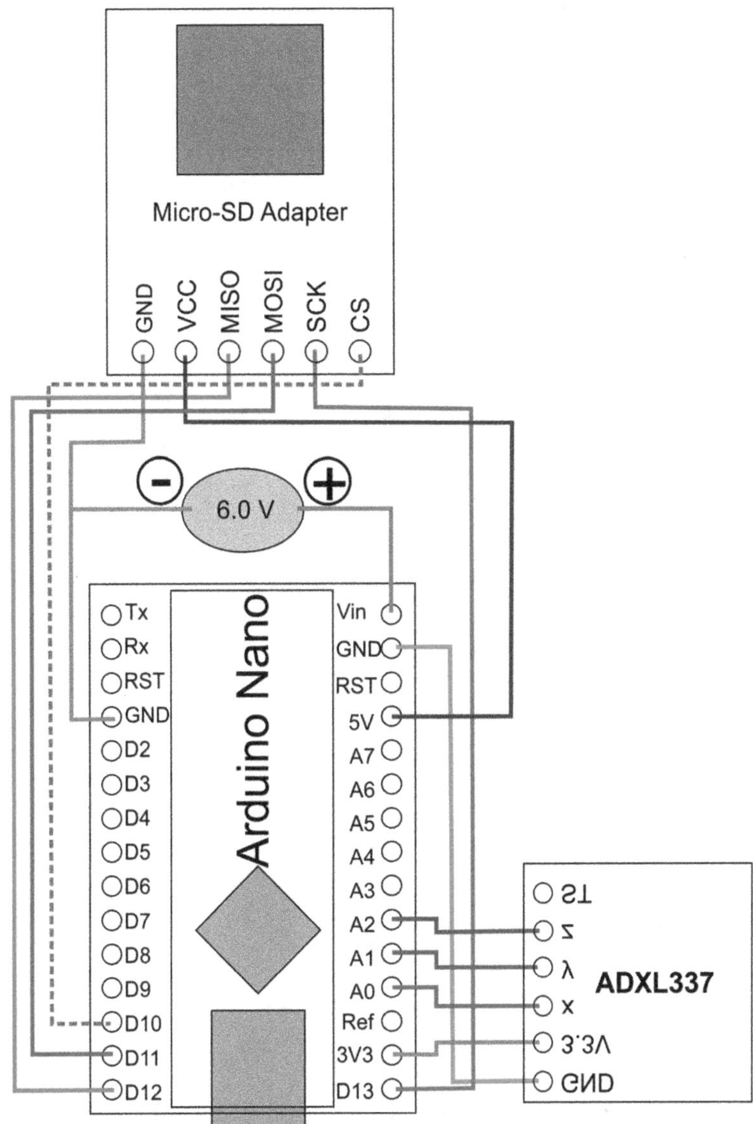

Fig. 4 Possible wiring diagram for an Arduino-based accelerometer module with microSD card. The sketch below represents one of the possible ways to program the microcontroller

arduino.cc/en/software (Fig. 4). The program (known as sketch) below is fully annotated and provides insight into adjustable features and alternative programming approaches (*see* **Note 3**).

A double slash (//) indicates comments, text between /* and */ can be comments or spare code; it is not compiled. The text below (between ****) can be copied into the IDE and uploaded to each microcontroller as is. Comments are provided for clarification of each step but do not affect the compiled program.

```
*****************************************************************************
#include <AccelStepper.h> //library required for subsequent code, must be
 //installed before program can be compiled.
#define stepPin 3 // stepPin, required for A4988
#define dirPin 4 //dirPin required for A4988, determines direction of rotation
#define MS1Pin 5 //MS pins 1-3 determine microstepping, see section below
#define MS2Pin 6 //MS2
#define MS3Pin 7 //MS3

AccelStepper stepper(1, stepPin, dirPin);

                    long RND; // RND holds value generated by the random() function
                    long time1; // variable, for time interval start
                    long time2; // variable for interval duration
                    float dspeed; // variable for motor speed
                    int rot = 1; // variable that determines clockwise or counter-
                    clockwise
                     // rotation
                    int Dir; // Variable that indicates change of rotation
                    void setup() { // setup section prepares the running conditions
                    Serial.begin(9600); // initialize communication at defined
                    baud rate, can be
                     // changed and is irrelevant for program execution
                    pinMode(MS1Pin, OUTPUT); // sets the first output pin for
                    microstepping
                    pinMode(MS2Pin, OUTPUT); // sets the second output pin for
                    microstepping
                    pinMode(MS3Pin, OUTPUT); // sets the third output pin for
                    microstepping
                    /* // full step
                    digitalWrite(MS1Pin, LOW); // 8 = 1 rpm
                    digitalWrite(MS2Pin, LOW);
                    digitalWrite(MS3Pin, LOW);
                    // 1/2 step
                    digitalWrite(MS1Pin, HIGH); // 16 = 1 rpm
                    digitalWrite(MS2Pin, LOW);
                    digitalWrite(MS3Pin, LOW);
                    // 1/4 step
                    digitalWrite(MS1Pin, LOW); //32 = 1 rpm
                    digitalWrite(MS2Pin, HIGH);
                    digitalWrite(MS3Pin, LOW);
                    // 1/8 step
                    digitalWrite(MS1Pin, HIGH); // 64 = 1 rpm
                    digitalWrite(MS2Pin, HIGH);
                    digitalWrite(MS3Pin, LOW);
                    */
                    digitalWrite(MS1Pin, HIGH); // 1/16 step - requires all MS pins
                    set high
```

```
digitalWrite(MS2Pin, HIGH);
digitalWrite(MS3Pin, HIGH);
stepper.setAcceleration(70); // determines acceleration value,
can be varied
} // end of the setup section

void loop(){ // beginning of loop, executed indefinitely
time1 = millis(); // sets time1 to processor time
if (time1 > time2){ // compares time1 to time2, which is zero
initially and
 // thus, executes the subsequent code; repeated at end
 // of each cycle
stepper.setCurrentPosition(0);
 RND = random(5,21); //selects value between 5 and 20
 time2 = time1 + (RND * 5000); // increments time2 by random
value between 25 and 100 s
 // 5000 milliseconds is the time base, can be adjusted
 RND = random(4,17); // selects value between 4 and 16, deter-
mines speed range
 dspeed = (RND *30); // 128 = 1 rpm @ 1/16 step; average = (10[=
mean of
 // range]*30)/128 = 2.34 rpm; can be different for each axis
 stepper.setMaxSpeed(dspeed); // sets new motor speed
 RND = random(1,11); //selects value between 1 and 10
 Dir = rot; // reads current rotation
 if (RND > 5){ // values > 5 result in positive rotation
 rot = 1;
 }
 else {
 rot = -1; //values <5 result in negative rotation
 }
 if (Dir != rot) { //change in direction of rotation if Dir ≠
rot
 stepper.setMaxSpeed(0); stops rotation before reversal
 delay (200); //wait 200 milliseconds before changing direc-
tion
 }
 stepper.moveTo(200000*rot); // step value (200000) must be
high enough to not be a
 // limiting factor - max steps = // 100{s}x480{speed} = 48000
} // end of parameter selection at the beginning of interval
 stepper.run(); // executes all parameters
} // end of loop, goes back to beginning
```

****

3.10 Accelerometer

The variability of the experimental load and the programmed conditions affect the movement of the system and despite careful programming it is essential to record the acceleration profile of an experiment. This task is possible with a separate accelerometer that is attached to the "payload." The following text describes yet another application of an Arduino-based system that records three-dimensional acceleration values to a microSD device so that data can be analyzed upon completion of the experiment. A second option is based on a commercially available system (RSL10 from ON semiconductors). The Arduino system is less expensive, weighs about 25 g and has on-board data recording; the RSL has ten different sensors including a 3D-accelerometer. Some of the additional sensors may be applicable to clinostat experiments (e.g., light meter, thermometer). This device is lighter (6.5 g) but more expensive and communicates via Bluetooth technology to a phone (with suitable app) or Bluetooth recording device, which could also be Arduino-based. Regardless of its nature, the recording device is an additional requirement and cost.

Whatever instrument is chosen, it must be calibrated. The easiest method requires recording raw data of Earth's 1-g acceleration in both directions of each dimension, or multiple measurements in all six directions. This configuration spans the dynamic range of ± 1 g for each axis. The obtained raw values can be mapped to represent true g-based acceleration data. The acquisition frequency determines the resolution of the spatially changing acceleration vector (*see* **Note 4**).

A suitable Arduino sketch is provided below. All serial.print statements are optional but facilitate troubleshooting; in contrast, the Data.print statements are essential as they write the measured values to the SD card. It is important to be aware of the limitations of the data transmission speed (baud rate). The output below contains about 40 bytes (320 bits) per write activity, which for 9600 baud limits transmission to 1 per about 35 milliseconds, i.e., the delay statement must be >35. However, higher baud rates are possible. Another variable is the required storage space. If 1000 bytes are written every second, then an eight-day long experiment requires $8 \times 24 \times 3600 \times 1000 = 0.65$ GB. However, since most Arduino-compatible microSD cards have up to 32 GB capacity, data storage should not be problematic.

```
******************************************************************************
#include <SPI.h> library required for serial communication
#include <SD.h> library required for microSD activities
int scale = 1000; accelerometer range expressed as +/- 1000 milli-g
boolean micro_is_5V = true;
File Data; declaration for the data file as Data
float SX, SY, SZ; Scaled values for each axis
void setup() { define conditions and parameters
```

```
Serial.begin(9600); Open serial communications and wait for port to open:
while (!Serial){ wait for serial port to connect.
}
Serial.print("Initializing SD card...");
pinMode(10, OUTPUT);
digitalWrite(10, HIGH); requirement for Arduino logic; value may be four for
some boards
if (!SD.begin(10)) {
Serial.println("initialization failed!");
while (1);
}
Serial.println("Initialization done.");
Data = SD.open("Accel.txt", FILE_WRITE); defines Accel.txt as name of data file

if (Data) {
Serial.print("Header Accel");
Data.println("Data are listed as x, y, z"); // header for all subsequent data
Data.close(); close the file:
} else {
Serial.println("Error opening File"); // error message if the file did not open
}

}
void loop() loop runs indefinitely
{
 Data = SD.open("Accel.txt", FILE_WRITE); opens data file "Accel.txt"
 int rawX = analogRead(A0); Raw accelerometer data for each axis
 int rawY = analogRead(A1);
 int rawZ = analogRead(A2);
 if (micro_is_5V) microcontroller runs @ 5V
 {
 SX = map(rawX, 298, 411, -scale, scale); the measured values 298, 411, will be
different for
 SY = map(rawY, 296, 409, -scale, scale); each sensor
 SZ = map(rawZ, 296, 430, -scale, scale);
 }
 else // microcontroller runs @ 3.3V
 {
 SX = map(rawX, 0, 1023, -scale, scale);
 SY = map(rawY, 0, 1023, -scale, scale);
 SZ = map(rawZ, 0, 1023, -scale, scale);
 }
 // Print out raw X,Y,Z accelerometer readings
Data.print("X: "); Data.print(SX); Data.print(" ");
Data.print("Y: "); Data.print(SY); Data.print(" ");
Data.print("Z: "); Data.print(SZ); Data.print(" ");
Data.print("Net g: "); Data.println(sqrt(SX*SX+SY*SY+SZ*SZ)/scale);
Data.close(); closes file, saves data after each writing cycle
```

```
delay(500); 500 milliseconds = 2 Hz sampling frequency
}

*******************************************************************************
```

3.11 Potential Experiments

The following points are meant to provide some incentives to utilize the random positioning machine as tool to study aspects related to gravisensing and general physiology, they are not to be misconstrued as experimental protocols. The proposed activities require careful programming of the controllers that goes beyond the scope of this introduction.

3.11.1 Randomized, Complete Rotation

Although clinostats are typically used as "micro-gravity simulators," the effect of mechanostimulation is an understudied area. The system could be used to investigate whether rotation (in any direction) enhances or diminishes mechano-sensitivity. In addition to rotation in the presence of inhibitors such as gadolinium or lanthanum, the effectiveness of auxin transport inhibitors would shed light on the relationship between induced or endogenous curvature, mechanostimulation, and activity and sensitivity of inhibitors.

3.11.2 Contrasting Complete Rotation with Fractional Rotations

The typical clinostat is based on 360 degrees of rotation. However, the programmability of the described system allows for detailed analyses of partial, pendulum-like rotations. The system can be programmed to perform partial rotations (e.g., $\pm$ 90°) avoiding stimulation either of the apical columella cells (during "vertical" clinorotation) or one side of the lateral statocytes (during the more typical lateral rotation around the longitudinal axis of the plant).

3.11.3 Asymmetrical Rotation

Modifying the angle such that the resultant between the positive and negative deflection from the gravity vector is unequal and not the plumb line, would provide a method to compare gravi-responsiveness after static and dynamic displacement. An offset by a defined angle would answer questions if the threshold of curvature [28] is sensitive to mechanostimulation and if plants can detect the resultant (mean) of moving amyloplasts. Fast rotation may cause centrifugation effects that can be examined by fast rotation in one direction followed by clinorotation at an angle to the original axis.

Although these suggestions address basic physiological questions, they are all amenable to genetic (transcription) studies.

3.11.4 Light Effects

The ability to change the illumination profile invites experiments on phytochrome and PAR-related effects during clinorotation. The adaptability of the sensing system to non-unilateral gravistimulation and light gradients could examine skewing effects of light gradients on roots or shoots.

4 Notes

1. Support frames can be printed as single piece or as two separate pieces (B) on smaller printers. Separate pieces are screwed together (M3 × 16 mm) and have the advantage of easier threading of wires through the internal channels (conduits).
2. All 3D printed materials can be obtained from a commercial supplier of printed on a personal printer. It is recommended to use a sufficiently sturdy material (PETG or ASA) with a 20% or greater infill setting.
3. The programming steps in Subheading 3.9 can be modified and refined to optimize the variability of the phase angle between the two axes. A constant phase angle will produce patterns (aka Lissajous figures) that indicate a non-random persistent pattern. Varying phase angles prevents such pattern formation. The code can be adjusted based on the chosen motors, gear ratios, and programmed rotational speeds.
4. The data acquisition rate for the accelerometer is variable but needs to be high enough to resolve the sum of the angular velocity of both axes. For example, if the outer frame rotates at 4 rpm and the inner frame at 3 rpm, then the total (i.e., 7 rpm) requires 250 readings per minute if a 10° resolution is desired ($7 \times 360/10 = 252$) or about 4.2 readings per second.

Acknowledgments

This chapter is the results of many years of continuous improvement of a variety of specialized clinostats that enabled research supported by NASA. Especially the effect of high gradient magnetic fields, mechanostimulation and longevity of stimulus retention were supported by NASA grants 80NSSC17K0344, NAG10-0190, NNX10AP91G.

References

1. Knight TA (1806) On the direction of the radicle and germen during the vegetation of seeds. Philos Trans R Soc Lond 99:108–120
2. Ciesielski T (1872) Untersuchungen über die Abwärtskrümmung der Wurzel. Beitraege zur Biology der Pflanzen 1:1–31
3. von Sachs J (1879) Über Ausschliessung der geotropischen und heliotropischen Krümmungen wärend des Wachsthums. Würzburger Arbeiten 2:209–225
4. Palmer JH (1973) Ethylene as a cause of transient petiole epinasty in *Helianthus annuus* during clinostat experiments. Physiol Plant 28 (1):188–193
5. Reinhardt D, Mandel T, Kuhlemeier C (2000) Auxin regulates the initiation and radial position of plant lateral organs. Plant Cell 12 (4):507–518
6. Hou GC, Mohamalawari DR, Blancaflor EB (2003) Enhanced gravitropism of roots with a

disrupted cap actin cytoskeleton. Plant Physiol 131(3):1360–1373

7. Shen-Miller J, Hinchman R, Gordon SA (1968) Thresholds for georesponse to acceleration in gravity-compensated Avena seedlings. Plant Physiol 43:338–344
8. Galland P, Finger H, Wallacher Y (2004) Gravitropism in phycomyces: threshold determination on a clinostat centrifuge. J Plant Physiol 161(6):733–739
9. Ma Z, Hasenstein KH (2006) The onset of gravisensitivity in the embryonic root of flax. Plant Physiol 140(1):159–166
10. John SP, Hasenstein KH (2011) Effects of mechanostimulation on gravitropism and signal persistence in flax roots. Plant Signal Behav 6:1–6
11. Shen-Miller J, Hinchman RR (1995) Nucleolar transformation in plants grown on clinostats. Protoplasma 185(3–4):194–204
12. Sobol, M.A., et al., Clinorotation influences rDNA and NopA100 localization in nucleoli. In: Space life sciences: gravity-related effects on plants and spaceflight and man-made environments on biological systems, 2005, pp. 1254–1262
13. Bouchern-Dubuisson E et al (2016) Functional alterations of root meristematic cells of *Arabidopsis thaliana* induced by a simulated microgravity environment. J Plant Physiol 207:30–41
14. Miyamoto K et al (2014) Analysis of apical hook formation in Alaska pea with a 3-D clinostat and agravitropic mutant ageotropum. Front Plant Sci 5
15. Yamazaki T et al (2012) Phenotypic characterization of *Aspergillus niger* and Candida albicans grown under simulated microgravity using a three-dimensional clinostat. Microbiol Immunol 56(7):441–446
16. Sawai S, Mogami Y, Baba SA (2007) Cell proliferation of Paramecium tetraurelia on a slow rotating clinostat. Adv Space Res 39 (7):1166–1170
17. Hader D-P, Lebert M, Richter P (1998) Gravitaxis and graviperception in Euglena gracillis. Adv Space Res 21:1277
18. Kessler, J.O., et al., Sedimenting particles and swimming microorganisms in a rotating fluid. In D. MontufarSolis, et al. (eds.) *Life sciences: microgravity research*, 1998. p. 1269–1275
19. Häder D-P et al (2005) Gravitational sensory transduction chain in flagellates. Adv Space Res 36:1182–1188
20. Ichigi J, Asashima M (2001) Dome formation and tubule morphogenesis by Xenopus kidney A6 cell cultures exposed to microgravity simulated with a 3D-clinostat and to hypergravity. In Vitro Cell Dev Biol-Anim 37 (1):31–44
21. Eguchi Y et al (2006) Cleavage and survival of xenopus embryos exposed to 8 T static magnetic fields in a rotating clinostat. Bioelectromagnetics 27(4):307–313
22. Yang HW, Bhat GK, Sridaran R (2002) Clinostat rotation induces apoptosis in luteal cells of the pregnant rat. Biol Reprod 66(3):770–777
23. Liu C et al (2020) Alteration of calcium signalling in cardiomyocyte induced by simulated microgravity and hypergravity. Cell Prolif 53 (3):e12783
24. Uchida T et al (2018) Reactive oxygen species upregulate expression of muscle atrophy-associated ubiquitin ligase Cbl-b in rat L6 skeletal muscle cells. Am J Phys Cell Phys 314(6): C721–C731
25. Hammond T, Allen P (2011) The Bonn criteria: minimal experimental parameter reporting for clinostat and random positioning machine experiments with cells and tissues. Microgravity Sci Technol 23(2):271–275
26. Hasenstein, K.H. and J.J.W.A. van Loon, Clinostats and other rotating systems—design, function, and limitation. In: Beysens DA, van Loon JJWA (Eds.), Generation and applications of extra-terrestrial environments on earth, 2015, River Publishers, pp. 147–156
27. Kiss JZ et al (2019) Comparison of microgravity analogs to spaceflight in studies of plant growth and development. Front Plant Sci 10:1577
28. Ajala C, Hasenstein KH (2019) Augmentation of root gravitropism by hypocotyl curvature in *Brassica rapa* seedlings. Plant Sci 285:214–223

Chapter 15

Arabidopsis Growth and Dissection on Polyethersulfone (PES) Membranes for Gravitropic Studies

Alexander Meyers, Nathan Scinto-Madonich, Sarah E. Wyatt, and Chris Wolverton

Abstract

Polyethersulfone (PES) membranes provide a versatile tool for gravity-related plant studies. Benefits of this system include straightforward setup, no need for specialized equipment, long-term seed viability between plating and hydration/growth, and adaptability to diverse protocols and downstream analyses. Methods outlined here include seed sterilization, planting, growth, and dissection that will transition directly into any RNA extraction protocol.

Key words Gravitropism, Polyethersulfone, Spaceflight

1 Introduction

When growing seedlings for phenotypic or transcriptomic analysis, reproducibility of the growth conditions is key. Plant growth on polyethersulfone (PES) membranes provides a simple platform that is especially useful in gravitropic studies. This is because the PES membrane material darkens upon hydration, providing excellent optical contrast for image analysis, while the solid backing allows for precise and rapid dissection and collection of organs for extraction. The system's merit lies in its simplicity, comprised of PES membranes, blotter paper, guar gum, Petri dishes of desired dimension, and nutrient solution (Figs. 1 and 2). Seeded plates maintain long-term viability, making this method ideal for high-throughput phenotyping or spaceflight experiments. For spaceflight applications, blotter paper can be soaked with nutrient media and dried, so only water is needed for hydration during flight. This system has been reliably used in spaceflight experiments [1, 2], but is also a useful method for ground-based studies. The PES growth protocol was developed and used extensively for the now-retired European Modular Cultivation System and has since been adapted for use in

Elison B. Blancaflor (ed.), *Plant Gravitropism: Methods and Protocols*, Methods in Molecular Biology, vol. 2368, https://doi.org/10.1007/978-1-0716-1677-2_15,

Fig. 1 Components of PES growth setup. From left to right: Larger half of Petri dish, blotter paper, PES membrane, smaller half of Petri dish, parafilm

Fig. 2 Fully assembled and sealed PES growth setup. Two lines of ten seeds were adhered to the membrane

the Veggie hardware, with versatility for use on other platforms. Outlined here are the methods for application in 100 mm square Petri dishes (*see* **Note 1**).

2 Materials

Sterile 100 mm square Petri dishes.

Metricel PES membranes: cut to 90 × 90 mm and autoclaved.

Whatman #17 chromatography paper cut to 90 × 90 mm and autoclaved.

Sterile filter paper, such as Whatman #3.

Guar gum powder.

Milli-Q water.
70% EtOH: 2 drops of Triton X-100 per 500 mL.
100% EtOH.
½ MS media: pH 5.8.
Sterile scintillation vials.
Sterile Pasteur pipette.
Sterile fine-pointed forceps, 2–3 pairs.
Sterile microscope slide.
Sterile 100 mm glass Petri dish.
Sterile deionized water.
Parafilm.
Sterile scalpel.
Omni Ruptor 12 Bead Mill (or similar).
2 mL ceramic bead tubes (Omni International Hard Tissue Homogenizing Mix or similar).
RNA extraction buffer.

3 Methods

3.1 Guar Gum Preparation

1. Add 250 mL milli-Q water in a 500 mL beaker with a stir bar.
2. With stir bar spinning close to max speed, quickly add all 2.5 g guar powder to make a 1% (w/v) solution.
3. Allow to mix until stir bar slows or stops completely.
4. Pour guar suspension into 500 mL media bottle and autoclave for 15 min at 120 °C.
5. Allow to cool. Sterile guar gum suspension can be stored at room temperature.

3.2 Seed Sterilization

1. Working in a laminar flow hood or other sterile cabinet, add seeds to scintillation vial.
2. Add two Pasteur pipette volumes of 70% EtOH with Triton X-100 to the vial and swirl for 5 min.
3. With a Pasteur pipette, aspirate off as much EtOH solution as possible and discard.
4. With clean Pasteur Pipette, add two pipette volumes of 95% EtOH to the scintillation vial and swirl for 1 min.
5. Aspirate off as much EtOH as possible and discard.
6. Repeat **step 4**. After the minute rinse is complete, aspirate EtOH and seeds onto a piece of sterile, dry filter paper.
7. Allow seeds to dry in hood for 30–60 min (*see* **Note 2**).

3.3 Plating Seeds

1. Working in a laminar flow hood or other sterile cabinet, invert the bottom of the glass Petri dish so the bottom is facing up. This will serve as a platform to work on (Fig. 3).

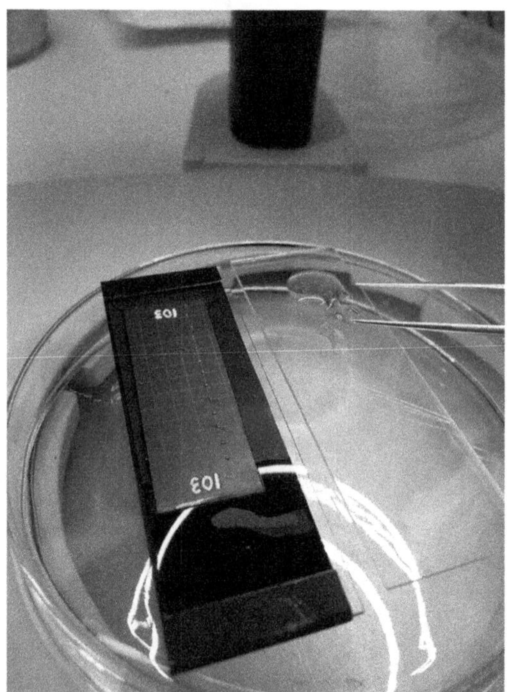

Fig. 3 Image of setup for seed planting on gridded PES membranes. Two microscope slides are shown here atop an upside-down glass Petri dish base, with the PES membrane on the slide on the left. The strip of guar gum is shown on the slide on the right with the forceps tip near an Arabidopsis seed

2. Use forceps to transfer the sterile PES membrane to the top of the glass Petri dish (*see* **Note 3**).
3. Use sterile DI water to lightly hydrate the membrane. This prevents the membrane from sliding while seeds are put into place.
4. Pipette a streak of guar gum across the top of a sterile microscope slide.
5. Dip one end of the forceps into the streak of guar gum and use the guar gum-covered forceps end to pick up several Arabidopsis seeds.
6. Dip the seeded forceps back into the guar gum streak, depositing the seeds into the guar gum streak to coat the seeds with guar gum.
7. Pick up one seed by dragging it out of the streak towards the bottom of the microscope slide. Use a "flicking" motion with one end of the forceps to pick up the seed. Excess guar gum may hinder germination; make sure there is not a large amount of guar gum covering the seed.
8. Use the forceps to lightly apply the seed to the membrane.

9. Repeat seed loading until the desired number of seeds has been attached to the membrane, e.g., 2 rows of 10 seeds each is optimal. Change forceps or rinse and resterilize them as the tips accumulate dry guar gum during planting.
10. Place the membrane in an open plastic Petri dish to dry in the laminar flow hood (*see* **Note 4**).

3.4 Hydration and Growth

1. Place a piece of sterile chromatography paper in a plastic 100 mm Petri dish (*see* **Note 5**).
2. With a Pasteur pipette, place a large drop of guar gum in the center of the top piece of blotter paper and quickly position the PES membrane on top of the blotter paper.
3. Allow blotter/PES membrane to dry completely in a laminar flow hood, about 1 h.
4. Pipette 5 mL of ½ MS solution onto the edge of the blotter paper and allow it to soak into the membrane (*see* **Notes 6** and **7**).
5. Wrap the dish in parafilm and place vertically under desired growth conditions.
6. After plant growth, if extracting RNA, use tongs to submerge the Petri dish in liquid nitrogen for 10 s or until completely frozen, then immediately store in −80 °C freezer. There is no need to remove the parafilm prior to this step.

3.5 Tissue Dissection

1. Prepare the stage of a dissecting microscope for RNA extraction. Treat the dissection as part of your RNA extraction protocol, keeping your area as RNase-free as possible (*see* **Note 8**).
2. Add the RNA extraction buffer to bead tubes and keep on ice (*see* **Note 9**).
3. Remove Petri dish from the freezer.
4. Remove the lid from the Petri dish. Flood the plants with RNA extraction buffer, enough to thaw the tissue but not so much that tissues will float on the plate during dissection (*see* **Note 10**).
5. With sterile scalpel, while observing through the dissecting microscope, cut root tips from seedlings (typically the apical 3 mm of the root). Collect tissue with sterile forceps and transfer to bead tubes. Thoroughly swish the forceps in the tube to ensure all tissue has been transferred. Inspect the forceps after to make sure no tissue remains on them, as you will use the same forceps for the next tissue.
6. Repeat **step 4** for collection of each of the other tissues: the rest of the root, then hypocotyl, and the cotyledons (*see* **Notes 11** and **12**).

7. Transfer the tubes with tissue to the bead mill. Seedlings at this stage are not particularly lignified, so usually 2 rounds of 30-s homogenization at a speed of 2.5 m/s is sufficient to lyse cells (*see* **Note 13**).
8. After homogenization briefly spin down bead tubes, remove the homogenate using a pipette, and continue with the preferred RNA extraction protocol.

4 Notes

1. This system works best when the Petri dish is used "upside-down"; the blotter paper is cut to the size of the lid (larger half) of the Petri dish. This allows the bottom (smaller part) of the Petri dish to hold the membrane in place. This also protects against membrane dehydration, as condensation that pools in the dish is wicked back up by the membrane. When a longer growth period (>5 days) is desired, PES membrane can be placed directly on ½ MS nutrient agar in a Petri dish and sealed with micropore tape. When using nutrient agar, it is best to use the usual Petri dish configuration (the smaller half as the bottom, the larger half as the lid). Seedling roots will typically reach the bottom of the plate in ~12 days.
2. This protocol is compatible with any seed sterilization method, but the protocol outlined here has been reliably used in preparation for spaceflight experiments with minimal loss in seed viability.
3. PES membrane is available gridded and non-gridded. The gridlines can help in planting, phenotyping, and dissection. Pre-sterile PES membrane is sold in 66 mm rounds. Larger sheets are available if using different sized Petri dishes, but the membrane must be cut to size. A rotary cutting tool works best when cutting PES membrane.
4. Seeded membranes can be stored long term in the sterile Petri dishes. Seeds stored in this manner have shown >90% germination rates for as long as 12 months of storage.
5. A 3-piece stack of filter paper, such as Whatman #3 or equivalent, can be used instead of chromatography paper, if needed.
6. This volume works best for square 100 mm Petri dishes with 60 mm membranes. For 100 mm dishes with 100 mm membranes, use 8 mL ½ MS. For 60 mm round Petri dishes with 60 mm membranes, 2–3 mL is adequate for hydration.
7. If desired, blotter paper can be soaked in ½ MS and allowed to dry, then rehydrated later with deionized water.

8. The typical protocol is for dissecting a 3–5-day-old seedlings into root tip, root, hypocotyl, and cotyledons. Generally, all like tissues are pooled from a single plate, and the four subsequent RNA extractions performed simultaneously.
9. Volume will depend on RNA extraction protocol. For Qiagen RNeasy kits, usually 450 μL RLT buffer is used.
10. Some extraction buffers will crystalize on the cold membrane. Crystalized buffer makes the seedlings difficult to dissect, so test the buffer by dropping some on the cold membrane prior to flooding. If the drop freezes, wait a few seconds and try again.
11. To facilitate dissection of the cotyledons at the petiole to retain the apical meristem within the hypocotyl tissue, grasp the hypocotyl with forceps and drag it down the membrane to bring the cotyledons above the hypocotyl. Both cotyledons can often be dissected from the seedling with a single cut.
12. Seedlings of this age often still have the seed coat attached to the tissue. The seed coat can be removed and discarded or included in the root or hypocotyl extraction, but the decision should be consistent across extractions.
13. Higher speeds and homogenization times have been used without substantial loss in RNA integrity but may require some optimization based on the bead mill, tissue, and downstream application.

Acknowledgments

Work in the authors' labs was supported by NASA Grant NNX15AG55G. The authors wish to thank the EMCS support team at NASA Ames Research Center for their work developing the specialized hardware and growth system from which this technique was modified.

References

1. Vandenbrink JP, Herranz R, Poehlman WL, Alex Feltus F, Villacampa A, Ciska M et al (2019) RNA-seq analyses of *Arabidopsis thaliana* seedlings after exposure to blue-light phototropic stimuli in microgravity. Am J Bot 106:1466–1476
2. Correll MJ, Edelmann RE, Hangarter RP, Mullen JL, Kiss JZ (2005) Ground-based studies of tropisms in hardware developed for the European Modular Cultivation System (EMCS). Adv Space Res 36:1203–1210

Chapter 16

Use of Reduced Gravity Simulators for Plant Biological Studies

Raúl Herranz, Miguel A. Valbuena, Aránzazu Manzano, Khaled Y. Kamal, Alicia Villacampa, Malgorzata Ciska, Jack J. W. A. van Loon, and F. Javier Medina

Abstract

Simulated microgravity and partial gravity research on Earth is a necessary complement to space research in real microgravity due to limitations of access to spaceflight. However, the use of ground-based facilities for reduced gravity simulation is far from simple. Microgravity simulation usually results in the need to consider secondary effects that appear in the generation of altered gravity. These secondary effects may interfere with gravity alteration in the changes observed in the biological processes under study. In addition to microgravity simulation, ground-based facilities are also capable of generating hypergravity or fractional gravity conditions whose effects on biological systems are worth being tested and compared with the results of microgravity exposure. Multiple technologies (2D clinorotation, random positioning machines, magnetic levitators, or centrifuges) and experimental hardware (different containers and substrates for seedlings or cell cultures) are available for these studies. Experimental requirements should be collectively and carefully considered in defining the optimal experimental design, taking into account that some environmental parameters, or life-support conditions, could be difficult to be provided in certain facilities. Using simulation facilities will allow us to anticipate, modify, or redefine the findings provided by the scarce available spaceflight opportunities.

Key words Clinostat, Random positioning machine (RPM), Magnetic levitation, Large Diameter Centrifuge (LDC), Seedlings, Cell suspension cultures

1 Introduction

1.1 Simulated Microgravity on Earth: Concept and Limitations

Simulated microgravity attempts to reduce the effective magnitude of the *g* vector to such a value that the weight load of an organism is below its biological gravisensing threshold. Real microgravity can only be achieved in freefall conditions. Experiments in these conditions can be performed in a durable and constant way on-board spaceflights like sounding rockets [1], satellites, space stations, parabolic flights (but only for times in the order of 20 s combined with hypergravity periods [2]) or in drop towers (for very short

Elison B. Blancaflor (ed.), *Plant Gravitropism: Methods and Protocols*, Methods in Molecular Biology, vol. 2368, https://doi.org/10.1007/978-1-0716-1677-2_16,

experiments, providing only 5–10 s of microgravity [3]) or the upcoming sub-orbital space flights [4] with a freefall time in the order of a couple of minutes. Particularly from the point of view of plant research, multiple biological questions demand longer exposures to the microgravity environment (days), restricting, in practice, the access to real microgravity conditions to orbital spaceflight experiments. In the last two decades, the International Space Station (ISS) has been a suitable platform, but the number of experiments performed on board of ISS has been necessarily limited. To somehow mitigate this serious limitation, ground-based facilities (GBFs) are available as microgravity simulators on Earth, in which the unilateral gravitational load is still present, but it may be compensated or averaged to near zero with time [5]. In the last decade, a significant number of plant biology experiments in both simulated and real microgravity conditions have been performed providing a comprehensive view on the differential effects caused by each of these facilities [6].

1.2 Simulating the Moon and Mars g-Levels on Earth

Since its full assembly, ISS has represented a major effort of international cooperation focused on scientific research in real microgravity. Plant space biology has greatly benefited from this unprecedented facility. All space agencies participating in ISS (NASA, ESA, JAXA, CSA, and Roskosmos) have contributed with relevant experiments to increase our knowledge on the alterations induced in plants by exposure to the space environment and on the adaptive response of these organisms [7–17]. However, in recent times, agencies are progressively refocusing their future activities from Low Earth Orbit (LEO) space to the Moon and beyond. LEO will be used by other countries interested in building National Space Stations and also by commercial parties, while main agencies are more and more interested in looking into nearby celestial bodies. The Moon will be visited again soon, according to NASA Gateway/Artemis plans to settle a woman there in less than 5 years, but a Mars settlement is also in the scope for the next two decades [18]. The *g*-load in these nearby objects is lower than the one on Earth and we are starting to unravel the consequences for the plants that have to grow in such a reduced gravity level environment. Both in the case of real spaceflight experiments using on-board centrifuges to provide a reduced *g*-level [10] and simulated conditions based on alternative uses of the random positioning machine [19], plant development, and specifically the early phases, decisive for the entire developmental plan appears particularly affected at the level of Moon gravity (one-sixth of the Earth gravity) while mild effects are observed at the Mars gravity (three-eighths of the Earth gravity). In this chapter, we will discuss different technological tools that can be chosen to perform this simulation and their expected limitations.

1.3 Mechanical Versus Magnetic Levitation Facilities

In order to neutralize the gravity force, we can choose between different technological approaches essentially based on two physical principles, namely mechanical and magnetic. None of them is absolutely optimal, and, consequently, the final choice will depend on the biological material and the experimental analyses to be performed.

Mechanical devices for simulation of microgravity include the most classical apparatuses designed with this purpose. These are the clinostats, dating from the nineteenth century. The operation of clinostats is based on redistributing the gravity vector in a circle (2D-clinostat) by means of mechanical devices that force the sample to rotate around a horizontal axis. A modern evolution of the concept of the clinostat is the random positioning machine (RPM), where the gravity vector is redistributed in a sphere and the sample is rotated around two orthogonal axes (3D). Depending on the rotational speed and the size of the sample, estimated by the distance from the center of the clinostat to the external edge of the sample container, we can obtain good quality microgravity simulation without too much residual gravity or shear forces as long as the sample is placed close to the rotation center [20]. A major advantage of this technology is readiness; there are commercially available models that can be fitted into a standard cultivation chamber providing multiple modes of operation that can be adapted to the experimental requirements of every user.

On the other hand, magnetic levitation technology uses the diamagnetic properties of mainly water, which is the major component of biological objects. A magnetic field applied to biological material can produce a diamagnetic force with the same magnitude as gravity and the opposite direction, capable of effectively compensating the weight of the sample, producing the levitation phenomenon [21, 22]. The diamagnetic effect is constant and acts at the atomic/molecular level, so it is not the result of averaging the forces in the system with time, but it is linked to the diamagnetic properties of most biological-related molecules. Actually, a residual internal force related to inhomogeneous organic solutions could appear as an unwanted secondary effect. The advantage of magnetic levitation is the stability of the compensated force (not randomized), together with the possibility of performing several parallel experiments simultaneously and in the same environmentally controlled magnetic bore, at different levels of effective gravity, including hypergravity, and, interestingly, 1 *g*, which represents an internal control reporting the effects of the magnetic field. The major disadvantage of magnetic levitation is the secondary effects of the high energy magnetic field required, together with the limitations in the access to this technology and the inhomogeneity of the levitation because of the difference in magnetic susceptibility of different elements. This type of equipment, ready for biological experiment hosting, is only available in around a dozen dedicated facilities in the world. In addition, some of the magnetic levitation facilities are quite expensive in terms of power supply requirements. Experiments should not be extended more than 3 or 4 h (due to refrigeration demands) and they may require the power supply of a small city (so they should be run during the night).

1.4 General Hardware Constraints: Comparison with Spaceflight Constraints

Spaceflight experiments are greatly constrained by hardware requirements (see the next chapter in this book). Particularly, in the case of plant biology, most of the experiments performed in the ISS were restricted to a few available hardware paradigms, mostly using the European Modular Cultivation System (EMCS), a facility that was located in the European "Columbus" module that, unfortunately, has been discontinued and is no longer available [23]. This means that future experimental designs should be compatible with the features—size and life-support systems—of the new facilities that NASA has developed looking into more agriculture friendly experiments. NASA has focused on demonstrating crop cultivation under microgravity conditions using the Veggie (Vegetable Production System) and, more recently, the APH (Advanced Plant Habitat) facilities on ISS [7, 8, 17], and there has been the Russian Lada module [24]. At the same time, ESA is considering to produce a new series of experiments based on the adaptation of hardware to the Biolab although final decisions and experimental plans are still pending. At the very moment, the ESA Biolab, together with the small Kubik incubator, are the only facilities incorporating the centrifuges that are required to expose the plants, in addition to microgravity, to a local control (1× Earth gravity, 1 *g*) and to fractional gravity, although China is also developing such a capacity [25]. To study crops/food plants in addition to model plant systems, an upgrade of the Biolab capabilities, and, consequently a refurbishment of the facility, is required (see Subheading 5), since the height of the cultivation chamber is limited by the configuration of the Biolab rotor [26]. These considerations affect the entire scientific community of this discipline due to the high degree of collaboration achieved by the ISS scientific users. ESA and NASA facilities have indeed been complementary and in the past, e.g., the EMCS has been used by both European and American researchers by means of collaborative agreements.

When using ground-based facilities not only the facility, but also the simulation paradigm, should be considered before the experimental design is executed. For example, the size of the container, and particularly the area in which the plant material is located, define the residual *g*-levels generated by the RPM operations (Fig. 1a). In the case of magnetic levitation, in addition to the topological constraints imposed by the magnet bore, as to the diameter and the level of effective gravity reached throughout it, the container should avoid metallic materials in its design and it should comply with requirements of temperature, video recording capabilities, and humidity (Fig. 1b).

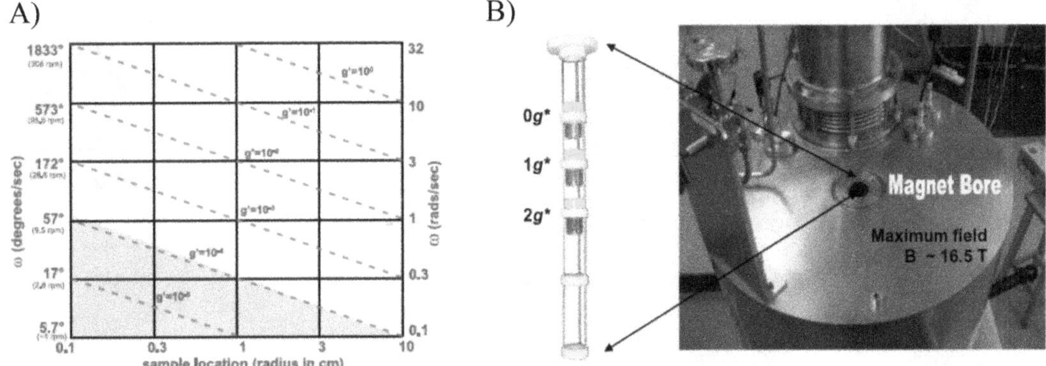

Fig. 1 (**a**) Residual gravity generated by an RPM, depending on rotational speed and distance of the sample to the rotation center [20]. (**b**) Image of a magnetic levitation facility at the University of Nottingham, UK, including the sample arrangement within the magnet bore for an experiment with seedlings [38]

2 Materials

2.1 Ground-Based Facilities to Simulate Microgravity

In this subsection, we provide a description of some of the ground-based facilities available in Europe (for a more comprehensive list, visit the ESA GBF Web page [27]) as an example of each type of available technology). Similar equipment with similar properties can be found elsewhere, but this list intends to provide the reader with a good scope of the available facilities to simplify the decision-making process when choosing a particular facility or experimental design for plant studies to simulate microgravity on Earth.

1. Classic 2D-clinostat. The simplest way to provide simulated microgravity is to use a two-dimensional (2-D), or one-axis, clinostat. This device has a single rotational axis, which runs perpendicular to the direction of the gravity vector. An example of such a facility is shown in Fig. 2a (clinostat provided in the frame of the United Nations Zero-Gravity Instrument Project ZGIP, UNOOSA). This facility, despite its conceptual simplicity, is quite adaptable to a wide range of experimental requirements due to flexible specifications (Table 1). Although in the literature there are numerous examples of the clinostat use for microgravity simulation, little agreement has been reached with regard to both the speed (ranging from 1–2 rpm to 30–60 rpm) and the orientation of the plants grown along the rotating axis (horizontal) or perpendicular (vertical clinorotation). An experiment performed in our laboratory [28] shows that the orientation (parallel or perpendicular) and the distance of the sample with respect to the rotation axis and the rotation speed greatly influence the magnitude of the gravity vector that is effectively sensed by the sample (Fig. 3). When fast clinorotation (60 or 90 rpm) is used, increasing the

Fig. 2 Examples of several ground-based facilities (GBFs) used in plant biology research. (**a**) 2D-clinostat provided by the UNOOSA Zero-gravity project initiative [53]. (**b**) 2D-Pipette clinostat available at DLR (Cologne, Germany). (**c** and **d**) Desktop and full-size RPMs available at DESC/ESTEC (Noordwijk, Netherlands). (**e**) Bitter (HFML, Nijmegen, Netherlands) and (**f**) Superconductive (Nottingham University, UK) magnetic levitation facilities. (**g**) Large Diameter (LDC) and (**h**) MidiCar centrifuges available at ESTEC-ESA, Noordwijk, Netherlands

Table 1
Comparison of the specifications of several clinostats

Clinostat model	2D-clinostat	Desktop RPM	Full-size RPM
Size (cm)	25 × 25 × 25	30 × 30 × 30	45 × 45 × 45
No. of rotational axes	1	2	2
Rotational speed	1–20 ± 0,5 rpm 20–90 ± 5 rpm	Depending on mode of operation	
Rotational axis angle	0–90°		
Experiment size	10 × 10 cm diameter	12 × 12 × 14 cm	45 × 45 × 30 cm
Experiment weight	0.5 kg	1.5 kg	10 kg
Connectivity	Not built-in	Switchable 12/15 V power line, RS232 (422) data bus (optical), fiberoptic video connection and camera	

A) Residual *g* force experienced under fast clinorotation

At 60 rpm At 90 rpm
4.15cm 2.5 0 1.95 4.15cm
0.17*g* 0.1*g* μ*g* 0.17*g* 0.37*g*
1 2 3 4 5 6 7 8 9
Centimeters
Effective *g* force

B) Clinostat 60 rpm experiment

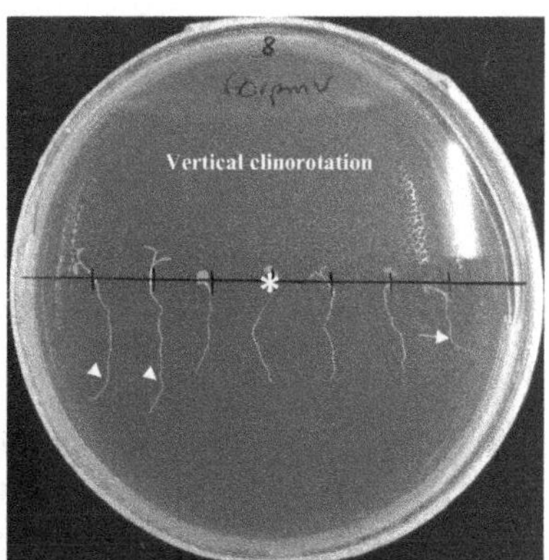

Fig. 3 Residual gravity under fast clinorotation. (**a**) Scheme of a Petri dish with a ruler indicating the distance from the rotation center and expected residual *g*-force in the line where the seeds were placed and the root tip at the end of the experiment. Note that even the sample in the center is easily growing out of the optimal simulated microgravity area. (**b**) *Arabidopsis thaliana* seedlings grown in photoperiod [16 h light/8 h darkness] and 1 *g* for 5 days, followed by 24 h of fast clinorotation (60 rpm) and the orientation of the plants grown perpendicular to the rotating axis (vertical clinorotation) in darkness. Roots grow straight due to the 1 *g* guided initial growth until they reach a length of around 2 cm and then show the curvature due to the last 24 h of clinorotation. The center of rotation is indicated by a star (*) and arrow heads indicate the root curvature points

distance between the sample and the rotation center, produces an increase of the centrifugal force acting on the sample substituting the gravity vector (aka residual *g*-level). It is possible to calculate the residual *g*-level in this case based on the rotation velocity and the distance of the sample from the rotation center. Indeed, it might be argued that it is possible to take advantage of this phenomenon to expose samples to reduced gravity simulation conditions (partial-*g*-levels) with a clinostat (Table 2), in a similar way to the use described in a modified RPM [19] although in such a partial-*g*-clinostat the *g* randomization and partial gravity generation always remain in the same plane. A future experiment should show whether this configuration is as effective as the clinostat centrifuge used by Galland and colleagues [29], or one can put the clinostat under an angle whose sinus value provides the simulated partial gravity value (see, e.g., [30]). The concept of fast clinorotation will be discussed further with more detail.

2. Pipette 2D-Clinostat. This facility is specifically dedicated to provide clinorotation conditions for microgravity simulation to suspensions of particles (cells, unicellular organisms, small aquatic organisms). It was developed by German Space Research Center (DLR, Cologne, Germany), which holds several instruments (Fig. 2b). Up to 10 samples inside 1 mL pipettes can be processed in parallel, and rotation speed varies from 60 to 90 rpm. Under the chosen experimental conditions (60 rpm, pipette diameter 4 mm), a maximal residual acceleration of 4×10^{-3} *g* is achieved at the border of the pipette, which decreases towards the center. Due to the relatively small size of the facility, it can be easily installed into a chamber with a defined temperature and atmosphere during clinorotation. Moreover, an interesting feature of this facility lies with the possibility of fixation of samples during rotation to prevent any experiment exposure to 1 *g* conditions before fixation.
3. Random Positioning Machine (RPM). This is the most advanced device for microgravity simulation regarding technology and novel operational modes. A computer interface with dedicated control software is available, and several operational modes can be defined by the user: real random (0.1–2 rad/s), centrifuge (0.1–20 rpm), clinostat (0.1–20 rpm), partial gravity (0.05–0.95 *g*, RPM Software paradigm for partial-*g* simulation [19], and freely programmable modes [31]. The main difference between the commercially available version (Desktop RPM from Airbus Defense and Space Netherlands (former Dutch Space, NV), Leiden, Netherlands, Fig. 2c), and the full-size version (Fig. 2d) is not the quality of simulated microgravity (similar in both cases, as can be inferred from the graph in Fig. 1a), but the size and weight of the sample which can be

Table 2
Experimental setups that should be tested to use a clinostat as a simplified version of the partial gravity RPM paradigms [19]

	Clinostat simplified versions of RPM partial-*g* paradigms						Verified RPM partial-*g* simulation paradigms [19]		
	Tilted clinorotation			Fast clinorotation			RPM^{SW}	RPM^{HW}	
	Residual force (*g*)	Tilted angle (radians)	Tilted angle (degrees)	Residual force (*g*)	Radius (cm)	Angular Speed (rpm)	Eccentricity	Radius (cm)	Angular Speed (rpm)
Micro*g*	0.00	0.0000	0	0.00	4.15	1.0	0	0	0
	0.10	0.0960	5.5	0.10	4.15	45.0			
MOON	0.17	0.1693	9.7	0.17	4.15	60.0	0.25	12	36
MARS	0.37	0.3840	22	0.37	4.15	90.0	0.53	12	53
	0.50	0.5236	30	0.50	4.35	100.0	0.66	16	53
	0.75	0.8465	48.5				0.87	12	75
Control 1 *g*	1.000	1.5708	90	1.00	9.00	100.0	–	16	75

Note that the use of each simulation paradigm must be tested both in terms of the mechanical forces acting and each biological system under study to avoid unexpected secondary effects [5]

loaded on the machine (Table 1). Particularly, the large RPM can accommodate an experiment including additional equipment, like a small microscope, onto the rotating platform, or even a built-in centrifuge to provide partial gravity into a simulated microgravity environment (RPM Hardware paradigm for partial-*g* simulation [19]). Several RPM models are available from the Dutch Experiment Support Center (DESC) facilities at the European Science and Technology Centre (ESTEC), a major ESA research center in Noordwijk, The Netherlands. The large RPM is installed within a temperature-controlled incubator ranging from +4 to +40 °C and also provides ambient light (generally not recommended because of phototropism interference with gravitropism).

4. Magnetic levitation. When choosing a diamagnetic levitation facility for plant microgravity research, duration of the experiment is a key, often limiting, factor. The Bitter technology, available, for example, in the High Field Magnetic Laboratory at Nijmegen University (Fig. 2e), provides intensities of magnetic fields higher than 16.5 T (as high as 33 T up to now), but the duration of an experiment is rarely longer than 5 h, due to power consumption and refrigeration demands. On the other hand, superconductive cryogenic technology, which is available, e.g., in the facility at Nottingham University (Fig. 2f), can run continuously for up to several weeks, but the experimental chamber cannot be larger than 4–5 cm in diameter. Further information, together with an introductory reading on the main concepts of ground-based facilities including the magnetic levitation technology, was published [5].
5. Centrifuges for hypergravity research. Hypergravity studies must complement microgravity and partial gravity experiments in order for a full understanding of the response of plants to mechanical stimuli. From the space research perspective, samples are subjected to hypergravity conditions during launch and the existence of planets with a higher level of gravity compared to Earth is a real fact. A number of centrifuges are available for biological research in different locations [32]. The use of lab-size centrifuges, such as the MidiCar (Fig. 2h), is convenient, but large facilities with a larger radius are necessary to avoid the undesired effect of shear forces (see [33] for a concise review on this topic). The Large Diameter Centrifuge (LDC), provided by the European Space Agency (ESA), is hosted by the Life and Physical Sciences Instrumentation and Life-Support (LIS) laboratory at ESTEC (Netherlands) [34]. This research tool is available for scientists working in the different fields where artificial gravity can play a key role. This centrifuge can indeed provide a stable hypergravity environment for fundamental research with a large capacity for accessory equipment (up to 80 kg, Table 3).

Table 3
Comparison on the specifications of several centrifuges

Centrifuge model	MidICar	LDC
Diameter (m)	0.3–0.4	8
Rotor arms	4	4 (6 gondola)
Acceleration	Up to 100 *g*	Up to 20 *g*
Rotational control	Not available	Central gondola
Experiment size	15 × 10 × 8 cm	50 × 50 × 75 cm
Experiment weight	0.5 kg	80 kg
Connectivity	Not built-in	230 V AC/6A socket Serial data connection (RS-232 Ethernet and USB connection) Video out analog input via NI-Rio controller module: input 16 × 0–10 V + 4 to be used for temperature sensors or alike
Environment control	Not built-in	Gas and (potable) water supply, temperature (room controlled) and static accelerometer sensors

Microgravity or reduced gravity can also be simulated making use of centrifuges. It is based on the premise that adaptations expected or seen going from a hypergravity level to a lower gravity level are similar as changes seen going from unit gravity to real microgravity or real partial gravity. It is called the Reduced Gravity Paradigm, RGP. For this one is not focusing on the absolute acceleration values, but rather on the responses generated due to the delta between two gravity levels. One starts such an experiment by adapting the plant sample to a higher gravity level, e.g., 2 *g* until it shows a stable, steady-state physiology. Then the 2 *g* is lowered to 1.5 or 1 *g* and the tissue will respond to this reduced gravity. The hypothesis is now that this adaptation from 2 to 1 *g* is similar to the adaptation from 1 to micro-*g*. The magnitudes of the responses might be different, but the directions of the response would be the same. This Reduced Gravity Paradigm is best used for measuring relatively fast responding phenomenon when reducing the acceleration load. See for more details [35].

2.2 Seedling Growth and Cell Cultures

1. Seed sterilization solution: 1.25% (v/v) sodium hypochlorite or Ethanol 70% (diluted from absolute ethanol) both supplemented with 1% (v/v) Triton X-100.
2. MS growth medium for *Arabidopsis* seedlings: 0.5% agar (*see* **Note 1**) with MS (Murashige and Skoog) plant culture medium. Prepare MS medium with 0.05% (*w/v*) MES (MES

hydrate) and 0.5% (*w/v*) MS medium supplemented with 1% (*w/v*) sucrose and pH 5.5–5.6 (adjusted with 1 M KOH). This medium was autoclaved at 110 °C.

3. MSS growth medium for *Arabidopsis* cell cultures: MS medium supplemented with 3% (*w/v*) sucrose and pH 5.8 (adjusted with 1 M NaOH). This medium is autoclaved at 110 °C and stored at 4 °C for no more than 2 weeks before use. Then, it is supplemented with 50 mg/L MS vitamins, 0.5 mg/L NAA (α-naphthalene acetic acid) and 0.05 mg/L kinetin (Kinetin) sterilized by filtration using Minisart® filter units.
4. Low-gelling agarose (2% (w/v); gelling below 26–30 °C; Sea-Plaque™ Agarose).
5. Pre-sterilized (by autoclave) 3MM paper, Nitrocellulose, 1% gum guar, and plastic Petri dishes.
6. Micropore™ tape is useful to close Petri dishes preserving humidity and gas exchange.
7. Double-sided scotch tape or Velcro® should be required for attaching the experiments to the altered gravity simulators.
8. Typical fixatives are 4% (w/v) formaldehyde (FA, EM grade, prepared from paraformaldehyde), for immunocytological studies, 3% (w/v) glutaraldehyde for ultrastructural microscopy analyses, or 90% acetone for GUS staining.
9. Prepare the fixatives in phosphate-buffered saline (PBS).

3 Methods

3.1 Use of Seedlings on Mechanical Facilities

The use of dedicated hardware is recommended in order to make more comparable the results obtained in different ground-based facilities. In preparation for flight studies, it is preferred to use flight modules in ground-based simulations as much as possible. It is also advisable to provide an illumination system built into the hardware, which is subjected to the same motion as the sample within the simulation facility, with the purpose of keeping constant the distance and the angle of incidence of the light on the sample (Fig. 4). This will assure a constant or controlled phototropic stimulus throughout the experiment that will allow a correct interpretation of results obtained in simulated altered gravity conditions. An array of LEDs providing light from the top of the culture chamber, operated with a programmable clock to provide day/night cycles in a stand-alone battery system is recommended for this purpose (Fig. 4).

Two alternative methods have been used to immobilize *Arabidopsis* seeds and seedlings within RPM/clinostat or LDC (centrifuge) facilities. The first one is to place the seeds on the surface of a Petri dish containing an agar solution of the culture medium, while

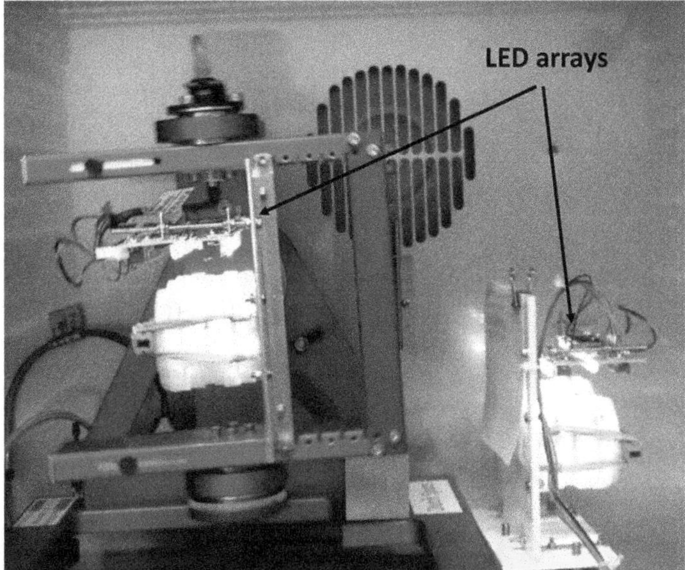

Fig. 4 Desktop RPM facility inside incubator along with 1 *g* control scaffold. The samples were grown in Petri dishes and the illumination was supplied from an array of LEDs located in association to samples, in order to guarantee a constant relative position of light and samples throughout the experiment [54]

the second one is to grow the plants onto a double layer of filter paper and a supporting sheet (nylon or nitrocellulose), in which nutrients are incorporated into the paper and seeds are affixed to the sheet. The use of gum guar as gluing agent may contribute to a better immobilization. This method has been used frequently in spaceflight experiments such as "GENARA-A" or "Seedling Growth" [15, 36] (Fig. 5). Since the results of experiments performed in ground-based facilities are going to be compared with both 1 *g* in-flight control experiments and spaceflight experiments, the method of choice, as well as the growth conditions in general, should be a compromise between the standard conditions used in the normal ground lab and the conditions used in space.

3.1.1 Use of Seedlings on Agar-Based Substrates [37]

1. Surface-sterilize seeds of *Arabidopsis thaliana*, in 1.25% (v/v) sodium hypochlorite and 1% (v/v) Triton X-100 for 10 in.
2. Wash the seed four times with distilled water (up to 2 h in total) before placing them on 9 cm diameter Petri dishes (*see* **Note 1**) containing 0.5% agar (*see* **Note 2**) with MS plant culture medium.
3. Keep seed at 4 °C (stratification in order to achieve a quick and synchronous germination) for 2 days until loading into the simulators. Induce seeds to germinate within the altered gravity facility by incubating them at 24 °C (*see* **Note 3**).

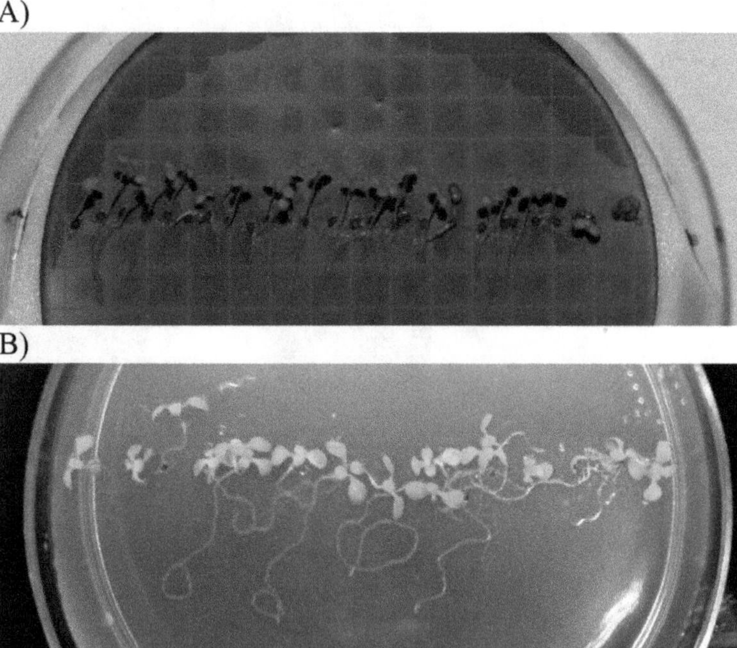

Fig. 5 Images of two experimental setups in Petri dishes with 4-day-old *Arabidopsis thaliana* seedlings grown in paper-based (**a**, 1 g control) or agar-based substrates (**b**, simulated microgravity showing random orientation of the seedlings)

4. A rotation and residual vibrational control experiment (1 *g*) should be placed in the rotation center of the centrifuge or in the RPM scaffold in order to detect secondary effects of the microgravity simulation device. Additionally, a 1 *g* external control (in a separate cultivation chamber) can be made, depending on the experiment design requirements. After several days of growth (*see* **Note 4**), samples should be quickly retrieved from the dishes, photographed and plunged into the fixative solution or frozen with liquid nitrogen until post-experiment analyses in the local lab. The time elapsed between stopping the gravity simulation and fixation should be minimal.

3.1.2 Use of Seedlings on Paper-Based Substrates [15]

1. Surface-sterilize seeds of *Arabidopsis thaliana* in 70% ethanol plus Triton X-100 (2 drops per 500 mL) for 5 min. Quickly wash the seeds three times with 100% (v/v) ethanol and immediately dry the seed on a pre-sterilized 3MM paper (*see* **Note 5**).
2. A row of seeds can be placed, one by one (*see* **Note 6**), into a line with 1% gum guar (*see* **Note 7**) in hardware containing one layer of 3MM paper with a nitrocellulose membrane on top (*see* **Note 8**). The experiment can be stored up to 3 months with this setup at room temperature and controlled, low-humidity conditions.

3. Inject the appropriate amount of MS plant culture medium (*see* **Note 9**) to activate the experiment. Allow seeds to germinate within the altered gravity facility by incubating them at 24 °C. Similar controls should be made as explained in Subheading 3.1.1.

3.2 Use of Seedlings on Magnetic Levitation Facilities [14]

The main constraints for the preparation of an experiment in the magnetic levitation facility are the small size and non-ferro-magnetic materials required in the experimental containers. It is also important to immobilize the samples in the precise area in which the desired effective gravitational load will be reached (normally 1–2 mm in height within the magnet bore). As an example, we have used a yellow cap cylinder tube (25 mm diameter) that fits into the scaffold that should be introduced into the magnet bore (Fig. 1b, [38]). It is also important to consider the number of samples that will be exposed simultaneously to the magnetic forces. The geometry of the magnet is important relative to the position of the sample in order to get suitable levels of effective gravity (g^*) and magnetic field intensity. For example, the parameters used in the experiment performed by our group at Nottingham University [38] were as follows: one sample was placed in the center of the magnet, where field intensity is maximum (16.5 T), but effective force is 1 g^* (no levitation), whereas 80 mm above the center we have the levitation point (0 g^*) but lower magnetic field intensity (11.5 T) and the opposite situation 80 mm below the center (2 g^* and 11.5 T). The 1 g^* position within the magnet provides an internal control, in which we can identify the effects of the magnetic field at the ground gravity level, but we have to perform an external 1 g control, with the same environmental conditions of temperature, light, etc.

1. Surface-sterilize seeds of *Arabidopsis thaliana* in 1.25% (v/v) sodium hypochlorite and 1% (v/v) Triton X-100 for 10 min and then rinse in sterile water.
2. For each sample (different gravitational loads and controls), place seeds on the surface of an agar slant containing 0.5% (w/v) agar with MS plant culture medium in a 25 mm diameter, 55 mm height plastic tube (*see* **Notes 1** and **5**). Around 20 seeds can be loaded into each tube and kept at 4 °C for 2 days in a refrigerator. Four experimental conditions can be investigated within four tubes.
3. After removal of the tubes from the refrigerator, position the first tube in the magnetic field such that the center of the tube is located at the 0 g^* point in the field. Henceforth, we refer to this tube as the 0 g^* tube. The effective gravitational force acting on water does not exceed 3×10^{-2} g for any of the seedlings with those settings.

4. Place a second group of seedlings in the magnetic field to enclose the 1 g^* point (center of the magnet). Place a third tube of seedlings to enclose the 2 g^* point in the field, while a control experiment (1 g) remains in a fourth tube outside the magnet in a temperature-controlled incubator.
5. Expose the seeds at 24 °C in the magnet and in the incubator, and allow them to germinate in the dark. The arrangement of seeds in the 1 g^* and 2 g^* tubes replicates the arrangement in the 0 g^* tube. The experiments in the 0 g^*, 1 g^*, 2 g^*, and 1 g tubes are run simultaneously.
6. After 2 or 4 days of growth in the dark, remove specimens promptly from the tubes, take photographs, and plunge the samples into a fixative solution or deep freeze. The elapsed time between removal of the first sample from the magnet and fixation of the last one should be as short as possible to avoid re-adaptation of the samples to the ground conditions.

3.3 Use of Callus Cell Cultures on Mechanical and Magnetic Levitation Facilities [39, 40]

A semisolid cell culture of *Arabidopsis* (callus) has often been used in simulated microgravity research [39–44]. It has the advantage of being immobilized but still remaining a relatively uniform population of actively proliferating cells. It is important to check that the cell callus is securely attached to the surface of the agar in which it has been placed since movements occurring during the simulation can invalidate the experiment (Fig. 6).

Callus semisolid cultures of *Arabidopsis thaliana* are prepared from suspension cultures [45]. In particular, we have used the MM2d line, described as highly suitable for cell cycle studies [45]. The best cultivation condition for the MM2d culture is without light, so all processing and experiments with this cellular line were done in the dark (*see* **Note 10**).

1. For the magnet experiments, prepare callus cultures in 40.8 mm high, 25 mm diameter tubes that are taped head-to-tail to form a column, divided into five levels (0 g^*, 0.1 g^*, 1 g^*, 1.9 g^*, and 2 g^*, for example) with an opaque non-magnetic cap at the top.
2. For experiments with mechanical facilities (LDC/RPM), prepare callus cultures in two regular 90 mm diameter Petri dishes. This material is especially unsuitable for partial gravity simulation experiments (*see* **Note 1**).
3. Grow callus cultures as a 1–2 mm thick layer on the surface of a 1 cm layer of 1% agar (*see* **Note 2**) with MSS medium. Due to this limited thickness, variations in the magnetic field and of the effective g-force are minimized.
4. For all devices and conditions, spread the suspension cultures on the agar surface one week before the start of the experiment and grow at 22 °C (*see* **Note 11**) to allow the callus to reach

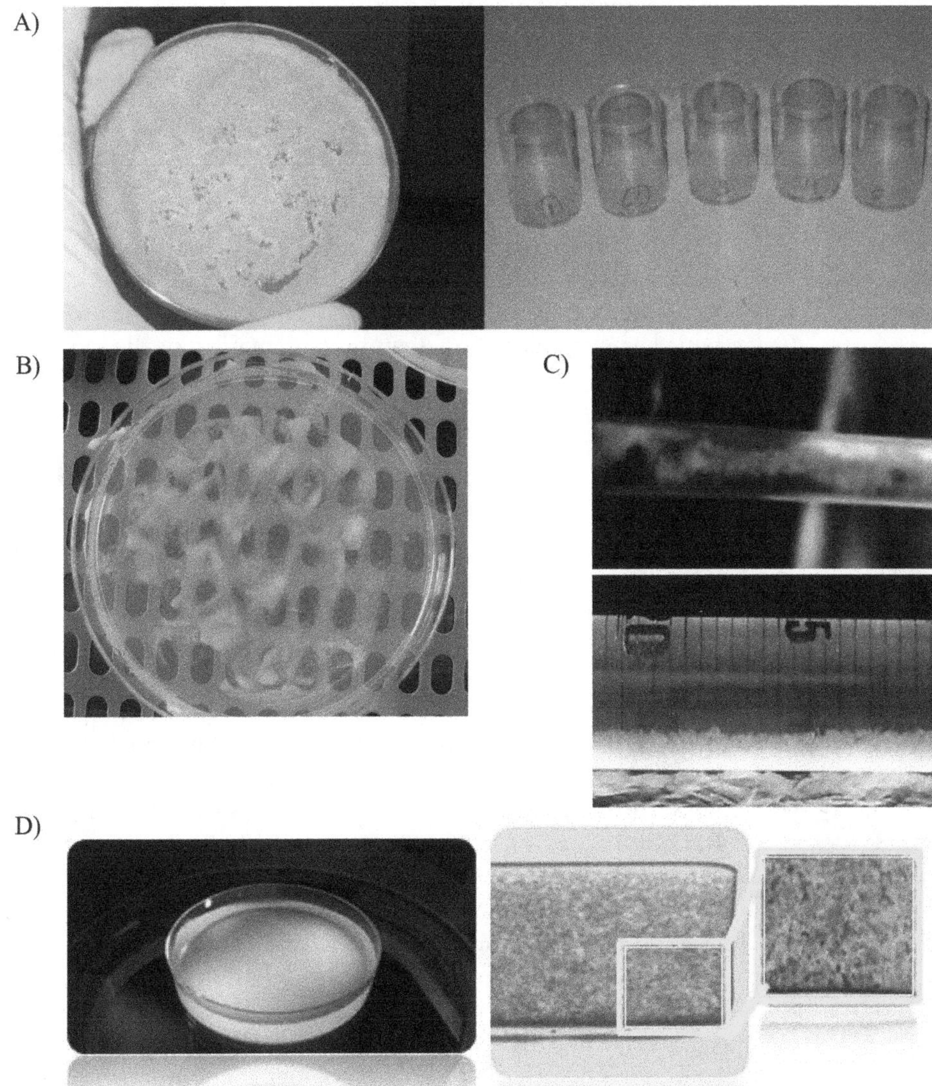

Fig. 6 Preparation of cell culture experiments and controls to be grown under altered gravity environments. (**a**) Callus culture prepared for exposure to both mechanical and magnetic GBFs [39]. (**b**) RPM rotation of a low density cell culture callus can lead to unexpected cell movements and, consequently, to a poor quality of simulated microgravity. (**c**) Suspension cell cultures under 2D-clinorotation (upper part) and under static 1 *g* control (lower part). See the cells have settled on the bottom of the pipet. Static conditions are not a proper 1 *g* control since cell suspensions used require shaking for survival. (**d**) Evenly suspended cell culture embedded in 2% (w/v) low-gelling agarose, consequently providing the required immobilization to be used in mechanical facilities. Detail on the distribution of cells in the agarose stack is provided on the right

maximum density (1–2 mm thick). Preserve the cell culture immediately after altered gravity treatment by quickly freezing in liquid nitrogen and storage at −80 °C until further processing.

3.4 Use of Cell Suspension Cultures on Mechanical and Magnetic Levitation Facilities

Historically, the 2D-clinostat has been used with some cell cultures without apparent constraints. In fact, a large collection of experiments have been done in both real and simulated microgravity conditions with human immune system cell cultures [46, 47]. However, it must be said that other cell culture systems, particularly plant cell cultures, rely on shaking for survival. Consequently, it is not a problem to maintain the cells under microgravity simulation, but it is mandatory to look for a proper 1 *g* control setup different from a static one [48] (Fig. 6c).

Similar constraints apply when exposing cell cultures to magnetic levitation facilities. Again, sedimentation of the samples is an open issue, although in this case sedimentation can occur even in the 0 *g** sample, with the understanding that we normally use the diamagnetic levitation point of water to calculate the effective forces acting in our system. In fact, we have already observed that sedimentation occurs inside a levitated droplet in a relatively short time. The alternative should be to increase the magnetic field intensity to try to prevent the cells' sedimentation, but this will induce the water droplet to try to escape from the magnet bore. In summary, according to our experience, using magnetic levitation is not a good choice for microgravity studies with cell suspensions [49].

The solution we propose to overcome this issue is to immobilize the cells in low-gelling agarose only during the simulation phase [50]. This solution is compatible with the maintenance of the advantages of cell suspension cultures before and after the simulation, for example, in studies of cell proliferation, the possibility of synchronizing cells in their progression throughout the cell cycle. It is possible to apply a drug treatment to the homogeneous cell culture for synchronizing it, just before the experimental phase [51], or to recover the cells from the low-gelling agarose in less than half an hour after the simulation to process the sample in solution. The procedure for immobilizing cells in low-gelling agarose to be used in the RPM for microgravity simulation experiments developed in our laboratory [50] is described in the steps below.

1. Subculture MM2d cultures (in dilutions 1:20) at the seventh day of growth in fresh MSS medium and keep cultures in a sterile 50 mL Falcon tube in darkness.
2. Dissolve low-gelling agarose (2% (w/v); gelling below 26–30 ° C) in MSS medium (*see* **Note 12**) in a sterile glass flask by boiling for 10 s in a microwave (*see* **Note 13**).
3. Allow the agarose to cool to 28–27 ° C.
4. Gently mix the agarose solution with an equal volume of the prepared cell suspension. This will result in a final concentration of 1% (w/v) agarose and a 1:40 dilution of cells in MSS medium.

5. Immediately pour 10 mL of the agarose-cell mixture into Petri dishes.
6. After the agarose is solidified, seal Petri dishes with Micropore™ tape. Carry out all steps at room temperature and under sterile conditions.
7. Keep the immobilized culture at 27 ° C in the dark, according to each experiment design.
8. Retrieval of embedded cells from agarose at the end of the experiment can differ on the basis of how they are fixed (for fixed samples follow to point 9, for frozen samples go to point 13).
9. To collect fixed samples, distribute 1 mL of 4% (w/v) formaldehyde (for immunocytological preservations) or 3% (w/v) glutaraldehyde (for ultrastructural microscopy analyses) to the surface of the plate containing the embedded cells, and incubate for 1 h. The fixative penetrates into the cells by free-diffusion through the agarose.
10. Wash fixed samples with 1 mL PBS buffer for 15 min after the chemical fixation to prevent over-fixation and cytoplasmic extraction.
11. Transfer the agarose containing the fixed cells to 15 mL Falcon tubes and dissolve by immersion in a water bath at 63 °C.
12. Centrifugation at 800–1000 × *g* for 5 min is enough to recover the pellet of fixed cells (without agarose) in order to use them in other protocols.
13. To collect frozen samples, pre-fix cells with 1 mL 1% (w/v) FA for 15 min to arrest any biological activity.
14. Quickly dissolve agarose using a 63 ° C water bath, extract cells by centrifugation, and directly freeze them by immersion in liquid nitrogen. The time to collect frozen samples should be minimized after the end of each experiment.

4 Notes

1. In the experiments of partial-*g* simulation, the placement of seeds in the Petri dish should be carefully designed, for the efficient use of partial gravity in the clinostat or RPM. As described extensively in [19], two RPM partial-*g* paradigms have been tested with plant seedlings producing similar results, but suggesting that RPM software version (that produce partial-*g* simulation in the 3D space by a dedicated algorithm, giving slight privilege to one axis with residual gravity) will have and ideal working range at 0.05–0.4 *g*, while the RPM hardware version may work better at >0.3 *g*-levels, including

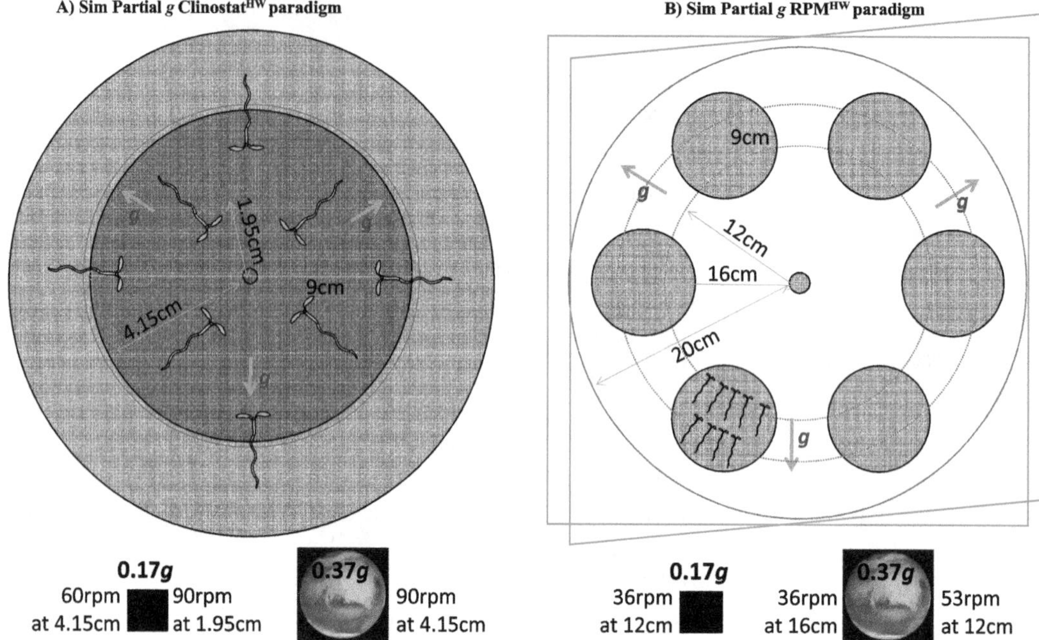

Fig. 7 Different setups to culture seedlings into Petri dishes to be exposed to partial gravity (Moon and Mars levels) using fast rotating clinostat (**a**) or the RPM (**b**), as calculated from Table 2

the simulated 1 *g* centrifuge control inside the real random RPM. To use the partial-*g* RPM software version (or a tilted clinostat assuming the sinus value provides the simulated partial gravity value [30]) the Petri dishes have to be properly oriented into the simulation machines but seedlings may grow anywhere in the plate (as close to the rotation center as possible for a better simulation, see Table 2). In the case that we try to take advantage of the fast rotating clinostat residual *g*-level, the seeds should be carefully placed in the precise spot that it is going to be exposed to the required *g*-level that has different value across the Petri dish. Note, however, that we recommend the use of the partial-*g* RPM hardware version. Figure 7 provides a couple of examples for the Moon and Mars simulation setups in both fast clinorotation and RPM hardware facilities.

2. Depending on the simulation technology and the experiment temperature and duration, a higher percentage of agar (up to 1%) should be used to prevent the appearance of water on the surface of the Petri dish.

3. Different illumination conditions can be used to prevent seedling etiolation. It is important to avoid the use of diffuse light; instead, LED illumination placed close to the samples should be used to take advantage of the synergistic effect of the main plant tropisms, light and gravity.

4. When using Petri dishes, it is not recommended that experiments last more than 14 days to prevent seedlings from reaching the walls of the container (thigmotropism could become an issue).
5. Sterilization procedures described in Protocol 1 and Protocol 2 are compatible. While the ethanol method is probably more effective, it may cause germination rates to drop if the exposure to ethanol is longer than 8 min. It is also important to dry the seeds in sterile conditions for at least 1 h to allow all traces of ethanol to evaporate.
6. It is important to carefully select the seeds one by one with a magnifying lens to exclude damaged or poorly developed seeds, and also to place all seeds in the same orientation. We recommend using two microscope slides, one with a narrow gum guar line and the other with a nitrocellulose membrane wetted by a few drops of sterile water, over a glass Petri dish cover. The selected seeds, having homogeneous size, a color neither too greenish nor too dark, and regular shape, are picked up from the gum guar with tweezers and dragged from the slide, avoiding taking an excess of gum guar and then put on the nitrocellulose membrane. Further details on the selection process for spaceflight qualifying seeds are here [52].
7. Gum guar 1% (w/v) must be prepared in a bottle and heated until boiling with a magnetic stirrer. Then the product is distributed in small vials and sterilized (by autoclave at 110 °C). Gum guar can diminish germination rates if used at too high density (taking the gum from the bottom from a non-shaken bottle) or after long exposures of the seeds under the magnifier (up to 15 min is reasonable for the selection process to be done safely).
8. Gum guar is used as an adhesive for attachment of both the 3MM paper to the nitrocellulose and the nitrocellulose to the seeds. It should be used at the minimum amount possible.
9. It is possible to use just distilled water to start the experiment by hydration in two scenarios: less than 4-day experiments (all nutrients are already in the seeds) or by using 3MM paper immersed in MS medium, dried and sterilized (by autoclave 120 °C) before use in Petri dishes.
10. The samples used were covered with aluminum foil during transportation and during the experiments at mechanical facilities; a PVC cap was used in the magnet experiments to prevent ambient light from entering into the magnet bore.
11. MM2d suspension cell culture grows at the optimal temperature of 27 °C. Depending on the comparability between experiments or facilities, it may be necessary to perform the experiment under suboptimal environmental conditions.

12. MSS medium should not carry vitamins if it is going to be boiled. Vitamins sterilized by filtration can be added later or provided 2× in the cell culture to be mixed 1:1 with the low-gelling agarose.
13. Low-gelling agarose needs to be dissolved at temperatures above 63 °C to be completely melted.

Acknowledgments

Most of the results and comments included in this book chapter have been the consequence of the authors' participation in several different "CORA-ESA Access to GBF" Projects of the European Space Agency, allowing the utilization of European facilities for altered gravity simulation, in close collaboration with the respective GBF managers. One of the co-authors of this paper (JvL) is the manager of the facilities hosted at DESC at ESA-ESTEC (Noordwijk, NL). Other facilities used are managed, respectively, by Dr. Hemmersbach (DLR), Dr. Pereda-Loth (Toulouse University), Dr. Hill (Nottingham University), and Dr. Christianen (Nijmegen University). We want to acknowledge Julio Martin Santos (3DOHMS) support in dedicated hardware design (3D printing and electronic components) to be used in our lab clinostats (obtained by UNZIP project grants from UNOOSA). Work performed in the authors' laboratory was financially supported by the Spanish Plan Estatal de Investigación Científica y Desarrollo Tecnológico, Grants #ESP2015-64323-R and #RTI2018-099309-B-I00 (co-funded by EU-ERDF) to F.J.M. and a grant from European Space Agency contract# 4000107455/12/NL/PA awarded to J.J.W.A.v.L.

References

1. European Space Acency E (2014) Sounding rockets. ESA User guide to low gravity platforms, Chapter 6: Sounding rockets. Noordwijk, Netherlands
2. Pletser V, Kumei Y (2015) Parabolic Flights. In: Beysens DA, van Loon JJ (eds) Generation and applications of extra-terrestrial environments on earth. Rivers Publishers, Aalborg, Denmark, pp 61–73
3. von Kampen P, Kaczmarczik U, Rath HJ (2006) The new Drop Tower catapult system. Acta Astronaut 59(1–5):278–283. https://doi.org/10.1016/j.actaastro.2006.02.041
4. Chang Y-W (2015) The first decade of commercial space tourism. Acta Astronaut 108:79–91
5. Herranz R, Anken R, Boonstra J, Braun M, Christianen PCM, Md G, Hauslage J, Hilbig R, Hill RJA, Lebert M, Medina FJ, Vagt N, Ullrich O, Loon JJWA, Hemmersbach R (2013) Ground-based facilities for simulation of microgravity, including terminology and organism-specific recommendations for their use. Astrobiology 13(1):1–17. https://doi.org/10.1089/ast.2012.0876
6. Kiss JZ, Wolverton C, Wyatt SE, Hasenstein KH, van Loon J (2019) Comparison of microgravity analogs to spaceflight in studies of plant growth and development. Front Plant Sci 10:1577. https://doi.org/10.3389/fpls.2019.01577

7. Massa GD, Wheeler RM, Morrow RC, Levine HG Growth chambers on the international space station for large plants. In: 8th International Symposium on Light in Horticulture, East Lansing, Michigan. KSC-E-DAA-TN29529, 2016
8. Zabel P, Bamsey M, Schubert D, Tajmar M (2016) Review and analysis of over 40 years of space plant growth systems. Life Sci Space Res 10:1–16. https://doi.org/10.1016/j.lssr.2016.06.004
9. Vandenbrink JP, Herranz R, Poehlman WL, Alex Feltus F, Villacampa A, Ciska M, Javier Medina F, Kiss JZ (2019) RNA-seq analyses of *Arabidopsis thaliana* seedlings after exposure to blue-light phototropic stimuli in microgravity. Am J Bot 106(11):1466–1476. https://doi.org/10.1002/ajb2.1384
10. Herranz R, Vandenbrink JP, Villacampa A, Manzano A, Poehlman WL, Feltus FA, Kiss JZ, Medina FJ (2019) RNAseq Analysis of the response of *Arabidopsis thaliana* to fractional gravity under blue-light stimulation during spaceflight. Front Plant Sci 10:1529. https://doi.org/10.3389/fpls.2019.01529
11. Valbuena MA, Manzano A, Vandenbrink JP, Pereda-Loth V, Carnero-Diaz E, Edelmann RE, Kiss JZ, Herranz R, Javier Medina F (2018) The combined effects of real or simulated microgravity and red-light photoactivation on plant root meristematic cells. Planta 248(3):691–704. https://doi.org/10.1007/s00425-018-2930-x
12. Vandenbrink JP, Herranz R, Medina FJ, Edelmann RE, Kiss JZ (2016) A novel blue-light phototropic response is revealed in roots of *Arabidopsis thaliana* in microgravity. Planta 244(6):1201–1215. https://doi.org/10.1007/s00425-016-2581-8
13. Bizet F, Pereda-Loth V, Chauvet H, Gerard J, Eche B, Girousse C, Courtade M, Perbal G, Legue V (2018) Both gravistimulation onset and removal trigger an increase of cytoplasmic free calcium in statocytes of roots grown in microgravity. Sci Rep 8(1):11442. https://doi.org/10.1038/s41598-018-29788-7
14. Karahara I, Suto T, Yamaguchi T, Yashiro U, Tamaoki D, Okamoto E, Yano S, Tanigaki F, Shimazu T, Kasahara H, Kasahara H, Yamada M, Hoson T, Soga K, Kamisaka S (2020) Vegetative and reproductive growth of Arabidopsis under microgravity conditions in space. J Plant Res. https://doi.org/10.1007/s10265-020-01200-4
15. Kiss JZ, Millar KD, Edelmann RE (2012) Phototropism of *Arabidopsis thaliana* in microgravity and fractional gravity on the International Space Station. Planta 236 (2):635–645. https://doi.org/10.1007/s00425-012-1633-y
16. Sychev V, Levinskikh M, Gostimsky S, Bingham G, Podolsky I (2007) Spaceflight effects on consecutive generations of peas grown onboard the Russian segment of the International Space Station. Acta Astronaut 60(4):426–432. https://doi.org/10.1016/j.actaastro.2006.09.009
17. Morrow R, Richter R, Tellez G, Monje O, Wheeler R, Massa G, Dufour N, Onate B A New plant habitat facility for the ISS. In: 2016, 46th International Conference on Environmental Systems
18. Deep Space Gateway and Transport: Concepts for Mars, Moon Exploration Unveiled. (2019). http://www.sci-news.com/space/deep-space-gateway-transport-mars-moon-exploration-04756.html. Accessed March 2020
19. Manzano A, Herranz R, den Toom LA, te Slaa S, Borst G, Visser M, Medina FJ, van Loon JJWA (2018) Novel, moon and mars, partial gravity simulation paradigms and their effects on the balance between cell growth and cell proliferation during early plant development. NPJ Microgravity 4. https://doi.org/10.1038/s41526-018-0041-4
20. Loon JJWA (2007) Some history and use of the random positioning machine, RPM, in gravity related research. Adv Space Res 39:5
21. Beaugnon E, Tournier R (1991) Levitation of organic materials. Nature 349:470
22. Valles JM Jr, Lin K, Denegre JM, Mowry KL (1997) Stable magnetic field gradient levitation of *Xenopus laevis*: toward low-gravity simulation. Biophys J 73(2):1130–1133. https://doi.org/10.1016/S0006-3495(97)78145-1
23. Kittang AI, Iversen TH, Fossum KR, Mazars C, Carnero-Diaz E, Boucheron-Dubuisson E, Le Disquet I, Legue V, Herranz R, Pereda-Loth V, Medina FJ (2014) Exploration of plant growth and development using the European Modular Cultivation System facility on the International Space Station. Plant Biol (Stuttg) 16(3):528–538. https://doi.org/10.1111/plb.12132
24. Baranova EN, Levinskikh MA, Gulevich AA (2019) Wheat space odyssey:"From Seed to Seed". Kernel morphology. Life 9(4):81
25. Wang S, Wang K, Zhou Y, Yan B, Li X, Zhang Y, Wu W, Wang A (2019) Development of the varying gravity rack (VGR) for the Chinese space station. Microgravity Sci Technol 31 (1):95–107
26. Wolff SA, Palma CF, Marcelis L, Kittang Jost AI, van Delden SH (2018) Testing new concepts for crop cultivation in space: effects of

rooting volume and nitrogen availability. Life 8 (4). https://doi.org/10.3390/life8040045

27. ESA GBF Web Page. (2020). http://www.esa.int/Our_Activities/Human_Spaceflight/Human_Spaceflight_Research/Ground_Based_Facilities. Accessed March 2020
28. Villacampa A, Sora L, Medina FJ, Ciska M Optimal clinorotation settings for microgravity simulation in *A. thaliana* seedlings. In: 69th International Astronautical Congress, Bremen (Germany), 2018. Curran Associates Inc., Red Hook NY, USA for International Astronautical Federation, pp 604–614
29. Galland P, Finger H, Wallacher Y (2004) Gravitropism in Phycomyces: threshold determination on a clinostat centrifuge. J Plant Physiol 161(6):733–739. https://doi.org/10.1078/0176-1617-01082
30. Dedolph R, Gordon S, Oemick D (1966) Geotropism in simulated low-gravity environments. Am J Bot S53(6 Part 1):530–533
31. Borst AG, van Loon JJWA (2009) Technology and developments for the random positioning machine, RPM. Microgravity Sci Technol 21 (4):287–292. https://doi.org/10.1007/s12217-008-9043-2
32. van Loon JJWA, Tanck E, van Nieuwenhoven FA, Snoeckx LHEH, de Jong HAA, Wubbels RJ (2005) A brief overview of animal hypergravity studies. J Gravit Phys 12(1):5–10
33. van Loon JJ, Folgering EH, Bouten CV, Veldhuijzen JP, Smit TH (2003) Inertial shear forces and the use of centrifuges in gravity research. What is the proper control? J Biomech Eng 125(3):342–346
34. van Loon JJ, Krausse J, Cunha H, Goncalves J, Almeida H, Schiller P The large diameter centrifuge, LDC, for life and physical sciences and technology. In: Life in Space for Life on Earth, Anger, France, 2008. vol SP-668. ESA/ISGP,
35. van Loon JJWA (2016) Centrifuges for microgravity simulation. The reduced gravity paradigm. Front Astron Space Sci 3:21. https://doi.org/10.3389/fspas.2016.00021
36. Mazars C, Briere C, Grat S, Pichereaux C, Rossignol M, Pereda-Loth V, Eche B, Boucheron-Dubuisson E, Le Disquet I, Medina FJ, Graziana A, Carnero-Diaz E (2014) Microgravity induces changes in microsome-associated proteins of Arabidopsis seedlings grown on board the international space station. PLoS One 9(3):e91814. https://doi.org/10.1371/journal.pone.0091814
37. Manzano AI, Herranz R, Van Loon J, Medina FJ (2012) A hypergravity environment induced by centrifugation alters plant cell proliferation and growth in an opposite way to microgravity. Micrograv Sci Technol 24(6):373–381. https://doi.org/10.1007/s12217-012-9301-1
38. Manzano AI, Larkin OJ, Dijkstra CE, Anthony P, Davey MR, Eaves L, Hill RJ, Herranz R, Medina FJ (2013) Meristematic cell proliferation and ribosome biogenesis are decoupled in diamagnetically levitated Arabidopsis seedlings. BMC Plant Biol 13(1):124. https://doi.org/10.1186/1471-2229-13-124
39. Manzano AI, van Loon JJWA, Christianen P, Gonzalez-Rubio JM, Medina FJ, Herranz R (2012) Gravitational and magnetic field variations synergize to reveal subtle variations in the global transcriptional state of Arabidopsis in vitro callus cultures. BMC Genomics 13:105. https://doi.org/10.1186/1471-2164-13-105
40. Herranz R, Manzano AI, Loon JJWA, Christianen PCM, Medina FJ (2013) Proteomic signature of Arabidopsis cell cultures exposed to magnetically induced hyper- and microgravity environments. Astrobiology 13(3):217–224. https://doi.org/10.1089/ast.2012.0883
41. Martzivanou M, Hampp R (2003) Hypergravity effects on the Arabidopsis transcriptome. Physiol Plant 118(2):221–231
42. Martzivanou M, Babbick M, Cogoli-Greuter-M, Hampp R (2006) Microgravity-related changes in gene expression after short-term exposure of *Arabidopsis thaliana* cell cultures. Protoplasma 229(2–4):155–162. https://doi.org/10.1007/s00709-006-0203-1
43. Barjaktarovic Z, Schutz W, Madlung J, Fladerer C, Nordheim A, Hampp R (2009) Changes in the effective gravitational field strength affect the state of phosphorylation of stress-related proteins in callus cultures of *Arabidopsis thaliana*. J Exp Bot 60(3):779–789. https://doi.org/10.1093/jxb/ern324
44. Barjaktarovic Z, Nordheim A, Lamkemeyer T, Fladerer C, Madlung J, Hampp R (2007) Time-course of changes in amounts of specific proteins upon exposure to hyper-g, 2-D clinorotation, and 3-D random positioning of Arabidopsis cell cultures. J Exp Bot 58 (15–16):4357–4363. https://doi.org/10.1093/jxb/erm302
45. Menges M, Murray JA (2006) Synchronization, transformation, and cryopreservation of suspension-cultured cells. Methods Mol Biol 323:45–61. https://doi.org/10.1385/1-59745-003-0:45

46. Cogoli A (1996) Biology under microgravity conditions in Spacelab International Microgravity Laboratory 2 (IML-2). Preface J Biotechnol 47(2–3):67–70
47. Cogoli A, Cogoli-Greuter M (1997) Activation and proliferation of lymphocytes and other mammalian cells in microgravity. Adv Space Biol Med 6:33–79
48. Kamal KY, Hemmersbach R, Medina FJ, Herranz R (2015) Proper selection of 1 g controls in simulated microgravity research as illustrated with clinorotated plant cell suspension cultures. Life Sci Space Res 5:47–52. https://doi.org/10.1016/j.lssr.2015.04.004
49. Kamal KY, Herranz R, van Loon JJ, Christianen PC, Medina FJ (2015) Evaluation of simulated microgravity environments induced by diamagnetic levitation of plant cell suspension cultures. Microgravity Sci Technol 28:309–317. https://doi.org/10.1007/s12217-015-9472-7
50. Kamal KY, van Loon JJWA, Medina FJ, Herranz R (2017) Embedding Arabidopsis plant cell suspensions in low-melting agarose facilitates altered gravity studies. Micrograv Sci Technol 29:115–119. https://doi.org/10.1007/s12217-016-9531-8
51. Kamal KY, Herranz R, van Loon JJWA, Medina FJ (2019) Cell cycle acceleration and changes in essential nuclear functions induced by simulated microgravity in a synchronized Arabidopsis cell culture. Plant Cell Environ 42 (2):480–494. https://doi.org/10.1111/pce.13422
52. Vandenbrink JP, Kiss JZ (2019) Preparation of a spaceflight experiment to study tropisms in arabidopsis seedlings on the International Space Station. Methods Mol Biol 1924:207–214. https://doi.org/10.1007/978-1-4939-9015-3_17
53. Kojima A, GarcíaYárnoz D, Pippo SD (2018) Access to space: a new approach by the united nations office for outer space affairs. Acta Astronaut 152:201–207
54. Manzano A, Pereda-Loth V, de Bures A, Sáez-Vásquez J, Herranz R, Medina FJ (2020) Light signals provide a mechanism of counteracting alterations caused by simulated microgravity in proliferating plant cells. Am J Bot. Under review

Chapter 17

Understanding the Mechanisms of Gravity Resistance in Plants

Kouichi Soga, Sachiko Yano, Motoshi Kamada, Shouhei Matsumoto, and Takayuki Hoson

Abstract

To understand gravity resistance in plants, it is necessary to analyze the changes induced when the magnitude of gravity in a growth environment is modified. Microgravity in space provides appropriate conditions for analyzing gravity resistance mechanisms. Experiments carried out in space involve a large number of constraints and are quite different from ground-based experiments. Here, we describe basic procedures for space-based experiments to study gravity resistance in plants. An appropriate cultivation chamber must be selected according to the growing period of the plants and the purpose of the experiment. After cultivation, the plant material is fixed with suitable fixatives in appropriate sample storage containers such as the Chemical Fixation Bag. The material is then analyzed with a variety of methods, depending on the purpose of the experiment. Plant material fixed with the RNAlater® solution can be used sequentially to determine the mechanical properties of the cell wall, RNA extraction (which is necessary for gene-expression analysis), estimate the enzyme activity of cell wall proteins, and measure the levels and compositions of cell wall polysaccharides. The plant material can also be used directly for microscopic observation of cellular components such as cortical microtubules.

Key words Arabidopsis, Cell wall, Chemical fixation bag, Cortical microtubules, Gravity resistance, Hypergravity, Microgravity, RNAlater®, Space experiment

1 Introduction

Mechanical resistance to gravitational acceleration (gravity resistance) is the principal graviresponse in plants besides gravitropism. Microgravity provides conditions that facilitate analysis of the mechanisms responsible for gravity resistance. The conditions produced by true microgravity are difficult to duplicate on Earth. The nature and mechanisms of gravity resistance have therefore been assessed primarily by examining hypergravity generated during centrifugation [1–3]. Hypergravity tends to suppress elongation growth but promotes lateral expansion of plant organs. Reorientation of cortical microtubules from transverse to longitudinal

Elison B. Blancaflor (ed.), *Plant Gravitropism: Methods and Protocols*, Methods in Molecular Biology, vol. 2368, https://doi.org/10.1007/978-1-0716-1677-2_17,

directions introduces modifications in growth anisotropy [4, 5]. Hypergravity also increases the rigidity of plant cell walls by altering the metabolism of cell wall constituents [6–8]. Mechanoreceptors are responsible for the plant's response to hypergravity [9, 10]. Thus, plants perceive gravitational acceleration by mechanoreceptors and resist it by modifying growth anisotropy and by increasing cell wall rigidity.

To confirm the contribution of cortical microtubules and cell walls to plant resistance under 1 *g* gravity, we analyzed changes in their properties in true microgravity aboard the Space Shuttle and the International Space Station (ISS) [11–17]. Experiments carried out in space involve a large number of constraints and differ substantially from ground-based experiments [18]. The numbers of samples and repetitions are severely limited in space-based experiments, as are the types of hardware available, amount of resources, techniques and expertise of the personnel, and available crew time. We describe the basic procedures for space experiments that can be used to study gravity resistance in plants.

2 Materials

2.1 Plant Cultivation

1. Seeds.
2. Cultivation chamber (e.g., Resist Tubule Chamber B, Plant Chamber (Fig. 1; *see* **Note 1**)).
3. Medium (e.g., Hatosheet (Oji Kinocloth, Fuji, Japan; *see* **Note 2**), rock wool).
4. Water.
5. 1% (w/v) Gum arabic (when necessary).
6. Incubator (e.g., Cell Biology Experiment Facility (CBEF); *see* **Note 3**).
7. Refrigerator (e.g., Minus Eighty-Degree Celsius Laboratory Freezer for ISS (MELFI), FROST; *see* **Note 4**).

2.2 Fixation and Storage of Plant Materials

1. Seedlings.
2. Chemical Fixation Bag (CFB; *see* **Note 5**).
3. RNAlater® (Ambion, Austin, Texas, USA).
4. Refrigerator (e.g., MELFI, FROST).
5. Freezer (e.g., MELFI, FROST).

2.3 Analysis of the Cell Wall Properties

1. Sample clamp (for tensile tester).
2. Tensile tester.
3. Liquid nitrogen.
4. Homogenizer (or mortar and pestle).

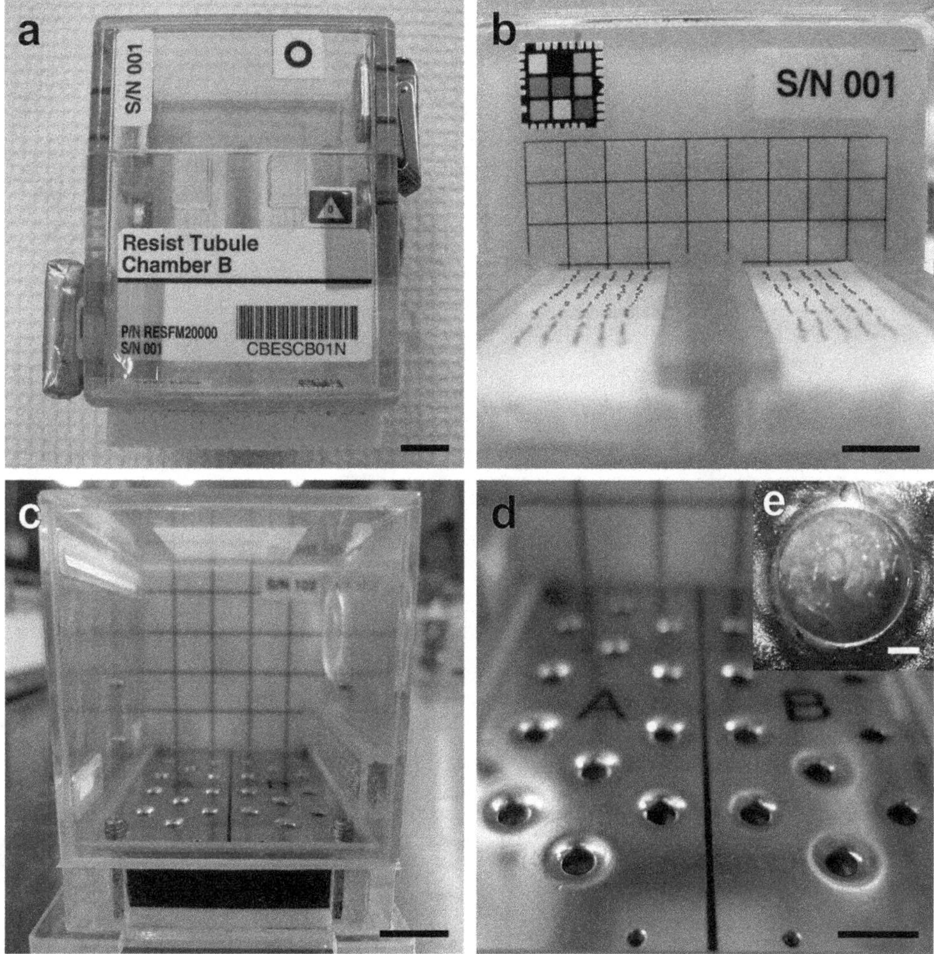

Fig. 1 The Resist Tubule Chamber B and the Plant Chamber. (**a**, **b**) The Resist Tubule Chamber B; interior dimensions are 56 mm in width, 46 mm in height, and 76 mm in depth. (**c**, **d**) The Plant Chamber; interior dimensions are 56 mm in width, 46 mm in height, and 48 mm in depth. (**e**) Arabidopsis seed fixed on rockwool medium. The bars in (**a–d**) and (**e**) are 10 mm and 1 mm, respectively

5. RNeasy Plant Mini Kit (Qiagen, Valencia, CA, USA).
6. RNase-Free DNase Set (Qiagen).
7. 10 mM sodium phosphate buffer, pH 6.0.
8. 1 M NaCl.
9. Vortex mixer.
10. Bench-top centrifuge.
11. Polypropylene mesh (32 μm).
12. Protein Assay Kit (Bio-Rad, Hercules, CA, USA).
13. Spectrophotometer.
14. Xyloglucans (Dainippon Sumitomo Pharma, Osaka, Japan).

15. 15% (w/v) Na_2SO_4.
16. Iodine solution: 0.5% (w/v) I_2 and 1% (w/v) KI.
17. 50 mM EDTA.
18. 24% (w/v) KOH containing 0.02% (w/v) $NaBH_4$.
19. Acetic acid.
20. Dialysis tube (or ultrafiltration device).
21. 72% (v/v) sulfuric acid.
22. 100% (v/v) sulfuric acid.
23. 5% (w/w) phenol.
24. Glucose.

2.4 Microscopic Analysis of Plant Materials

1. Seeds (e.g., fluorescent protein-expressing Arabidopsis line).
2. Culture medium (e.g., Hatosheet).
3. Cultivation/Observation Chamber (*see* **Note 6**).
4. Syringe.
5. Water.
6. Light-proof bag.
7. Incubator (e.g., CBEF).
8. Refrigerator (e.g., MELFI, FROST).
9. Fluorescence microscope (*see* **Note 6**).

3 Methods

3.1 Plant Cultivation

Different chambers are needed according to the growing period of the experimental plant and the purpose of the experiment. The Resist Tubule Chamber B, which was developed for the Resist Tubule experiment using seedlings of Arabidopsis, is suitable for cultivating plants, etiolated seedlings in particular, for approximately 1 week (Fig. 1a, b). The Plant Chamber was developed for the Space Seed experiment using adult Arabidopsis plants, which can be grown for about 2 months, allowing for an examination of a complete life cycle [19].

1. Prepare the culture medium. Hatosheet medium and rock wool medium are optimal growth media for the Resist Tubule Chamber B and the Plant Chamber, respectively.
2. When using Arabidopsis in the Resist Tubule Chamber B, place the seeds on moistened Hatosheet medium, and dry them immediately. The seeds remain attached to the Hatosheet because of the mucilage, a hydrophilic polysaccharide extruded from the seeds. When using plants that do not extrude mucilage, the seeds must be fixed to the medium with an aqueous

solution of gum arabic. For the Plant Chamber, drop a small amount of 1% (w/v) aqueous solution of gum arabic on rock wool medium, then place the seeds, and dry them immediately, even when using Arabidopsis (*see* **Note 7**). The gum arabic ensures the seeds remain attached to the rock wool.

3. Set the medium, along with the seeds, into the chamber. Then, set the Resist Tubule Chamber B to the Video-Measurement Unit (V-MEU) and the Plant Chamber to the Plant Experiment Unit (PEU), respectively. Store the chamber at 4 or 25 °C until launch (*see* **Note 8**).
4. Set the V-MEU or PEU in the CBEF in the Kibo Module after arriving at the ISS. Supply water to the seeds and begin cultivation.
5. After incubation, fix the plant material immediately using a suitable method as described in Subheading 3.2.

3.2 Fixation and Storage of Plant Materials

1. For chemical fixation of seedlings with reagents such as the RNAlater® solution, use the appropriate sample storage containers (e.g., CFB), as shown Fig. 2a, b, and described in **Note 5**.
2. Grow the plants for an appropriate period, as described in Subheading 3.1. When using the CFB, put the plants into the upper space of the inner bag, and seal the bag with a clip to prevent the RNAlater® solution from leaking. Put the inner bag into the middle bag, and then put both bags into the outer bag.
3. Roll one edge of the CFB and push the pouch inside the inner bag until the pouch breaks (Fig. 2b). Shake the CFB gently to mix the RNAlater® with the seedlings, to facilitate permeation of RNAlater® into the plants.
4. Keep the CFB in the refrigerator at 2–4 °C for 1–4 days to allow the RNAlater® solution to permeate deeper into the plants. Then transfer the CFB to a freezer for long-term storage at −80 °C. The plant material can be stored in this way for at least several months.

3.3 Analysis of the Cell Wall Properties

The amount of plant material available during space experiments is extremely limited, and it is essential that various parameters are analyzed using identical plant materials. Plant material fixed with the RNAlater® can be used for this purpose (Fig. 3).

1. Remove the plant material frozen in the RNAlater® solution and measure growth parameters such as length and diameter.
2. To measure the mechanical properties of their cell walls, first thaw the frozen plant material, and then cut it into segments of appropriate lengths if it is too long. Clamp the middle portions

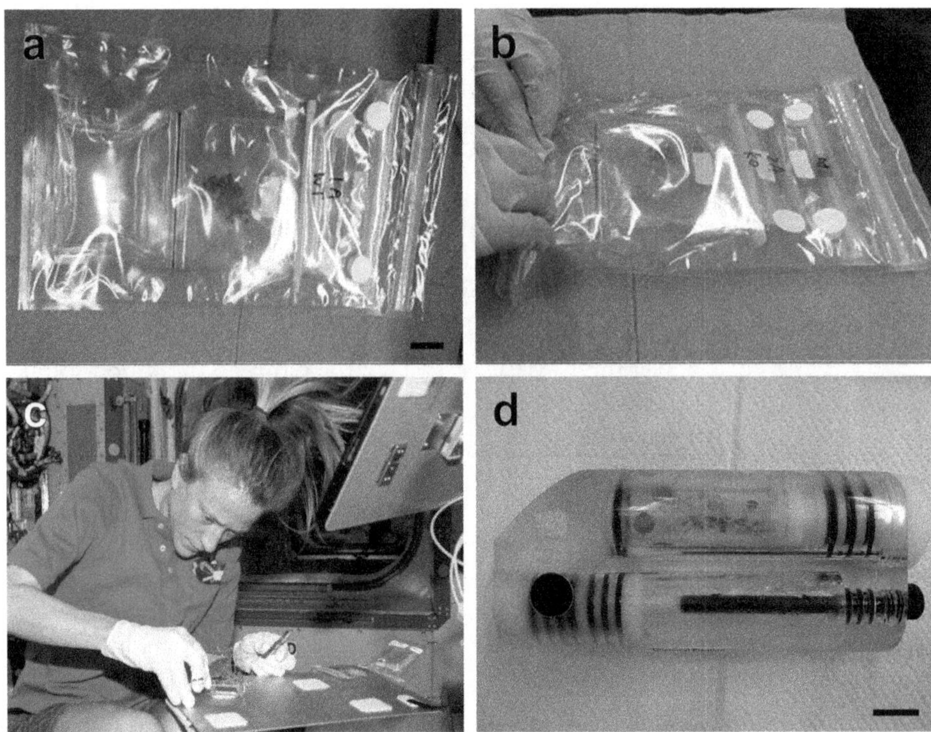

Fig. 2 The Chemical Fixation Bag (CFB) and the Chemical Fixation Apparatus (CFA). (**a**, **b**) The CFB. (**c**) NASA astronaut Karen Nyberg operates the Resist Tubule experiment using the CFB. (**d**) The CFA. Bars in (**a**) and (**d**) are 20 mm

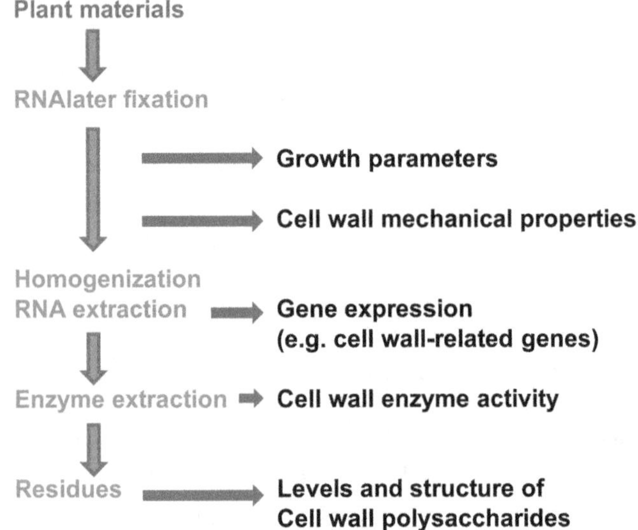

Fig. 3 Sequential analysis of cell wall properties of plant material fixed with RNAlater® solution

of these segments and measure the mechanical properties of their cell walls with a tensile tester, according to the manufacturer's instructions (*see* **Note 9**).

3. After measuring the mechanical properties of the cell walls, freeze the segments immediately with liquid nitrogen. Homogenize the frozen segments with a mortar and pestle. Then isolate total RNA using the RNeasy Plant Mini Kit. This step includes a DNA elimination step (RNase-Free DNase Set) (*see* **Note 10**). Analyze gene expression with real-time polymerase chain reaction amplification, a microarray analysis, or RNA sequencing using extracted RNA.
4. Collect the cell wall homogenate from the RNA extraction column of the RNeasy Plant Mini Kit (*see* **Note 11**) and incubate it in a 10 mM sodium phosphate buffer (pH 6.0) containing 1 M NaCl. Extract the cell wall proteins at 4 °C for 6–24 h. Collect the cell wall proteins by centrifugation using a benchtop centrifuge or by filtration through a polypropylene mesh (32 μm).
5. Measure the protein content in the extract with the Protein Assay Kit or by measuring absorbance at 280 nm. The extract can be used to measure various enzyme activities (*see* **Note 12**). To measure xyloglucan-degrading activity, mix 1 μg of cell wall proteins with 20 μg of xyloglucans in 10 mM sodium phosphate buffer (pH 6.0) and incubate at 37 °C for 24 h. After incubation, terminate the reaction by boiling. Determine the amount of xyloglucans with an iodine-staining method after mixing with 15% (w/v) Na_2SO_4 and express the activity of the xyloglucan-degrading enzymes in terms of the decrease in absorbance at 640 nm.
6. After extracting cell wall proteins, incubate the remaining cell wall materials thrice (15 min each) with 50 mM EDTA at 95 °C to obtain the pectic substances. Then extract the hemicellulose fraction thrice (12 h each) at 25 °C with 24% (w/v) KOH containing 0.02% (w/v) $NaBH_4$ (*see* **Note 13**). Neutralize the hemicellulosic materials with acetic acid and then dialyze against water. Dissolve the remaining alkali-insoluble material (cellulose) in 72% (v/v) sulfuric acid for 1 h at 25 °C and dilute with a 29-fold volume of water.
7. Determine the total sugar content in each of the cell wall fractions using a phenol-sulfuric acid method and express the results as glucose equivalents (*see* **Note 14**).

3.4 Microscopic Analysis of Plant Materials

In some experiments, plant materials are used directly after cultivation for microscopic observation of cellular components, such as cortical microtubules. Live cell imaging should be performed in the medium in which the seeds germinated [20]. We therefore

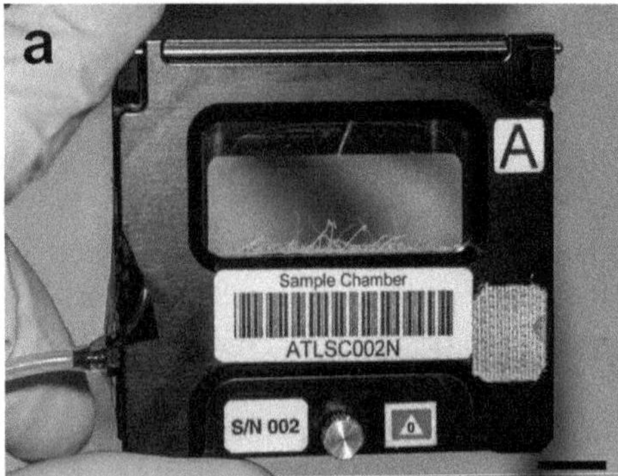

Fig. 4 The Cultivation/Observation Chamber. (**a**) The Cultivation/Observation Chamber; the interior dimensions are 38 mm in width, 16 mm in height, and 0.5 mm in depth. (**b**) JAXA astronaut Koichi Wakata operates the Aniso Tubule experiment using the chamber. The bar in (**a**) is 10 mm

designed a special chamber, a Cultivation/Observation Chamber (Fig. 4a), for the Resist Tubule and the Aniso Tubule experiments using green fluorescent protein-expressing Arabidopsis lines [16, 21] that can be used during both space and ground-based experiments.

1. Prepare the culture medium. The Hatosheet medium is the optimal growth medium for the Cultivation/Observation Chamber.
2. Place the seeds of the fluorescent protein-expressing Arabidopsis line on the edge of a moistened Hatosheet medium and dry them immediately. The seeds remain attached to the Hatosheet because of the mucilage.
3. Set the medium, along with the seeds, in the chamber. Store the chamber at 4 or 25 °C until launch.
4. Add water to the medium using a syringe. Put the chamber in a light-proof bag and refrigerate for 2–4 days, then expose the chamber to weak white light to induce germination. Put the chamber in the light-proof bag and grow the plants in the dark.
5. After cultivating the seedlings for the appropriate time inside the incubator, remove the chamber from the bag. Confirm seedling growth by observing them through the window and add water to fill the cultivation space between the glass sheets.
6. For microscopic observation, reduce the depth of the cultivation space by tightening the screw on the chamber.
7. Transfer the chamber to a microscope stage (Fig. 5) using a chamber holder.
8. To image the all seedlings in the chamber in a bright field, acquire tile images and make a larger montage by stitching together individual tile images (Fig. 6a).

Fig. 5 The JAXA fluorescence microscope. (**a**) The JAXA fluorescence microscope; a commercially available fluorescence microscope (DMI6000B, Leica Microsystems) partially modified to fit the ISS. (**b**) The Cultivation/Observation Chamber can be set directly the microscope stage

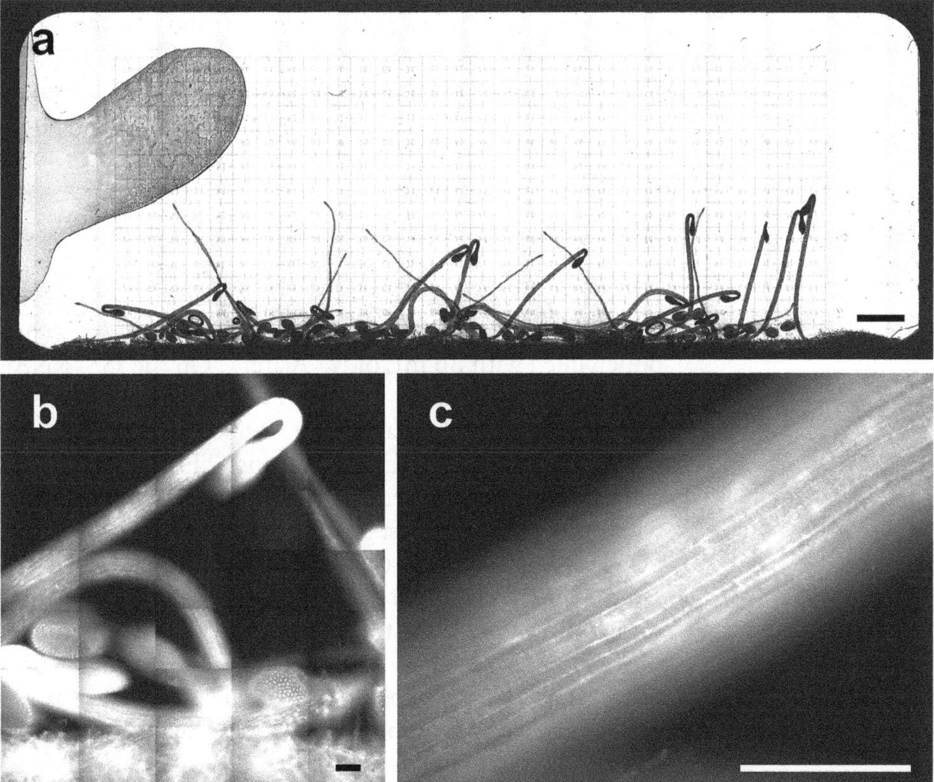

Fig. 6 Microscopic images of Arabidopsis seedlings grown in space. (**a**) Brightfield images of the entire cultivation space. (**b**, **c**) Fluorescent images of arrays of cortical microtubules adjacent to the outer tangential wall of epidermal cells in hypocotyls. The bars in (**a**), (**b**), and (**c**) are 2 mm, 200 μm, and 200 μm, respectively

9. Determine a place to observe based on the bright field image and acquire a fluorescence image (Fig. 6b, c).

4 Notes

1. Both the Resist Tubule Chamber B and the Plant chamber were developed by the Japan Aerospace Exploration Agency (JAXA) for plant cultivation in space. The interior dimensions of the Resist Tubule Chamber B are 56 mm in width, 46 mm in height, and 76 mm in depth. Those of the Plant Chamber are 56 mm in width, 46 mm in height, and 48 mm in depth.
2. The Hatosheet is a nonwoven fabric composed primarily of cellulose.
3. The CBFE is an incubator developed by JAXA for life science experiments involving microorganisms, cells, tissues, small animals, and plants. It has two working areas; a microgravity compartment and an artificial gravity compartment on a turntable (centrifuge).
4. Both the MELFI and the FROST can be used as a refrigerator or as a freezer. The MELFI has four compartments, and each compartment can be set separately to −80 ° C, −26 °C, and 4 °C. In the FROST, temperature can be varied from −65 °C to room temperature (it is not equipped with a heater). Generally, the refrigerator is kept at 4 ° C and the freezer is operated at −35 ° C, respectively.
5. The CFB was developed by JAXA for chemical fixation of plant materials under the influence of microgravity (Fig. 2a–c). The CFB comprises three bags (a zipper-sealed outer bag, a middle bag, and a clip-sealed inner bag). The inner bag contains a pouch filled with 30 mL of RNAlater®. On the ISS, RNAlater® is a toxic hazard level 1 (THL1) reagent, and therefore be contained in double-sealed containers. The CFB bag provides for two levels of containment against THL1 reagents. In this case, the seals must tolerate a depressurized environment, even when the module pressure is decreased to zero during emergencies. The safety of the CFB was confirmed by the JAXA safety panel for the Kibo Module and used for the first time during for the Resist Tubule experiment conducted in November 2013 (Fig. 2c). The CFB can also be used for other space-based experiments involving plants and small organisms. For the more toxic THL reagents, an additional level of containment is needed. As an alternative, JAXA has developed another fixation tool, the Chemical Fixation Apparatus (CFA), which provides for three levels of containment against reagents categorized with greater THL, such as a 4% paraformaldehyde solution (Fig. 2d).

6. The Cultivation/Observation Chamber was developed by JAXA for live imaging of plant cells under microgravity conditions (in the ISS module Kibo), using a JAXA fluorescence microscope (DMI6000B, Leica Microsystems, partially modified to fit the ISS). The chamber comprises two metal frames joined by a hinge. Each frame has a glass plate. One of the glass plates is a slide glass, and the other is a cover glass (e.g., Matsunami class NO. 1S, with a thickness of 0.17 mm). The interior dimensions of the chamber are 38 mm in width, 16 mm in height, and 0.5 mm in depth.
7. The optimum rock wool thickness for the Plant Chamber is 10 mm, and it takes time to dry moistened rock wool. The seeds are therefore attached by gum arabic.
8. The V-MEU is suitable for observing plants and has both a light-emitting diode source of illumination and a video camera. Water must be supplied to the plants manually by an astronaut. The PEU is an automatic system for plant cultivation. Cultivation begins with the initial water supply and plants are then supplied automatically with nutrients and water. Plants are regularly observed with a video camera by an automated system.
9. The mechanical properties of the cell wall are generally measured by performing a sequential analysis, using the stress-strain and stress-relaxation methods [11–14]. The mechanical properties of the cell wall in RNAlater®-fixed inflorescence stems of Arabidopsis were comparable to those of ordinary methanol-killed stems [22].
10. The amounts of RNA and gene-expression levels of the material observed following measurement of cell wall extensibility were similar to those observed in the case of the material used directly for RNA extraction [22].
11. A part of the cell wall components may flow through the first column of the RNeasy Plant Mini Kit.
12. The 1 M NaCl extract can be used to directly measure the activities of cell wall enzymes after dilution. Selecting suitable substrates and methods of measurement based on the type of target enzymes is recommended. The activity of the xyloglucan-degrading enzymes was well maintained, even after fixation in RNAlater® solution and after extraction of RNA [22].
13. Hemicellulose can be divided into hemicellulose I and hemicellulose II by extracting it with 4% (w/v) KOH thrice (12 h each) before 24% (w/v) KOH extraction.

14. The amount of cell wall polysaccharides was not significantly affected by the RNAlater® fixation, RNA extraction, or cell wall protein extraction [22].

References

1. Hoson T, Soga K (2003) New aspects of gravity responses in plant cells. Int Rev Cytol 229:209–244
2. Hoson T, Saito Y, Soga K, Wakabayashi K (2005) Signal perception, transduction, and response in gravity resistance. Another graviresponse in plants. Adv Space Res 36:1196–1202
3. Soga K (2013) Resistance of plants to gravitational force. J Plant Res 126:589–596
4. Soga K, Wakabayashi K, Kamisaka S, Hoson T (2006) Hypergravity induces reorientation of cortical microtubules and modifies growth anisotropy in azuki bean epicotyls. Planta 224:1485–1494
5. Matsumoto S, Kumasaki S, Soga K, Wakabayashi K, Hashimoto T, Hoson T (2010) Gravity-induced modifications to development in hypocotyls of Arabidopsis tubulin mutants. Plant Physiol 152:918–926
6. Soga K, Wakabayashi K, Hoson T, Kamisaka S (1999) Hypergravity increases the molecular mass of xyloglucans by decreasing xyloglucan-degrading activity in azuki bean epicotyls. Plant Cell Physiol 40:581–585
7. Soga K, Arai K, Wakabayashi K, Kamisaka S, Hoson T (2007) Modifications of xyloglucan metabolism in azuki bean epicotyls under hypergravity conditions. Adv Space Res 39:1204–1209
8. Wakabayashi K, Nakano S, Soga K, Hoson T (2009) Cell wall-bound peroxidase activity and lignin formation in azuki bean epicotyls grown under hypergravity conditions. J Plant Physiol 166:947–954
9. Soga K, Wakabayashi K, Kamisaka S, Hoson T (2004) Graviperception in growth inhibition of plant shoots under hypergravity conditions produced by centrifugation is independent of that in gravitropism and may involve mechanoreceptors. Planta 218:1054–1061
10. Hattori T, Otomi Y, Nakajima Y, Soga K, Wakabayashi K, Iida H, Hoson T (2020) MCA1 and MCA2 are involved in the response to hypergravity in Arabidopsis hypocotyls. Plants 9:590
11. Hoson T, Soga K, Mori R, Saiki M, Nakamura Y, Wakabayashi K, Kamisaka S (2002) Stimulation of elongation growth and cell wall loosening in rice coleoptiles under microgravity conditions in space. Plant Cell Physiol 43:1067–1071
12. Soga K, Wakabayashi K, Kamisaka S, Hoson T (2002) Stimulation of elongation growth and xyloglucan breakdown in *Arabidopsis* hypocotyls under microgravity conditions in space. Planta 215:1040–1046
13. Hoson T, Matsumoto S, Soga K, Wakabayashi K, Hashimoto T, Sonobe S, Muranaka T, Kamisaka S, Kamada M, Omori K, Ishioka N, Shimazu T (2009) Growth and cell wall properties in hypocotyls of Arabidopsis *tua6* mutant under microgravity conditions in space. Biol Sci Space 23:71–76
14. Hoson T, Soga K, Wakabayashi K, Hashimoto T, Karahara I, Yano S, Tanigaki F, Shimazu T, Kasahara H, Masuda D, Kamisaka S (2014) Growth stimulation in inflorescences of an *Arabidopsis* tubulin mutant under microgravity conditions in space. Plant Biol 16 (S1):91–96
15. Wakabayashi K, Soga K, Hoson T, Kotake T, Yamazaki T, Higashibata A, Ishioka N, Shimazu T, Fukui K, Osada I, Kasahara H, Kamada M (2015) Suppression of hydroxycinnamate network formation in cell walls of rice shoots grown under microgravity conditions in space. PLoS One 10:e0137992
16. Soga K, Yamazaki C, Kamada M, Tanigawa N, Kasahara H, Yano S, Kojo KH, Kutsuna N, Kato T, Hashimoto T, Kotake T, Wakabayashi K, Hoson T (2018) Modification of growth anisotropy and cortical microtubule dynamics in Arabidopsis hypocotyls grown under microgravity conditions in space. Physiol Plant 162:135–144
17. Wakabayashi K, Soga K, Hoson T, Kotake T, Yamazaki T, Ishioka N, Shimazu T, Kamada M (2020) Microgravity affects the level of matrix polysaccharide 1,3:1,4-β-glucans in cell walls of rice shoots by increasing the expression level of a gene involved in their breakdown. Astrobiology 20:820–829
18. Hoson T, Takahashi A, Nikawa T, Fukui K, Ogawa S, Higashitani A (2011) Toward future space experiments for life sciences. Biol Sci Space 25:21–24
19. Karahara I, Suto T, Yamaguchi T, Yashiro U, Tamaoki D, Okamoto E, Yano S, Tanigaki F, Shimazu T, Kasahara H, Kasahara H,

Yamada M, Hoson T, Soga K, Kamisaka S (2020) Vegetative and reproductive growth of Arabidopsis under microgravity conditions in space. J Plant Res 133:571–585

20. Dyachok J, Yoo C-M, Palanichelvam K, Blancaflor EB (2010) Sample preparation for fluorescence imaging of the cytoskeleton in fixed and living plant roots. In: Gavin RH (ed) Cytoskeleton methods and protocols, methods in molecular biology, vol 586. Humana Press, Totowa, pp 157–169
21. Hoson T, Akamatsu H, Soga K, Wakabayashi K, Hashimoto H, Yamashita M, Hasegawa K, Yano S, Omori K, Ishioka N, Matsumoto S, Kasahara H, Shimazu T, Baba SA, Hashimoto T (2012) Objectives, outlines, and preparation for the resist tubule space experiment to understand the mechanism of gravity resistance in plants. Trans JSASS Aerospace Tech Japan 10:Tp1–Tp5
22. Matsumoto S, Kumasaki S, Higuchi S, Inoue Y, Kirihata K, Fujie M, Soga K, Wakabayashi K, Hoson T (2008) Development of an efficient procedure for Resist Wall space experiment. Abstract of 37th COSPAR Scientific Assembly, Montreal, F11-0029-08

Chapter 18

NASA's Ground-Based Microgravity Simulation Facility

Ye Zhang, Jeffery T. Richards, Jessica L. Hellein, Christina M. Johnson, Julia Woodall, Tait Sorenson, Srujana Neelam, Anna Maria J. Ruby, and Howard G. Levine

Abstract

Since opportunities to conduct experiments in space are scarce, various microgravity simulators and analogs have been widely used in space biology ground studies. Even though microgravity simulators do not produce all of the biological effects observed in the true microgravity environment, they provide alternative test platforms that are effective, affordable, and readily available to facilitate microgravity research. The Microgravity Simulation Support Facility (MSSF) at the National Aeronautics and Space Administration (NASA) John F. Kennedy Space Center (KSC) has been established for conducting short duration experiments, typically less than 1 month, utilizing a variety of microgravity simulation devices for research at different gravity levels. The simulators include, but are not limited to, 2D Clinostats, 3D Clinostats, Random Positioning Machines, and Rotating Wall Vessels. In this chapter, we will discuss current MSSF capabilities, development concepts, and the physical characteristics of these microgravity simulators.

Key words Microgravity simulation, Space biology, MSSF, Plant research, Facility

1 Introduction

All Earth-based life has developed under the force of gravity, which affects growth, development, and morphology at all levels of biological organization [1]. The effects of gravity on physical phenomena (e.g., buoyancy, convection, and sedimentation) are central to these effects on biological processes [2].

There are a number of critical NASA-relevant questions relating to the effects of gravity that need to be answered for humans to establish a continuous human presence on the Moon, Mars, and survive long duration conditions of microgravity. Many of these questions relate to the effects of microgravity and partial gravity conditions (Lunar 0.166 g; Martian 0.38 g) on humans and those life forms (e.g., plants and selected animals) that will be essential for our long-term survival in space and on other planetary surfaces. For the foreseeable future, there will continue to be a limited number of

Elison B. Blancaflor (ed.), *Plant Gravitropism: Methods and Protocols*, Methods in Molecular Biology, vol. 2368, https://doi.org/10.1007/978-1-0716-1677-2_18,

opportunities to study the effects of gravity in space, which require significant resource expenditures to conduct. Additionally, the opportunity to perform replicated space experiments may only be realized sporadically and require long delays to achieve.

One approach taken to study the effects of altered conditions of gravity on Earth (with more regularity and a greatly reduced cost) has been to constantly change the direction of the gravity vector by using rotating devices of various designs (e.g., clinostats). Clinostats do not eliminate the force of gravity, but rather randomize the direction of gravity with respect to the sample over time, i.e., samples are rotated to prevent a consistent perception of the gravitational acceleration vector. True microgravity cannot be achieved with clinostats, but "functional near weightlessness" may be achieved at gravity levels below the known acceleration sensitivities of relevant biological processes [3]. There are a variety of clinostat designs and other microgravity analog devices that have been extensively reviewed [3–10].

Simulated microgravity conditions produce some, but not all of the biological effects observed in the true microgravity environment. While microgravity simulators provide test beds that are effective, affordable, and readily available to facilitate microgravity research, it must always be considered that each type of simulator has artifacts (e.g., centrifugal accelerations, vibrations) that can result in a misinterpretation of responses if not carefully considered. Therefore, for each simulator, the physical parameters, operational principles, and their specific impact on the biological processes and specimens of different sizes need to be critically evaluated.

In Europe and elsewhere, there has been a recognition of the need for ground-based facilities to enable testing to address gravity-related issues prior to incurring costs associated with spaceflight experiments [10, 11]. In this vein, at John F. Kennedy Space Center (KSC), the National Aeronautics and Space Administration (NASA) has invested in the establishment of a Microgravity Simulation Support Facility (MSSF) that provides the science community in the United States the opportunity to conduct microgravity simulation studies that might otherwise be beyond the cost limits for individual investigators. At MSSF, engineering and science support is provided to meet individual experiment requirements and to share lessons learned. This chapter describes the current capabilities provided by the KSC MSSF, along with a selected review of relevant past studies and recommendations for future investigations.

2 Research Retrospective

As opportunities for conducting experiments in space are rare, more studies have been reported using ground-based analogs that simulate microgravity. A literature search in PubMed (https://

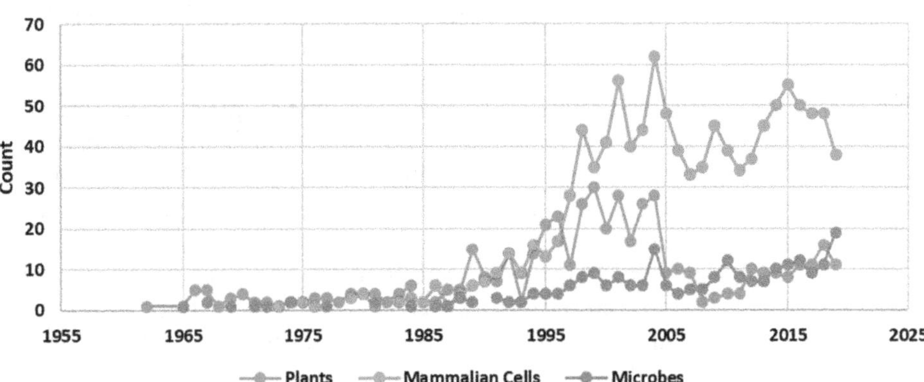

Fig. 1 A retrospective of life science research using simulated microgravity. Search keywords: *Plants* (477 entries) by combinations of "clinostat," "simulated microgravity," "microgravity analog," and "plant"; *Mammalian Cells* (1073 entries) by combinations of "clinostat," "simulated microgravity," "microgravity analog," and "mammalian cells"; and *Microbes* (271 entries) by combinations of "clinostat," "simulated microgravity," "microgravity analog," "microbial," "microbe," "biofilm," "bacteria," "fungi," and "virus"

pubmed.ncbi.nlm.nih.gov/) revealed more than 5000 peer-reviewed articles have been published on studies using simulated microgravity conditions since 1961. Approximately 2000 of these studies investigated mammalian cell cultures, plants, and microbial cell cultures (Fig. 1). While NASA's Bioastronautics Program [12] sponsored research using microgravity simulators for decades, the invention of Rotating Wall Vessels (https://synthecon.com) in the 1970s was a technological breakthrough that significantly increased the number of ground-based simulated microgravity research studies, primarily utilizing mammalian cells [13, 14]. The Synthecon website has cited more than 700 publications using their Rotating Cell Culture (RCC) systems (https://synthecon.com/pages/research_publications_rotary_cell_culture_system_8.asp). Compared to the plant and mammalian cell biology fields, fewer microbiology studies have been conducted using microgravity simulation devices, but in recent years this number has been steadily increasing. The number of plant research publications have also been increasing since 2008.

We conducted a review of these 2000+ articles to gain insights into the trends of life sciences research using microgravity simulation devices. Inclusion criteria were defined as studies in English that utilized two-dimensional (2D) Clinostats, three-dimensional (3D) Clinostats, Random Positioning Machines (RPM), and/or Rotating Wall Vessels. Studies that paired simulated microgravity experiments with true microgravity platforms, such as those flown on the Space Shuttle, International Space Station (ISS), parabolic flight, and sounding rocket missions, were given in-depth analysis. Also emphasized were those articles that combined simulated microgravity with other study variables relevant to spaceflight, such as high carbon dioxide (CO_2), ionizing radiation, and partial gravity states.

2.1 Plant Research

The first study of a plant's response to gravity was a hypergravity experiment conducted in 1806 by Sir Thomas Andrew Knight, when he grew bean plants on a rotating water wheel [15]. In 1868, the term "geotropism" was introduced for the first time by Alert Bernhard Frank to describe the process of plant response to gravity [16]. Of the early gravitational plant biology researchers, many used rudimentary clinostats. By 1902, the statolith model for plant gravity sensing was beginning to form, with the writings of Haberlandt [17].

In general, the goal of a plant clinostat study is to balance out the influence of the directional g-vector to near zero. This can be achieved on a 2D clinostat by slowly rotating a plant along a horizontal axis. As the plant rotates, the influence of gravity is negated, thus simulating microgravity by cancelling the unilateral gravitational influence on the plant. While 3D clinostats have been in use for biology research since 1924, the first report of a 3D clinostat for plant research was made in 1963 by Scano [18]. 3D clinostats were not reported for growing plants again until the early works of Chapman in 1980, and Hoston in 1993 [19, 20].

Our PubMed search found more than 230 original research articles on simulated microgravity research with plants. These articles cover a wide range of species spanning a range of photosynthetic eukaryotic organisms commonly described as plants including algae, pteridophytes, and flowering plants such as *Triticum* and *Arabidopsis thaliana*. *Arabidopsis thaliana* and *Pisum sativum* are two of the most studied species (accounting for almost one third of these 230 articles). Forty three of these studies were performed using microgravity simulation devices in parallel comparisons with spaceflight experiments, mostly on the ISS or other low Earth Orbit vehicles, plus several using sounding rocket and parabolic flight. Interestingly, the majority of these studies were published prior to 2005, with 6 articles after 2005, and only 2 after 2010. Few studies have been reported on investigating the combined effects of simulated microgravity with other deep space environmental risk factors, such as ionizing radiation and lack of a magnetic field, as well as others related to confined spaceflight environments, such as high CO_2 levels, exposure to volatile organic compounds (VOCs), and plant pathogen or opportunistic pathogen infections. Furthermore, advanced molecular and imaging analytical technologies have only been applied to more recent investigations. Most of these studies utilized 2D clinostats with low speed rotation, with RPM usage only emerging after 2000.

For many of these studies, a plant is initially grown with a standard Earth gravity vector, and once it is established, the plant is placed on a clinostat for analyzing the effects of simulated microgravity. Some studies have also been conducted to investigate plants initiated from seeds onward. However, seed-to-seed studies have rarely been conducted. It is important to note that experimental parameters, e.g., the age of the plant, and the duration of the study, vary dramatically between plant studies, which makes in-depth comparisons between existing studies difficult.

2.2 Non-Plant Research

Studies examining mammalian cell biology under simulated microgravity conditions gained traction beginning in the 1970s. One of the first of these cell biology investigations assessed the ability of human lymphocytes to undergo activation following exposure to simulated microgravity on a fast-rotating 2D clinostat [21]. While plant research continued to use the 2D clinostat, the cell biology community shifted towards the NASA-developed rotary cell culture system upon its commercialization by Synthecon Corporation in the 1990s.

Cellular and molecular responses to simulated microgravity have been investigated in many cell types, both primary cells and immortalized cell lines, with the most common types being osteoclasts and osteoblasts, skeletal muscle cells, and T lymphocytes. These cell types reflect systems considered to be the most profoundly affected by spaceflight (bone, muscle, immune) and remain active areas of human spaceflight research. Moreover, the simulated microgravity environment has been shown to be a suitable environment for growing co-cultures and spheroids [22–24], making it an attractive platform for researching 3D cultures that better recapitulate the cellular organization observed in the human body. Beginning in the mid-90s through present day, cancer cell cultures have become increasingly studied under simulated microgravity conditions, particularly on RPMs and rotating wall vessels [25–28].

Over 20 articles reported performing simulated microgravity experiments in parallel with true microgravity research platforms such as ISS, parabolic flights, or sounding rockets. For example, Camberos et al. (2019) compared a 7-day culture of cardiac progenitor cells within a 2D clinostat at 4 rpm to the culture of the same cells for 30 days on the ISS and found that in both simulated and true microgravity conditions, there was a down-regulation of a critical cardiac repair gene [29]. Although at different timepoints, the results indicated that in this instance a simulated microgravity device was able to partially recapitulate cellular effects from spaceflight. Another study by Krüger et al. (2019) investigated the formation of 3D endothelial cell structures (spheroids) after 1, 5, or 12 days on an RPM, on the ISS, and within ground controls [30]. This investigation observed the formation of 3D endothelial structures in both simulated microgravity and true microgravity groups, which further corroborates the value of simulated microgravity devices as a useful tool in studying spaceflight associated changes to cell biology.

In addition to altered gravity states, radiation exposure is one of the critical risks associated with future long duration exploration missions to the Moon and Mars. To date, only a few studies have assessed the combined effects of simulated microgravity and radiation on mammalian cells, mostly using low-Linear Energy Transfer (LET) radiation sources such as X-rays, gamma rays, or entrance plateau protons. These studies mostly concentrated on radiation induced chromosomal aberrations and cells responsible for ossification and immune function [31–36]. In 2019, Hada's group reported their investigation on chromosome aberrations in fibroblast cells exposed simultaneously to simulated microgravity using a 3D clinostat and high-LET Carbon ion (C-ion) irradiation [37]. The combined treatment showed synergistic changes in the expression of genes as well as increased chromosomal aberrations, which were partially in agreement with several low-LET studies [33–36]. The aforementioned integrated research areas as well as others, e.g., host–microbe interactions, are generally underrepresented, but will remain active as novel and deep space relevant approaches in the space life science community.

3 Gaps in Ground-Based Microgravity Simulation Research

Based on this literature review, several major knowledge gaps were identified for ground-based microgravity simulation research:

(a) Integrated approaches to conduct microgravity simulation research in parallel with other simulated or true microgravity platforms with both phenotypic and molecular analyses.

(b) Combined effects of simulated microgravity and other space-related environmental risk factors, e.g., ionizing radiation.

(c) Interactions among different organisms, species, or cell types using co-culture systems, or 3D tissue organoids for investigating microgravity effects at the tissue level and host responses.

(d) Multi-generational studies under simulated microgravity.

(e) State-of-the-art molecular and imaging analysis approaches in addition to phenotypic observations.

4 KSC Ground-Based Microgravity Simulation Support Capabilities

The MSSF was first established in 2017 to support ground-based life sciences research funded by the NASA Space Biology Program. Since that time the MSSF has supported more than 20 projects and hosted visiting scientists, postdoctoral fellows, and student interns funded by various NASA programs and partnerships. These projects range from ground-based microgravity studies, radiation studies, Low Earth Orbit (LEO) studies, parabolic flight studies, and student outreach activities covering a wide variety biological organisms including microbes, plants, and mammalian cells.

Table 1 lists the simulation devices currently available in the MSSF. These devices can be accommodated within controlled environment chambers, including tissue culture incubators and plant growth chambers, allowing investigators to customize and monitor environmental conditions such as temperature, humidity, CO_2, and light exposure. Additionally, tissue culture facilities, cellular and molecular analysis tools, and advanced confocal fluorescence microscopy are also available for investigators to conduct their research projects. Cellular and molecular analysis tools provided in the MSSF include, but are not limited to, flow cytometry, a real-time PCR system, microplate readers for absorbance and/or fluorescence signals, a Nanodrop spectrophotometer, and a fluorescence stereo-microscope. These capabilities allow live samples to be analyzed right after an experiment on a microgravity simulation device.

Each simulation device has its own pros and cons, not only for the way "microgravity" is simulated, but also for its ability to accommodate different specimen types due to their individual unique constraints. The selection of an appropriate simulation method (device and rotation speed for many cases) is driven by the science, the biology, and the experimental objectives. However, it requires more than an appropriate simulation method for conducting a reliable microgravity simulation experiment and subsequent data interpretation. One of the major constraints that affects the quality of a microgravity simulation experiment is the number of replicates, often limited by the size of the rotating plates. Theoretically, for every type of simulation device, the most centrally located point has the best *g*-vector cancellation. As specimens move away from the center point the perceived gravity level changes. The threshold for a biological organism to respond to a directional "*g*-vector" is determined by its gravity-sensing mechanism and the magnitude and duration of the accelerations. In addition, careful planning, adequate logistic support, and high fidelity 1-g controls are some of the most important determinant factors for the success of a microgravity simulation experiment in reducing confounding variables and mitigating non-device-related artifacts.

Several strategies being implemented in the MSSF to reduce the controllable artifacts are: (1) using multiple microgravity research platforms with high fidelity controls; (2) minimizing the artifacts mechanically; (3) understanding the localized physical nature around the specimens through fluidic dynamics modeling; and (4) developing customized science modules based on individual experimental requirements.

Table 1
List of microgravity simulation devices in the MSSF

Type of Simulation Devices	Quantity	Max Science Carrier Size	Weight	Rotation Speed	On-stage Power Supply	Information
Large Slow Rotating 2D Clinostat	1	A standard mid-deck locker or ISS EXPRESS locker.	Up to 30 kgs	Up to 3 rpm along the horizontal axis with 0.5 rpm increments.	Provided	KSC engineers have designed a slow rotating clinostat that allows researchers to subject biological specimens within an ISS EXPRESS locker, or other large containers, to simulated microgravity conditions. Power is provided by built-in computers, lamps, fans, and auxiliary equipment. Adapters can be developed to accept various hardware configurations.
Petri Plate 2D Clinostat	2	Up to 8- 10cm x 10cm petri plates, or customized holders.	NA	1-20 rpm with 1 rpm increments.	NA	A 2D clinostat was developed by KSC engineers, accommodating 10cm x 10cm petri plates and other containers specialized for particular life science model organisms.
Rotating Wall Vessel (RWV) Bioreactor	2	4 mL to 500 mL culture vessels, including autoclavable Stem Cells Culture Vessels, High Aspect Ratio Vessels (HARVs), Slow Turning Lateral Vessels (STLVs), disposable vessels, and customized sample containers.	NA	0-20 rpm along the horizontal axis.	NA	Customized Synthecon Rotary Cell Culture systems (RCCs) can be operated vertically or horizontally. Each vessel can be operated separately at a desired rotation speed. Well-designed rotation conditions allow specimens to remain in constant free fall simulating near weightlessness conditions and minimizes various mechanic stresses, such as interactions between specimens and vessels.
Eisco 2D Clinostat	6	Support any size or shape of pot up to 7.5 cm in diameter with maximum baseline root zone depths of 10 cm. Holders are also available for 100 mm petri dishes and tubes. The units can accommodate customized sample holders as well. Axis angle can be adjusted from 0-90°.	NA	2 rpm	NA	These commercially available Eisco 2D clinostats enable horizontal to vertical clinorotation studies with large plants and other biological samples.
3D Clinostat: Random Positioning Machine (RPM)	4	Flasks, well plates, petri dishes, and rotating wall vessels. Modifications to the inner platform can allow the use customized hardware with various configurations.	Up to 1.5kgs	0-20 rpm	Provided	The Airbus RPM 2.0 (www.airbusDS.nl) has two independently rotating frames that rotate with different speeds and different directions. An experimental apparatus containing research specimens is placed within the inner of two counter-rotating platforms. Based upon operating configurations, treatments of simulated microgravity (<10-3 g) or partial gravity can be achieved.
RPM with Microscope Module	2	Chamber slide, Nunc™ Lab-Tek™ Flask on Slide, or ibidi µ-Slide.	NA	0-20 rpm	NA	A RPM Microscope Module was developed at KSC to enable time-lapse imaging capability. Magnification: 750X – 900X Imaging Capability: Up to 5.0 MP vivid color resolution (2592x1944), live video (30fps), or time-lapse video. Single Imaging Mode: bright-field, dark-field, or fluorescence. Other Capabilities: Wi-fi adaptor and real-time image transfer.
3-D Clinostat: GRAVITE	2	Flasks, well plates, petri dishes, and rotating wall vessels. Modifications to the inner platform can allow use of various hardware configurations.	Up to 1.5kgs	1-4 rpm with 1 rpm increments for microgravity simulation.	NA	The GRAVITE from the Space Bio-Laboratories (http://www.spacebio-lab.com/ENG-index.html) contains two independently rotating frames that are mounted and rotate perpendicular to each other. An experimental apparatus containing research specimens is placed within the inner of the two counter-rotating platforms. Based upon operating configurations, treatments of simulated microgravity ($<10^{-3}$ g), partial gravity, or hypergravity (to 2-3 g) can be achieved.
3D Clinostat: SciSpinner	1	60mm petri dish; Light provided: IR and RGB LEDs; and Imaging capability: : 3280 x 2464 px for stills, and 1080P at 30fps / 480p at 90fps for video.	NA	0.3-15 rpm with decimal grade precise speed control.	NA	SciSpinner was originally developed by a Space Biology investigator team and manufactured by Cosē Instruments. Most parts are 3D printed.

4.1 Using Multiple Microgravity Research Platforms and High Fidelity Controls

Using multiple microgravity research platforms, whenever possible, is an effective way to conduct ground-based microgravity studies. All platforms have their own sets of unique characteristics. This raises a fundamental question: What is the true microgravity effect on biological organisms? For research conducted on the ISS, we often cannot distinguish true microgravity effects from associated spaceflight effects. This may contribute to why it is not a simple matter to repeat an ISS finding using a ground simulation device. In addition, the unique device-specific method for achieving "weightlessness" may generate sufficient artifacts to affect the biological responses to simulated "weightlessness." Using multiple platforms with high fidelity experimental settings in parallel comparisons can, in some instances, reduce the device-specific artifacts and reveal some true microgravity-associated biological changes.

To achieve high fidelity experimental settings, flexibility in the design of the rotating stages is required. Customized science holders for sample containers (e.g., flasks, petri dishes, chamber slides) are frequently fabricated by the MSSF staff to meet experimental needs. These sample holders can be 3D printed and are often designed based upon sample containers that have been flown for ISS studies, such as ibidi µ-Slides, 100 mm petri plates, 60 mm petri dishes, BioCell cell culture chambers, microplates, and more. As an example, a bracket was developed for holding multiple 2 U hardware (e.g., Biological Research in Canisters (BRIC)-like fight hardware) on a large 2D clinostat (Fig. 2).

Biofilm formation under simulated microgravity has been an area of interest in fundamental space biology, and the topic of several research investigators' teams supported by the MSSF. For one team, the requirement was to use 50 mL tube sample containers for both the RPM and the Gravite 3D clinostat [38]. For another team also studying biofilms, the High Aspect Ratio Vessel (HARV) was preferred. Given that there are no easy attachment points on the RPM or Gravite to secure tubes, two sets of tube holders were fabricated, one 9 tube holder for the RPM and two 5 tube holders (one for each side of the baseplate for 10 tubes in total) for the Gravite (Fig. 3). The designs minimize distances around the 3D center of rotation to reduce as much as possible disproportional centrifugal forces between the tubes and comparable conditions between two simulation devices. For the study using HARV, minor modifications of a 3D-printed baseplate allowed the attachment of the HARV vessels at the center of rotation within the inner-ring on the RPM. With modified HARV vessels reducing the length of the fill ports, two 50 mL HARV vessels can fit (one on each side of the baseplate). In another implementation, eight 10 mL HARV vessels can be accommodated. These highly flexible and customized designs allow for different sample containers to be placed on simulators at optimal positions comparable with other microgravity platforms.

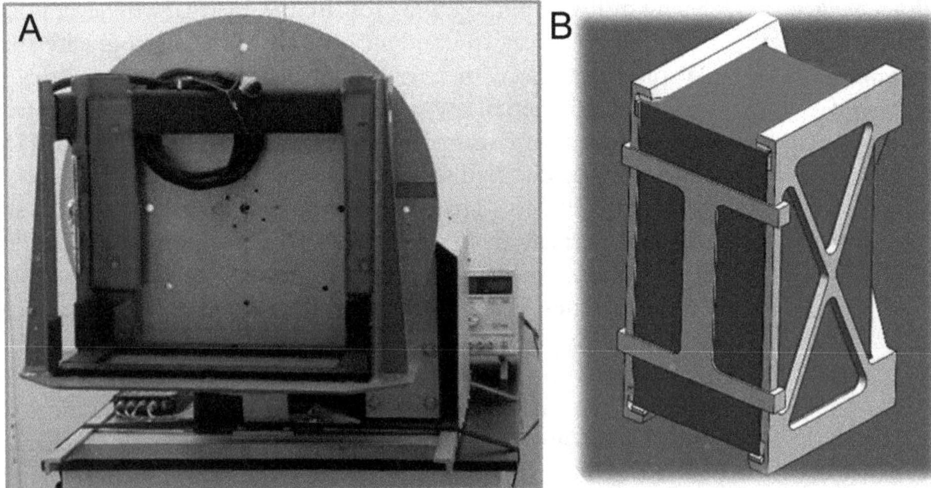

Fig. 2 The design of the bracket for a 2 U container, showing (**a**) the large 2D clinostat without the bracket; and (**b**) the CAD rendering of the bracket that can be installed at the center of the large 2D clinostat

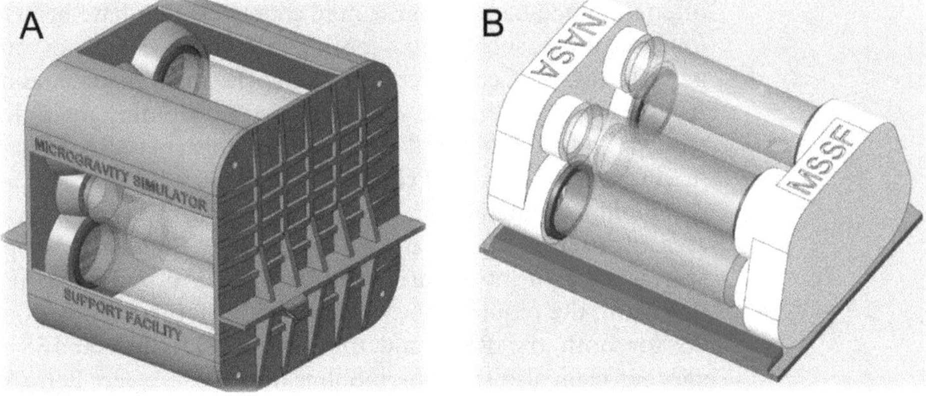

Fig. 3 Tube holders designed for biofilm coupon experiment on (**a**) RPM and (**b**) Gravite

4.2 Minimizing Mechanical and Design Artifacts

Artifacts can be generated in multiple ways from the devices, sample holders, or even the samples themselves. As a best practice, for anything to be added onto the rotating plate, the center of mass (with samples) should be located at the physical center of the rotating plate(s). Counterweights may be needed at appropriate positions on the stage based on holder/sample layout. Sample containers with liquid media need to be conditioned and completely filled up without visible bubbles. Furthermore, vibration is a major source of unwanted acceleration.

Vibrations from environmental chambers or cell culture incubators can be attenuated by adding vibration-damping materials under simulation devices. If vibrations are generated from the device while operational, then maintenance or even modifications are required. Periodic measurements, regular maintenance, and close observations are important for preventing unfavorable vibrations during operations. The customized sample holders may also generate some relative movement against the simulation device and the samples. Detailed analysis is often required to locate the vibration source. For example, for a microscope module designed for the RPM, a four-leg scope holder did a better job at reducing vibration with improved image stability compared to a two-leg holder version. Analysis of the four-leg scope holder still found residual movements skewed in the Y direction, suggesting the conjunction of the sample slide with the holder connecting with the RPM frame might have relative movement at the micron scale. Even this small movement caused unstable images when viewed in time-lapse, but these could be eliminated using post-imaging correction. More importantly, any vibrations, if not well characterized and controlled, may generate artifacts that impact the reliability of the microgravity simulation.

4.3 Computational Fluid Dynamic Models

While the use of simulated microgravity methods has extensive precedence in the scientific community, a more complete understanding of the mechanical environment around the biological samples created by different simulation techniques can allow some discrimination between the intended effects of microgravity simulation and the unintended physical artifacts inherent to simulation techniques. This approach can improve the simulation and data interpretation significantly. Therefore, we are developing Computational Fluid Dynamic (CFD) models of simulated microgravity conditions using ANSYS CFX software to better understand the physical stresses that the biological samples are exposed to in these analogs.

One example of this effort is to improve our understanding of fluidic shear forces on suspended cells within rotating wall vessels. Multiphase modeling capabilities of the ANSYS CFX software tool were applied with an experimental flake (black mica power, ~5μm) suspension to simulate suspended cells within a 50 mL High Aspect Ratio Vessel (HARV) rotating wall vessel (Fig. 4a). The models analyzed fluid flow and suspended solid–fluid interaction in a vessel rotating at 15 rpm (revolutions per minute) under simulated microgravity conditions. The analysis of particle image velocimetry using MATLAB was then conducted for model validation. The measurements of interphase momentum transfer between the fluid and solid phases revealed the effects of fluid flow on cell suspension systems in simulated microgravity. The rotational and stationary velocity were then calculated along the timeline of rotation (Fig. 4b, c). When a single cell suspension was introduced into this system, no significant velocity changes were detected. However, the condition may be dramatically altered if large-sized 3D tissue organoids or other large-sized objects are introduced to the

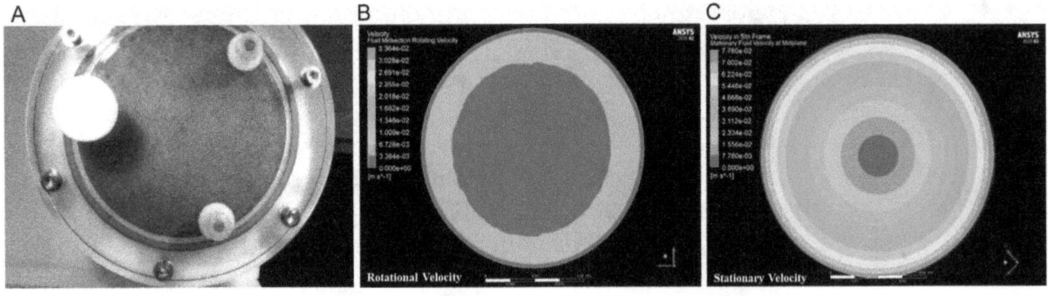

Fig. 4 Fluid dynamic modeling using (**a**) 50 mL HARV with experimental flakes in water rotating at 15 rpm for particle tracking; and contour plots of (**b**) Rotational velocity, and (**c**) Stationary velocity

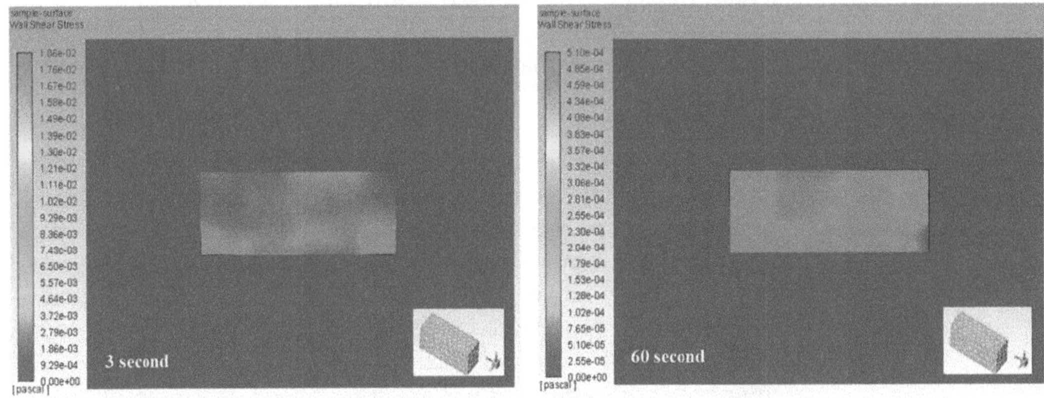

Fig. 5 Analysis of the potential shear stress imposed upon adherent cells on the slide chamber surface after 3 s and 60 s exposures to simulated microgravity on a 3D clinostat

system, e.g., scaffolds or fish eggs [39]. Other factors will be further analyzed with more variables and parameters, such as gas exchange, convection, and circulated fluidic perfusion.

Fluidic analysis was also employed to understand the potential shear stress imposed upon the adherent cells on a 9.0 cm^2 slide surface of a Nunc™ Lab-Tek™ chamber flask on the 3D Gravite clinostat using the same software. In this experiment, epithelial cells were cultured on a fibronectin-coated slide surface, and the center of the slide was placed at the center of rotation. At 2 rpm rotation speed, the cell sample's wall shear stress was at a maximum at $1.82e^{-02}$ pascals after 3 s and then converged to $5.10e^{-04}$ pascals after 60 s (Fig. 5). The fibronectin coating on the slide provided additional structural support to the cells.

This type of analysis is important for understanding what biological samples experience on a microgravity simulation device more precisely, assisting problem-solving and data interpretation. Fluidic dynamic analysis is not only beneficial for samples with fluid filled sample containers [40], but also for plant research where there is "free" water within the root zone. Fluidic modeling for plant research is inherently much more complicated, with both micro-environment simulation targeting root tips and macro-environment simulation on the whole root zone of the plant habitat.

4.4 Capabilities for Designing Experiment Unique Equipment Add-ons

Although there is increasing demand for the use of 3D clinostats and RPMs for biological research, these devices provide only a simple baseplate on which researchers can attach their experiments. In most instances, there are specific requirements for experiment unique equipment (EUE) that can attach to these simulators to support plant, microbial, or cell culture experiments. With this pressing need, we have used our 3D printing and machining capabilities in the KSC Prototype Shop to perform customized design and development according to each individual unique set of science requirements. Examples of the EUE that we have developed at MSSF, particularly those can be used for plant research, are presented below.

4.4.1 General Considerations for Designing Hardware to Accommodate Plant Research

A key consideration for designing hardware to accommodate plant research on microgravity-simulated devices is the provision of light. Phototropism and gravitropism are two interactive mechanisms in plants affecting plant growth and development. For 2D clinostats (including RCC systems), controlled light intensities can be provided by environmental chambers and independent light sources to provide comparable directional light to both samples at 1-g static and under simulated microgravity. However, for 3D clinostats and RPMs, on stage power supply is required for implementing light emitting diode (LED) light bands to provide directional light that moves with the specimens.

The RPM has pins mounted on the inner rotational axes allowing for power to feed small devices while the system is running. As such, a set of Agrivolutin white LEDs can be attached to the hardware. These LEDs are capable of providing modulated intensities greater than 300μmol m^{-2} s^{-1} photosynthetic photon flux density (PPFD) of broad spectrum lighting. Combinations of red green blue LEDs are also used based on the science requirements, especially if the goal is to compare results with an experiment in the Vegetable Production System on the ISS. Separate systems are also designed with LED lighting for supporting 1-g static controls simultaneously under the same intensity of light as those held in place on the RPM.

4.4.2 Customized Plant Growth Systems

Unique plant growth modules with white LED lighting have been designed to fit within the inner rotational ring of the RPM to support specialized plant growth studies. One example is to support a circadian rhythm study. Two custom modules were developed to meet the unique requirements of this investigation. Each module allows dual light cycles to be delivered separately to

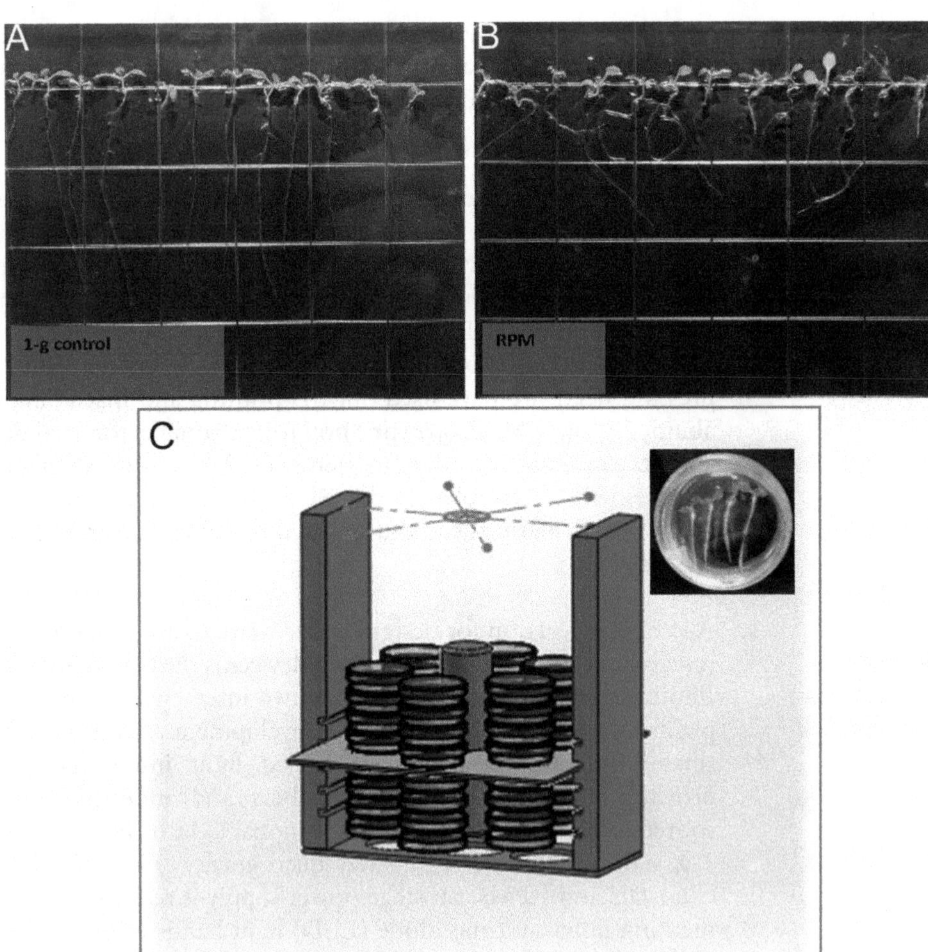

Fig. 6 Customized Plant Growth Systems with (**a**) *Arabidopsis* seedlings in a Petri plate after 6 days of growth under static conditions on an RPM; (**b**) *Arabidopsis* seedlings in a petri plate after 6 days of growth under rotation conditions on an RPM; and (**c**) The time release module CAD design to support 35 mm petri dishes with seedlings

two sets of 100 mm^2 Petri plates with *Arabidopsis thaliana* seeds/ seedlings on an RPM [41]. Figure 6a shows *Arabidopsis* seedlings in a Petri plate after 6 days of growth on an RPM. The other module, shown in Fig. 6b, mounts 12 stacks of 35 mm^2 Petri plates with a central core of LED lights. Each stack of 6 Petri plates is designed to be easily removed at various time points without stopping the RPM rotation, thereby preventing interruption of the microgravity simulation on the other stacks during a light cycle [42].

Fig. 7 Plant growth chamber developed for plant early development studies. (**a**) Microgreen growth; and (**b**) Chamber CAD design

4.4.3 Microgreen Plant Growth Chamber

This module created for the RPM shown in Fig. 7 is designed to support two sets of plant trays for microgreens or young plants, which fit the designated height constraints [43]. The design is flexible and expandable to other size plant trays. White LED light strips are powered through the pin connector positioned on the inner rotational ring and provide a controlled photoperiod programmed through an attached Arduino board. The chamber contains two Bluetooth cameras mounted above the growth area to take time-lapse pictures as the chamber is rotating. The plant trays sit at the center of rotation to minimize artifacts in microgravity simulations. For microgreens and young plants with specially curated growth substrate/media, active water delivery is not necessary for short duration studies.

For supporting experiments using taller and older plants, the two set design can be adjusted to a one-set design; however, the plant tray has to be moved away from the center. In addition, an active water delivery system, such as a system controlled by a syringe pump, is required as an add-on. For microgravity research involving large-sized plants, a large 2D or 3D clinostat is recommended. However, off-center artifacts are always of concern for these types of research. Therefore, experimental objectives and potential science gains need to be carefully weighed with technical considerations and disadvantages in order to guarantee meaningful data interpretation and to draw reliable conclusions.

4.4.4 Live Cell Imagery under Simulated Microgravity Conditions

Real-time imaging during simulations are of particular interest, as it allows study of basic cell functions such as cell division, cell migration, proliferation progress, and other potential biological time-lapse analyses under simulated microgravity conditions. We developed a live cell microscopic imaging module for the RPM, which can also be modified for other 3D clinostats (Fig. 8a, b)

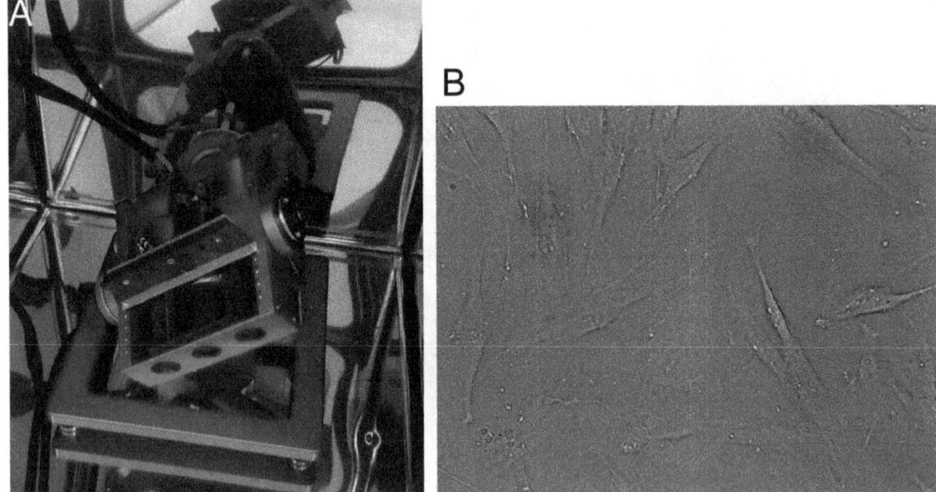

Fig. 8 Live cell microscopic imaging module for the RPM. (**a**) Microscopic imaging module integration on an RPM in an incubator; (**b**) An example of original image obtained during a time-lapse experiment

[44, 45]. All structural components of the module are 3D printed to support live bright-field and fluorescent imaging using a digital microscope. Additionally, this module contains a sample carrier holder for either a chamber slide, a Nunc™ Lab-Tek™ Flask on Slide, or an Ibidi μ-Slide. Maximum magnification can reach 750×–900×, the imaging capability is up to 5.0 MP vivid color resolution (2592 × 1944), and live (30fps) or time-lapse video is provided. Single imaging mode includes bright-field (including a backlight), dark-field, or fluorescence capabilities. A Wi-Fi adaptor is incorporated for real-time image transfer.

5 Future Work to Support Science Beyond Low Earth Orbit

Life on Earth has developed under 1 g gravity and is well protected from space radiation by Earth's magnetic field and atmosphere. The ISS environment provides true microgravity conditions, but is still within the Earth's magnetic field, and thus receives effective protection from radiation. However, for missions beyond LEO, such as those in Lunar orbit, on the Lunar surface, or during Mars transit missions, life will experience the absence of Earth's protective magnetic field, increased space radiation, micro to reduced gravity, and enclosed environments with significant space and resource constraints. For the past several decades, microgravity simulation devices have been developed and widely utilized for ground microgravity research, particularly focusing on LEO and changes associated with living in microgravity. Studies on combined effects of microgravity and other space-related risk factors are extremely under-represented in space biology investigations and the published literature. With NASA's vision of future deep space missions beyond LEO, the responses of biological organisms to the deep space environment, especially for those species potentially carried onboard these missions, are of particular interest. Because opportunities for conducting deep space biological science are rare, ground studies using microgravity simulation devices at micro g and partial g levels in combination with ionizing radiation exposures relevant to deep space radiation, as well as other risk factors, are promising analog approaches for continuing research beyond LEO. These approaches require next level capabilities of ground microgravity simulation based on the extensive experience gained over decades and the large body of knowledge from research onboard the ISS.

Acknowledgments

The KSC MSSF facility is supported by NASA's Biological and Physical Science Division within the NASA Science Mission Directorate. We greatly appreciate the LASSO MSSF science support team, especially Jason Fischer and Stephanie E. Richards for logistic support, Jacob J. Torres, Jonathan R. Gleeson, Lawrence L. Koss, as well as Randall I. Wade and Michael A. Lane from NASA KSC, and Bill Wells from the Bionetics Corporation for hardware design support. We also thank Caesar Udave, Audrey Lee, Antonina Tsinman, Jessica L. Hellein, Julia Woodall, Tait Sorenson, and Emily N. Keith, for their dedication and excellent internship work to improve MSSF operations. Srujana Neelam contributed greatly to the establishment of the MSSF when she was a NASA postdoctoral fellow at KSC through the Universities Space Research Association. Appreciation is also given to Gioia D. Massa and Raymond M. Wheeler for their critical review of this book chapter.

References

1. Vandenbrink JP, Kiss JZ, Herranz R, Medina FJ (2014) Light and gravity signals synergize in modulating plant development. Front Plant Sci 5:563. https://doi.org/10.3389/fpls.2014.00563
2. Braun M, Böhmer M, Häder D-P, Hemmersbach R, Palme K (2018) Gravitational biology I: Gravity sensing and graviorientation in microorganisms and plants. Springer International Publishing, Cham, Switzerland. https://www.springer.com/la/book/9783319938936
3. Briegleb W (1992) Some qualitative and quantitative aspects of the fast-rotating clinostat as a research tool. ASGSB Bull 5:23–30
4. Häder D-P, Rosum A, Schafer J, Hemmersbach R (1995) Gravitaxis in the flagellate *Euglena gracilis* is controlled by an active gravireceptor. J Plant Physiol 146:474–480
5. Hoson T, Kamisaka S, Masuda Y, Yamashita M, Buchen B (1997) Evaluation of the three-dimensional clinostat as a simulator of weightlessness. Planta 203:S187–S197. https://doi.org/10.1007/PL00008108
6. Klaus DM, Schatz A, Neubert J, Höfer M, Todd P (1997) *Escherichia coli* growth kinetics: a definition of "functional weightlessness" and a comparison of clinostat and space flight results. Naturwissenschaften 143:449–455

7. Klaus DM (2001) Clinostats and bioreactors. Gravit Space Biol Bull 14:55–64
8. van Loon JJWA (2007) Some history and use of the random positioning machine, RPM, in gravity related research. Adv Space Res 39:1161–1165. https://doi.org/10.1016/j.asr.2007.02.016
9. Hasenstein KH (2009) Plant responses to gravity—insights and extrapolations from ground studies. Gravit Space Biol 22:21–32
10. Herranz R, Anken R, Boonstra J, Braun M, Christianen PCM, de Geest M, Hauslage J, Hilbig R, Hill RJA, Lebert M, Medina FJ, Vagt N, Ullrich O, van Loon J, Hemmersbach R (2013) Ground-based facilities for simulation of microgravity, including terminology and organism-specific recommendations for their use. Astrobiology 13:1–17. https://doi.org/10.1089/ast.2012.0876
11. Brungs S, Egli M, Wuest S, Christianen P, van Loon J, Ngo-Anh TJ, Hemmersbach R (2016) Facilities for simulation of microgravity in the ESA ground-based facility programme. Microgravity Sci Technol 28(3):191–203. https://doi.org/10.1007/s12217-015-9471-8
12. Bioastronautics roadmap: a risk reduction strategy for human space exploration https://humanresearchroadmap.nasa.gov/Documents/BioastroRoadmap.pdf
13. Goodwin TJ, Prewett TL, Wolf DA, Spaulding GF (1993) Reduced shear stress: a major component in the ability of mammalian tissues to form three-dimensional assemblies in simulated microgravity. J Cell Biochem 51(3):301–311
14. Duray PH, Hatfill SJ, Pellis NR (1997) Tissue culture in microgravity. Sci Med 4:46–55
15. Knight TA (1806) V. On the direction of the radicle and germen during the vegetation of seeds. By Thomas Andrew knight, Esq. FRS In a letter to the right Hon. Sir Joseph banks, KBPRS. Philos Trans R Soc Lond 96:99–108
16. White CA (1900) The structure and signification of certain botanical terms. Science 12 (289):62–64
17. Haberlandt, Gottlieb (1902) Ueber die Statolithenfunktion der Stärkekörner. Borntraeger
18. Scano, A (1963) Effeti di una variazione continua del campo gravitazionale sullo svoluppo ed accrescimento di Lathyrus Odororatus. Communication at 6th International and 12th European Congress on Aeronautical and Space Medicine. Rome, October 1963
19. Chapman DK, Venditti AL, Brown AH (1980) Gravity functions of circumnutation by hypocotyls of *Helianthus annuus* in simulated hypogravity. Plant Physiol 65(3):533–536. https://doi.org/10.1104/pp.65.3.533
20. Hoson T, Kamisaka S, Miyamoto K, Ueda J, Yamashita M, Masuda Y (1993 Dec) Vegetative growth of higher plants on a three-dimensional clinostat. Microgravity Sci Technol 6 (4):278–281
21. Cogoli A, Valluchi-Morf M, Mueller M, Briegleb W (1980) Effect of hypogravity on human lymphocyte activation. Aviat Space Environ Med 51(1):29–34
22. Dietz C, Infanger M, Romswinkel A, Strube F, Kraus A (2019) Apoptosis induction and alteration of cell adherence in human lung cancer cells under simulated microgravity. Int J Mol Sci 20(14):3601. https://doi.org/10.3390/ijms20143601
23. Strube F, Infanger M, Dietz C, Romswinkel A, Kraus A (2019) Short-term effects of simulated microgravity on morphology and gene expression in human breast cancer cells. Physiol Int 106(4):311–322
24. Romswinkel A, Infanger M, Dietz C, Strube F, Kraus A (2019) The role of C-X-C Chemokine Receptor Type 4 (CXCR4) in cell adherence and spheroid formation of human Ewing's sarcoma cells under simulated microgravity. Int J Mol Sci 20(23):6073. https://doi.org/10.3390/ijms20236073
25. Arun RP, Sivanesan D, Vidyasekar P, Verma RS (2017) PTEN/FOXO3/AKT pathway regulates cell death and mediates morphogenetic differentiation of colorectal cancer cells under simulated microgravity. Sci Rep 7(1):5952. Published 2017 Jul 20. https://doi.org/10.1038/s41598-017-06416-4
26. Chung LW, Zhau HE, Wu TT (1997) Development of human prostate cancer models for chemoprevention and experimental therapeutics studies. J Cell Biochem Suppl 28–29:174–181
27. Pisanu ME, Noto A, De Vitis C, Masiello MG, Coluccia P, Proietti S, Giovagnoli MR, Ricci A, Giarnieri E, Cucina A, Ciliberto G, Bizzarri M, Mancini R (2014) Lung cancer stem cell lose their stemness default state after exposure to microgravity. Biomed Res Int 2014:470253. https://doi.org/10.1155/2014/470253
28. Ma X, Pietsch J, Wehland M, Schulz H, Saar K, Hübner N, Bauer J, Braun M, Schwarzwälder A, Segerer J, Birlem M, Horn A, Hemmersbach R, Waßer K, Grosse J, Infanger M, Grimm D (2014) Differential gene expression profile and altered cytokine secretion of thyroid cancer cells in space. FASEB J 28(2):813–835. https://doi.org/10.1096/fj.13-243287

29. Camberos V, Baio J, Bailey L, Hasaniya N, Lopez LV, Kearns-Jonker M (2019) Effects of spaceflight and simulated microgravity on YAP1 expression in cardiovascular progenitors: implications for cell-based repair. Int J Mol Sci 20(11):2742. https://doi.org/10.3390/ijms20112742
30. Krüger M, Pietsch J, Bauer J, Kopp S, Carvalho DTO, Baatout S, Moreels M, Melnik D, Wehland M, Egli M, Jayashree S, Kobberø SD, Corydon TJ, Nebuloni S, Gass S, Evert M, Infanger M, Grimm D (2019) Growth of endothelial cells in space and in simulated microgravity—a comparison on the secretory level. Cell Physiol Biochem 52 (5):1039–1060. https://doi.org/10.33594/000000071
31. Shanmugarajan S, Zhang Y, Moreno-Villanueva M, Clanton R, Rohde LH, Ramesh GT, Sibonga JD, Wu H (2017) Combined effects of simulated microgravity and radiation exposure on osteoclast cell fusion. Int J Mol Sci 18(11):2443. https://doi.org/10.3390/ijms18112443
32. Moreno-Villanueva M, Feiveson AH, Krieger S, Brinda AK, von Scheven G, Bürkle A, Crucian B, Wu H (2018) Synergistic effects of weightlessness, isoproterenol, and radiation on DNA damage response and cytokine production in immune cells. Int J Mol Sci 19 (11):3689. https://doi.org/10.3390/ijms19113689
33. Hada M, Ikeda H, Rhone JR, Beitman AJ, Plante I, Souda H, Yoshida Y, Held KD, Fujiwara K, Saganti PB, Takahashi A (2019) Increased chromosome aberrations in cells exposed simultaneously to simulated microgravity and radiation. Int J Mol Sci 20:43. https://doi.org/10.3390/ijms20010043
34. Canova S, Fiorasi F, Mognato M, Grifalconi M, Reddi E, Russo A, Celotti L (2005) "Modeled microgravity" affects cell response to ionizing radiation and increases genomic damage. Radiat Res 163:191–199
35. Mosesso P, Schuber M, Seibt D, Schmitz C, Fiore M, Schinoppi A, Penna S, Palitti F (2001) X-ray-induced chromosome aberrations in human lymphocytes *in vitro* are potentiated under simulated microgravity conditions (Clinostat). Phys Med 17:S264–S266
36. Manti L, Durante M, Cirrone GAP, Grossi G, Lattuada M, Pugliese M, Sabini MG, Scampoli P, Valastro L, Gialanella G (2005) Modeled microgravity does not modify the yield of chromosome aberrations induced by high-energy protons in human lymphocytes. Int J Radiat Biol 81:147–155. https://doi.org/10.1080/09553000500091188
37. Mognato M, Celotti L (2005) Modeled microgravity affects cell survival and HPRT mutant frequency, but not the expression of DNA repair genes in human lymphocytes irradiated with ionizing radiation. Mutat Res 578:417–429. https://doi.org/10.1016/j.mrfmmm.2005.06.011
38. Gleeson JR, Richards JT, Ceth P, Zhang Y (2020) Culture tube holder for both the gravite and RPM microgravity simulators. NASA NTR No. 1596561636
39. Hammond TG, Hammond JM (2001) Optimized suspension culture: the rotating-wall vessel. Am J Phys Renal Phys 281:F12–F25
40. Wuest SL, Stern P, Casartelli E, Egli M (2017) Fluid dynamics appearing during simulated microgravity using random positioning machines. PLoS One 12(1):e0170826. https://doi.org/10.1371/journal.pone.0170826
41. Torres JJ, Richards JT, Zhang Y (2019) Petri dish plant growth platform Aka J^4 Arabidopsis Independent Lighting (JAIL) system. NASA NTR No. 1564677590 (KSC-14263)
42. Torres JJ, Richards JT, Doherty C, Tolsma J, Zhang Y (2020) Plant and microbial growth system with in motion sample removal. NASA NTR No. 1597332009
43. Gleeson JR, Richards JT, Torres JJ, Johnson CM, Massa GD, Koss LL, Zhang Y (2020) RPM Microgravity simulator EUE hardware to support plant and microbial growth. NASA NTR No 1596555114
44. Neelam S, Lee A, Lane MA, Udave C, Levine HG, Zhang Y (2021) Development of modules to support real-time microscopic imaging of living organisms on ground-based microgravity simulators. Appl Sci 11:3122. https://doi.org/10.3390/app11073122
45. Neelam S, Lee A, Udave C, and Zhang Y (2018) Modules to support live microscopic imaging and samples on ground-based microgravity simulator devices. NASA NTR No. 1538140007 (KSC-14221)

Chapter 19

Evaluating the Effects of the Circadian Clock and Time of Day on Plant Gravitropic Responses

Joseph S. Tolsma, Jacob J. Torres, Jeffrey T. Richards, Imara Y. Perera, and Colleen J. Doherty

Abstract

Circadian rhythms are regular oscillations of an organism's physiology with a period of approximately 24 h. In the model plant *Arabidopsis thaliana*, circadian rhythms regulate a suite of physiological processes, including transcription, photosynthesis, growth, and flowering. The circadian clock and external rhythmic factors have extensive control of the underlying biochemistry and physiology. Therefore, it is critical to consider the time of day when performing gravitropism experiments, even if the circadian clock is not a focus of study. We describe the critical factors and methods to be considered and methods to investigate the possible circadian regulation of gravitropic responses.

Key words Circadian, Gravistimulation, Time of day, Diel, Simulated microgravity

1 Introduction

1.1 The Importance of Time in Experimental Design

On Earth, the ability to track time's progression enables the anticipation of recurring events in the environment, such as the sun's daily rising or seasonal weather changes. The fact that almost all organisms possess an internal molecular timekeeping mechanism indicates the importance of measuring time and monitoring daily and seasonal changes. An internal clock tuned appropriately to the environment optimizes fitness and provides a competitive advantage [1–7]. In many species, proper timing of physiological activities such as energy acquisition is so important that almost all aspects of metabolism and growth are moderated by this molecular time-keeper known as the circadian clock. Plants, dependent on the recurring arrival of the sun for their energy, are no exception. The circadian clock modulates most aspects of plant growth and physiology, from energy acquisition, hormone signaling [8–15], primary metabolism [16–22], and responses to biotic and abiotic stresses

Elison B. Blancaflor (ed.), *Plant Gravitropism: Methods and Protocols*, Methods in Molecular Biology, vol. 2368, https://doi.org/10.1007/978-1-0716-1677-2_19,

[12, 23–30]. Therefore, the timing of stress exposure or an environmental cue, such as gravistimulation, can affect that response.

Experiments to measure the circadian and diel effects on plant responses require attention to light, temperature, nutrients, and time. Here, we will overview some of the critical aspects important for experimental design when considering the role of the time of day and the circadian clock in gravitropic responses.

1.2 The Time of Day Variation in Control Conditions

The core circadian clock functions through a series of transcriptional-translational feedback loops in plants, mammals, yeast, fish, and invertebrate systems [31–35]. This regulatory system produces endogenous, self-sustaining oscillations that are tuned to the local environmental cycles. In both plant and mammalian systems, transcription factors are components of the central oscillator. These components not only contribute to the self-sustaining ~24 h oscillations, but they also provide the connections between the clock and the clock-controlled output. These core components act as master regulators, targeting clock-controlled genes, including other transcription factors, initiating a transcriptional cascade. In *Arabidopsis thaliana*, about 25% of the genome is targeted directly by core clock components [36–40]. Beyond these direct targets, a number of the targets are themselves transcription factors expanding the clock network. About 30% of the transcriptome shows a rhythmic expression pattern driven by the circadian clock [8, 16, 41]. However, these clock-controlled genes are only a portion of the effects of time of day. In addition to the circadian controlled targets, many genes show rhythmic patterns of expression in response to the daily cycles of light and temperature. In a meta-analysis of transcriptional responses to light and temperature cycles, Michael et al. identified that over 80% of the plant transcriptome exhibits rhythmic expression [42]. And while the most comprehensive characterization of these daily molecular rhythms has been done on transcript expression levels, RNA stability, splicing, translation, protein levels, and metabolite levels all show diel variation [43]. Consequently, the molecular landscape is different for experiments performed at any two times of the day. Thus, the time of day the investigation is conducted can have significant consequences.

At the transcriptional level, this changing landscape impacts our perception of the response. Most analysis methods identify genes of interest based on a difference between control and treatment, e.g., the significance of the fold change. However, if a gene has a rhythmic expression pattern in the control condition, the ability to detect the response will be affected. Therefore, the time of day an experiment is performed will alter the genes identified as responsive to the experimental stimulus. For example, a transcript that is repressed by gravistimulation will be much harder to detect as differentially responsive when the basal transcriptional cycle is at

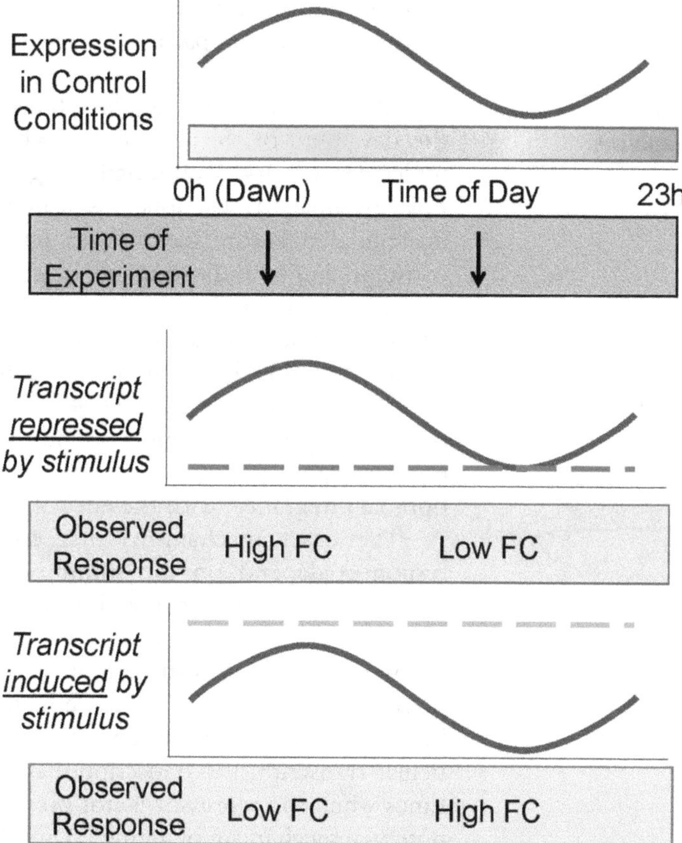

Fig. 1 Rhythmic patterns of expression in control conditions influence the perceived response to stresses, such as gravitropism. The majority of genes in Arabidopsis show variation in their expression levels throughout the day, as indicated in the top panel. If a gene is repressed in response to the stimulus, the fold change (FC) between control and stress will be evident if the experiment occurs when the gene is at its maximal expression level. However, the expression difference will be undetectable if the treatment happens when the gene is already off or very lowly expressed. Therefore, this gene would not be considered responsive to the stimulus at this time of day. Likewise, a gene induced by the stress will be identified as differentially expressed when the experiment is done at the minimum expression level in control conditions. But detecting a change in expression may be more difficult if the experiment is performed when the gene is already highly expressed

its nadir (Fig. 1). Likewise, induced transcripts may be harder to identify as differentially expressed between control and treatment when they are at their maximum expression level when the treatment is applied. Thus, even if a transcript consistently responds to a stimulus by accumulating to the same absolute level at all times of day, the significance of the change in expression will depend on the time of day the experiment is performed if the transcript has a rhythmic expression pattern in control conditions.

1.3 Circadian Gating of Responses

Not all transcripts respond consistently to a stimulus at all times of the day. One of the circadian clock functions is to ensure that molecular activity is temporally organized to optimize efficiency. For example, transcripts associated with photosynthesis are up-regulated before dawn. This temporal regulation persists in constant light, indicating it is circadian regulated [16, 41–43]. Not only are light and energy-related activities restricted to particular times, but almost all aspects of plant growth, including water usage [44], hormone activity [45], UV-B response [46], and temperature response [23, 24, 47], show this "gating" effect. By controlling *when* a plant responds to external changes, gating enables plants to temporally partition responses to achieve the optimal integration with the environment.

The exact mechanisms that enable the gating of specific responses depend on the stimulus. For example, many of the genes with rhythmic expression are transcriptional activators or repressors. If a repressor is absent when the stress is applied, a responsive transcript can be induced. However, if the repressor is present, gating will result if the induction in response to the stimulus cannot overcome the repressor's presence. Thus, for that particular transcript, the transcriptional response will be gated to the times when the repressor is not present. This example is only one potential mechanism of gating. Daily changes in hormones, metabolite, protein levels, and chromatin occupancy all can result in periods of permissive molecular and physiological responses and inhibitory periods.

1.4 What Time of the Day Should an Experiment Be Performed?

A frequent question is, "What time of day should I do my experiment?" This is not a trivial question to answer. This area is relatively uncharted because the effects of the time of the day on many plant physiological processes have not been studied systematically (Fig. 2). In an ideal situation, performing the experiment multiple times of the day will provide a comprehensive picture of the effect of a stimulus on the underlying networks. Repeating an experiment at different times of day will produce richer results and valuable insights. For example, you would not detect a gene repressed by gravity unless you sample during the time of the day when that gene is expressed.

Unfortunately, sampling at multiple times of the day is not usually an economically viable option. If at least two times of day are feasible, this will provide an initial understanding of how variable the response is throughout the day. Many genes and physiological responses show peaks around dawn and dusk. In the plant, this is the time of daily minimum and maximum photosynthate availability [19]. Therefore, initial time points around dawn and

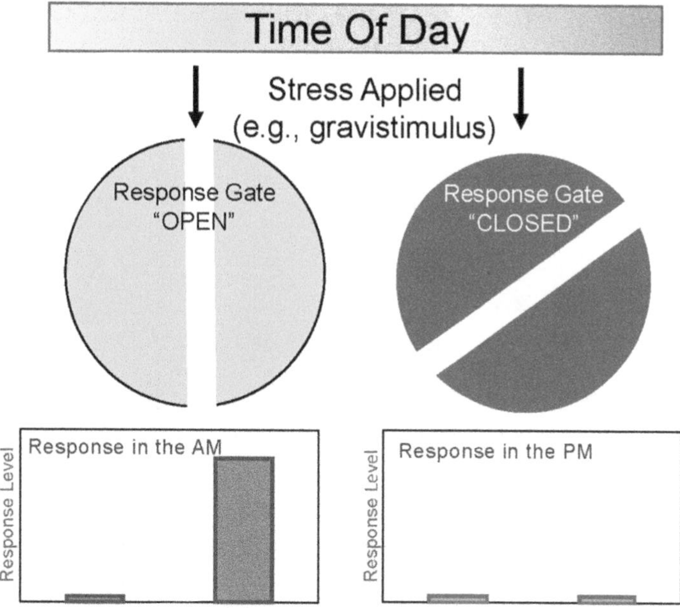

Fig. 2 Some responses are gated by the time of day. In a gated response, the same stimulus produces different responses depending on the time of day the stimulus is presented. In this example, the response is robust in the morning but is absent at night

dusk can help gauge the variability in responses due to the time of day. Based on this information, the value of testing additional time points throughout the day can be evaluated.

Frequently, budget constraints and analysis costs limit researchers to examining the responses at only one time of day. However, it is challenging to determine the optimal time point to perform an experiment *a priori* without some knowledge of the output.

If the response is transcriptionally regulated and some candidate genes are known, these can guide in determining the time of day to do the experiment. The candidate genes can be examined in the DIURNAL database (http://diurnal.mocklerlab.org/). DIURNAL contains microarray-based transcriptome analysis from 24 h time-series experiments. For Arabidopsis, there are 14 time series performed on "wild-type" plants. These experiments span various conditions: different photoperiods, developmental stages, and cycles of light and temperature. The abbreviations in DIURNAL are shorthand for the environmental chamber settings in which the plants were grown. The 14 experiments on wild-type plants include three investigations in short-day and one in long-day conditions. The remaining ten experiments are performed in 12 h:12 h day/night cycles. These experiments are annotated based on whether the lights (first two letters) or the temperatures (last two letters) are cycling or are in constant conditions. For example, the time series annotated LDHC has cycling light

(LD) and cycling temperature (HC) while the experiment LLHC has only cycling temperature. For the four experiments performed in constant conditions, the first two letters before the underscore indicate if they were collected in continuous light (LL) or dark (DD). So, LL_LDHC was an experiment on plants grown in cycling light and cycling temperature and then moved to constant light (and constant temperature) before the samples were collected. Most of these experiments were collected every 2 h for 48 h. However, the experiments Col_SD and LER_SD were only collected for 24 h, and the data is double plotted. Enter the candidate genes into the search box and select the most similar experiments to the planned analysis conditions. The first thing to examine is the consistency of expression of the candidate genes across various experimental conditions. Suppose a candidate gene is peaking at a similar time across all experiments selected. In that case, this indicates the peak and trough times will be robust, and the appropriate time to perform the experiment can be determined based on this information. Comparing across different conditions, for example, plants grown in photocycles alone or combined photocycles and thermocycles can also be used to determine the best environmental conditions to evaluate the response of interest.

Unfortunately, most of the time, little is known about the results until the experiment is performed. If this is the case for your investigation, the most important thing is consistency. Pick a time of day that is logical for the experimental design (e.g., can be completed during regular working hours), knowing that you may miss responses that are unique to other times. Ensure that all data collection and sample harvest occur in a narrow time window. The experiments can be performed in batches across multiple days if necessary to minimize the time spread of the data collection. The timing should be measured from dawn and not wall time. When recording your data, ensure to include the day length because the photoperiod also affects timing and basal transcriptional levels [48].

1.5 Comparing Results Between Experiments

It is frequently of interest to compare results across labs or repeat an experiment to identify robust responses. A response can be very consistent, yet still not be detected at all times of the day. Therefore, information on the time of day an experiment is performed must be available to enable time-corrected analysis between experiments. This information should be recorded as the time since the last "time cue" or Zeitgeber the plant received. Light is a strong Zeitgeber, so documenting the experimental time as the time since dawn is essential. Another option is to provide both the dawn and experimental times so that this information can be calculated. However, surprisingly few experiments in GEO (https://www.ncbi.nlm.nih.gov/geo/) or GeneLab (https://genelab.nasa.gov/) provide this information. Although plants maintain circadian rhythms with a period of approximately 24 h, they are highly responsive to the environment and "dynamically plastic" (Reviewed [49]). Thus, even for two experiments performed at the same time relative to the dawn, the circadian phase can be out of sync if they are in different photoperiods or levels of light intensity or nutrients. Therefore, recording growth and harvesting protocols ensures that the circadian phase (the position of the internal clock when the experiment is performed) can be correctly compared between experiments.

The photoperiod has substantial impacts on gene expression. Complicating things further, these effects are gene dependent. There are dawn-tracking genes and dusk-tracking genes [49–51]. This means that for some genes, their expression at 4 h after dawn will be the same in short-day and long-day, but for other genes, their expression will vary based on the photoperiod. To reiterate that transcription is not the only aspect affected, the diel expression variations interact with photoperiod-driven changes in translational rates to coordinate seasonal control of protein levels [52]. The light intensity and quality (wavelength distribution) also affect the circadian period. Increased intensity shortens the period in Arabidopsis grown in continuous conditions [53] although the effects in diel conditions have not thoroughly been evaluated. Media composition is also a critical factor that affects the circadian phase [54]. Consistency in media preparation is essential for replicate experiments, and this information should be provided in the metadata. Therefore, in addition to recording the time of dawn and the experiment, photoperiod, media composition, light quality, and light intensity information is essential metadata that should be included when documenting experiments.

Although the circadian clock is temperature compensated, the clock is *not* temperature independent. The expected biochemical effects of temperature are an approximate doubling of enzymatic activity for every 10 °C increase in temperature from room temperature. These effects are reduced in the circadian oscillator. Temperature compensation means that while one would expect the period to be substantially shorter than 24 h in warmer temperatures and longer in colder temperatures, it changes little within a specific physiological range [55]. However, outside this range, the circadian clock is affected by temperature [56]. Moreover, although the oscillator shows circadian compensation, the outputs are not necessarily compensated. To complicate things further, temperature itself is a zeitgeber, or entrainment cue, for the circadian clock. Even in the absence of light cycles, rhythms can be entrained by warm days and cool nights, also known as thermocycles [57]. Therefore, monitoring changes in temperature, particularly when the chamber is opened and lights are turned on, is critical.

Although it may seem impossible to control all the factors that can alter the circadian phase, it can be done with careful monitoring and consideration of details that affect the circadian clock. Following these measures will improve overall consistency across experiments, and sharing the metadata related to these factors will improve comparisons of experiments between laboratories.

Here, we outline the methods to analyze for time of day or circadian effects on plant gravitropism. We provide detailed protocols and considerations of the critical points for querying time of day effects in two scenarios. First, we provide methods to assay root responses to gravistimulation as the initial experimental system. Next, we incorporate these considerations into simulated microgravity experiments using the random positioning machine (RPM) housed in the microgravity simulation lab at the Kennedy Space Center (KSC). These points can be extended to any assay to consider the potential effects of the time of the day.

2 Materials

2.1 Plant Growth

1. Seeds.
2. Laminar flow hood.
3. Micropore surgical tape.
4. Ethanol solutions (75% and 95%) for seed surface sterilization.
5. Square Petri dishes.
6. Growth chamber.
7. Murashige and Skoog (MS) nutrients.
8. Plant growth agar.

2.2 Time Course and IR Imaging

1. Raspberry Pi computer (Raspberry Pi Foundation, https://www.raspberrypi.org/products/).
2. Raspberry Pi NoIR camera (Raspberry Pi Foundation, https://www.raspberrypi.org/products/). This camera has no Infrared (IR) filter.
3. 3D printed Petri dish holders.
4. 3D printed camera mounts.
5. Aluminum foil.
6. Green filter paper Roscolux Supergel (#389 or as appropriate for selected species, Rosco https://us.rosco.com/en/products/catalog/supergel).
7. Headlamp.

3 Methods

3.1 Plant Growth Basics

1. Prepare plant growth media so it consists of 0.6% (w/v) plant agar and ½ strength MS nutrients in deionized water. Adjust to pH 6 before adding the agar (*see* **Note 1**).
2. Sterilize agar solution using an autoclave for 45 min on a liquid cycle.
3. Add 35 mL of agar to 10 cm, square Petri dishes. Allow to cool and solidify.

3.2 Seed Surface Sterilization and Plating

1. Aliquot seeds to be sterilized into 1.5 mL microcentrifuge tubes.
2. Add ~1 mL of 75% EtOH and one drop of Tween 20. Shake continuously for 2 min.
3. Allow the seeds to settle then aspirate the ethanol solution.
4. Add ~1 mL of the 95% EtOH solution to the seeds. Shake continuously for 2 min.
5. Allow the seeds to settle, then aspirate the ethanol solution.
6. If storing seeds long term, dispense seeds onto sterile filter paper and allow the ethanol to evaporate for 2–4 h. Place filter paper into a Petri dish and store it in a cool, dry place for no longer than 4 months.
7. If using seeds immediately, wash with sterile distilled water at least five times and store with ~1 mL of sterile, distilled water. Store seeds for 3–5 days at 4 °C in the dark to stratify (*see* **Note 1**).
8. Using a micropipette or sterilized, long-nose Pasteur Pipette, place individual seeds in a row on the agar surface near a plate's top. Use the grid on the Petri dish as a guide for seed placement.
9. Seeds can be placed along the first grid line with ~3 seeds per square.
10. Seal the plate with micropore surgical tape and orient the plate vertically in a growth chamber.
11. Grow plants under 12:12 light/dark cycles and constant temperature for 7 days. Circadian experiments are generally conducted in 12:12 light:dark periods to allow consistent sampling rate between day and night.

3.3 Raspberry Pi Computer Set-up

Cost-effective time course imaging can be implemented using Raspberry Pi computers and IR cameras. Programming these computers for time course imaging is relatively easy and can be done with little coding experience. A basic code can be found in the GitHub repository linked below. Utilizing 3D printed Petri dish

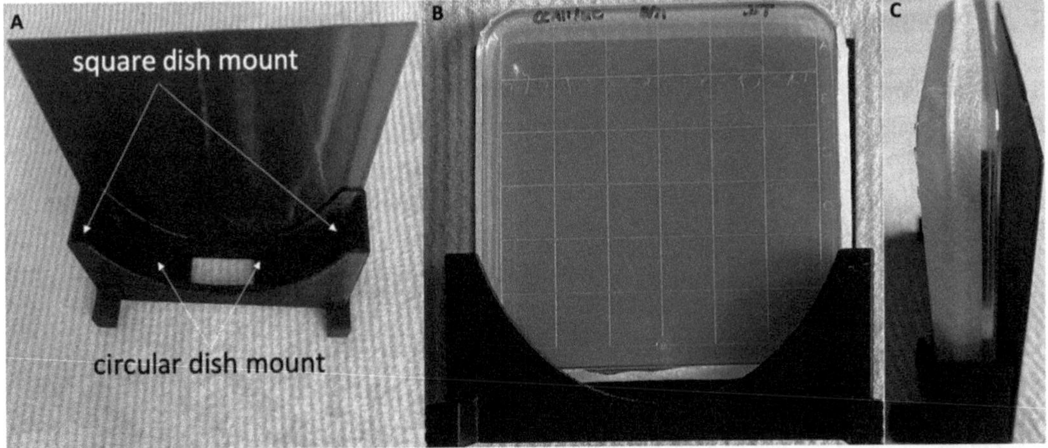

Fig. 3 Image detailing 3D printed plate holder, which provides stability and black background for both square and circular Petri dishes. (**a**) Empty plate holder without background. (**b**) Front-view of plate holder with a square plate and black background. (**c**) Side-view of plate holder with a square plate and black background

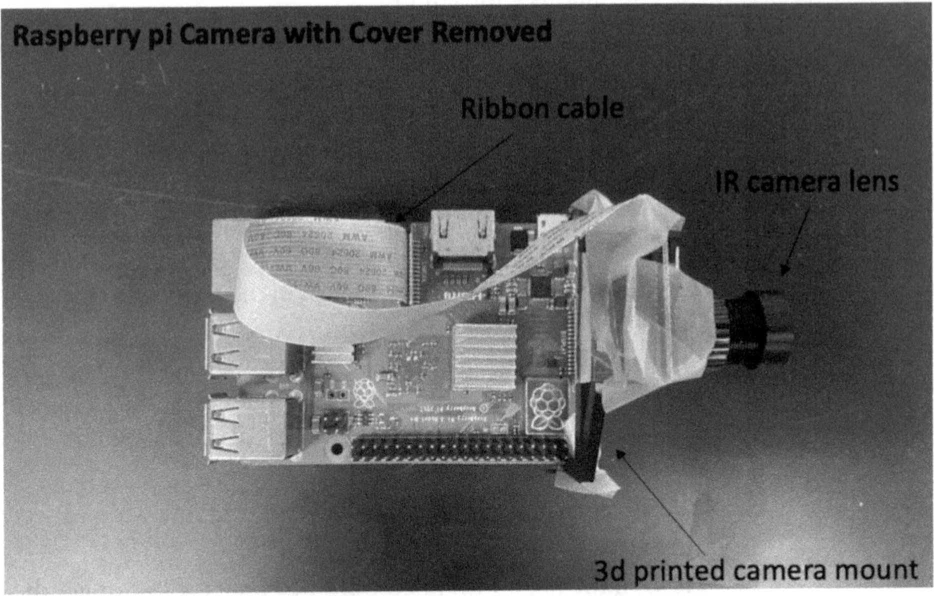

Fig. 4 Raspberry Pi camera set-up detailing ribbon cable connections and camera mount. This image is taken without a cover to show internals and connections

holders allow for clearer visualization of root and shoot structure. A black, non-reflective filament is most appropriate (Fig. 3). Code for use with popular 3D printers and splicers can be found at this GitHub repository: https://github.com/DohertyLab/JosephStantenTolsma-pi_imaging.

1. Use Raspberry Pi computer models 3–4 following the manufacturer's instructions.

2. A microSD card with at least 32GB of space and sufficient writing speed for imaging must be used to run the operating system (OS).
3. Easy installation of the Raspberry Pi OS can be done by following the instructions and downloading [57] or purchasing an SD card with the OS pre-installed.
4. The Raspberry Pi NoIR camera should be used. Remove the IR light source before connecting the camera to the Raspberry Pi computer using a ribbon cable, ensuring metal contacts are correctly oriented (Fig. 4) (*see* **Note 2**).
5. Mount the camera to the Pi computer using 3D printed mounts (e.g., https://www.thingiverse.com/thing:128617).
6. Cron jobs, time-based schedulers used in Unix operating systems, can be used to image and store files. A simple example can be found in the link below: https://github.com/DohertyLab/JosephStantenTolsma-pi_imaging.
7. Cover the ethernet and power LED indicators with aluminum foil so that no light is visible.
8. In the chamber, any number of Pi cameras can be installed.
9. IR lights should be installed separately and overhead to avoid glare in the images captured. IR lights connected to a spare Pi NoIR camera and Raspberry Pi will automatically illuminate when darkness is detected (*see* **Note 3**).
10. The Raspberry Pi computer used for IR light illumination can simultaneously record temperature data using the code can be found in the link below: https://github.com/DohertyLab/JosephStantenTolsma-pi_imaging.

3.4 Gravistimulation Time Course

1. Gravistimulation experiments performed over a 24-h time course to measure circadian effects require a minimum of 4 time points although higher resolution is preferred (*see* **Note 4**).
2. Rotate vertically oriented plates by 90° every 2 h and measure the responses over a 24 h period [58, 59] (*see* **Notes 4** and **5**).
3. Turn individual plates every 2 h so that a different set of plates is turned at ZT0, 2, 4,....20, 22 (Fig. 5).

3.5 Measuring Root Growth Response

1. Measure root growth and bending angles with default tools in Image J analysis software (https://imagej.net).
2. Calibrate length measurements using the analysis package and gridlines on the Petri dishes (Fig. 6).
3. Measure the root growth rate by monitoring change in root length every 4 h and dividing length over time elapsed.

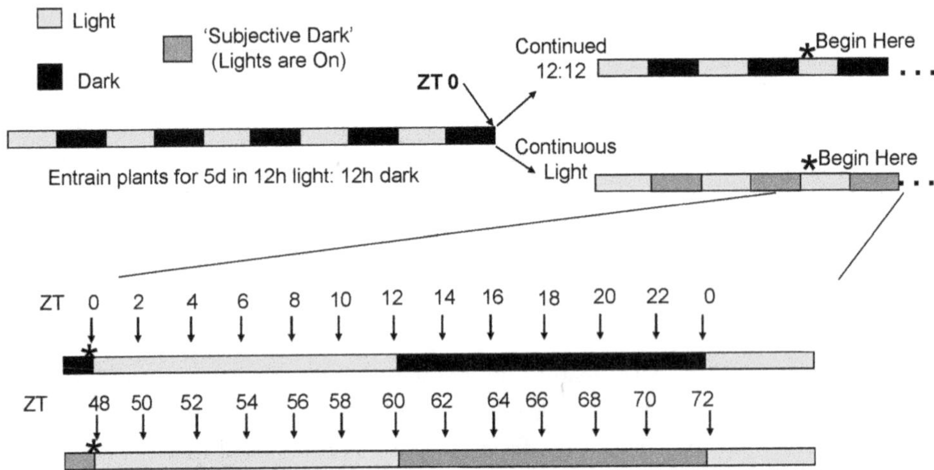

Fig. 5 Schematic detailing lighting conditions and timing of plate turning. Yellow indicates daytime/lights on, black indicates nighttime/lights off, and gray indicates subjective night (lights on but plants entrained to night conditions). * indicates when the experiment starts with plants maintained in 12 h light:12 h dark conditions (top rows) or moved to continuous light (bottom rows). Arrows indicate the time at which plates should be turned to complete a 24-h time course. Note that the time is measured from the last cue to the circadian clock (dawn), so in continuous light, the count does not reset at subjective dawn

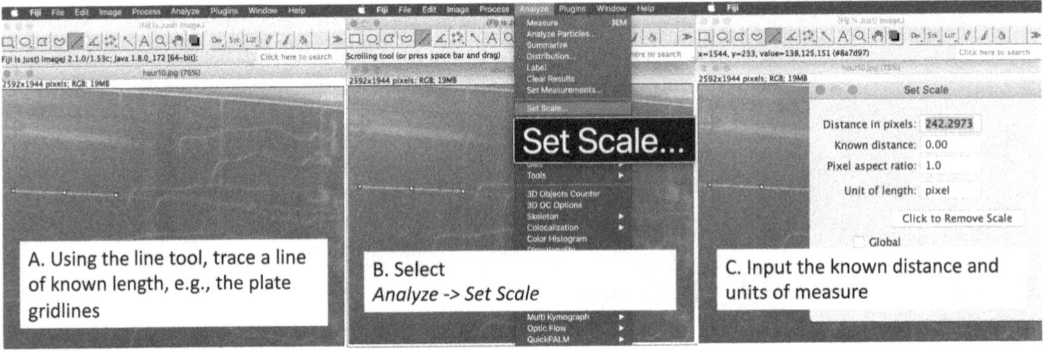

Fig. 6 Steps for setting the scale using ImageJ. ImageJ measures by counting pixels in an image and calibrating with a known scale. (**a**) Use the line tool to trace a line of known length. For example, the grid lines on the plate. (**b**) Select *Analyze- > Set Scale*. (**c**) Input the known distance and units into the menu that appears

4. Measure the root bending angle using the angle tool in Image J by measuring from the root: shoot junction to the root tip location pre-bend and extending to the root tip location after turning (Fig. 7)

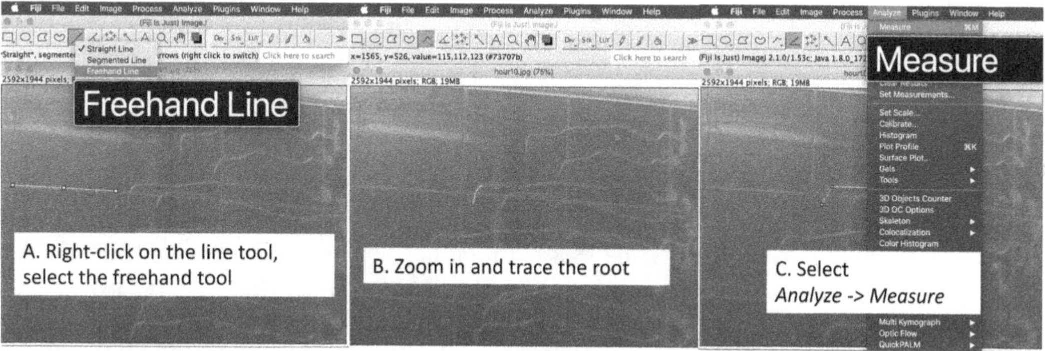

Fig. 7 Measuring root length and angles. After the scale has been set: (**a**) Right-click on the line tool to select the freehand tool. (**b**) Zoom in and trace the root. (**c**) Select *Analyze- > Measure* (or Ctrl+M) to save the length measurement in a table. To measure the angles, select the angle tool and save the measurements in the same manner (Ctrl+M)

3.6 Evaluating for Time of Day or Circadian-Specific Responses

If significant differences are observed between the responses at different times of day, this will indicate a time of day effect that is important to consider.

To determine if the observed time of day effects are under the control of the circadian clock, the first step is to perform the experiment in constant conditions. The plants should be grown in two stages: entrainment and release to constant conditions. First, plants are grown in light:dark cycles to entrain the circadian cycle. After entrainment, plants should be released to continuous conditions (light and temperature) for two full days before performing experiments. This entrainment and release experiment helps to differentiate responses regulated by the circadian clock from responses regulated by other timekeeping mechanisms or modulated in real-time in response to external stimuli. If the variation persists in constant conditions, this indicates that it is controlled, at least in part, by the circadian clock. If the variation is lost in circadian conditions, this can indicate that the time of day differences may be due to environmental factors such as light or temperature. A lack of persistence in constant light conditions does not necessarily indicate that the response is not under circadian control. Many circadian-regulated outputs are masked in constant light or constant dark conditions [38, 60].

1. The experiment set-up should be followed precisely as above for media preparation, seed surface sterilization, and Raspberry Pi set-up (*see* **Note 6**).
2. Plants are grown initially in 12:12 light/dark cycles, as in the first experiment, for 5 days. Then the plants are released to constant light and temperature conditions.

4. After 2 days in constant light, the same experimental time course should be performed to turn different plates at ZT48, 50, 52..., 70. (The ZT times indicate the time from the last Zeitgeber in constant conditions, dawn) (Fig. 5) (*see* **Note 7**).
5. A second way to evaluate the circadian clock's role is to examine the response in genotypes with a disrupted circadian clock. Commonly used genotypes include CCA1 overexpressing plants [61], *prr7* mutants [62], and *elf3* mutants [63]. CCA1 overexpressing plants are arrhythmic; therefore, analyzing the response here is straightforward. However, these plants have multiple pleiotropic phenotypes, and the secondary effects from the altered physiology must be considered. Plants with mutations in *prr7* show reduced variation to different entrainment signals as *prr7* has a role as a regulator of the plasticity of the circadian clock [51]. Working with this mutant can be valuable for examining if the stress impacts the circadian clock function, but may be less helpful for determining if the time of day effects are circadian controlled. Genotypes with mutations in e*lf3* show normal rhythms in entrained environments but become arrhythmic in constant light conditions. Therefore, comparing the response between wild-type and *elf-3* mutants in constant light will clarify the circadian clock's role in the response.

3.7 Simultaneously Sampling Multiple Time points on the RPM

While root bending assays monitor the response to the direction of the gravity stimulus, monitoring the response to microgravity requires a different analysis method. Here, we describe extending these studies to monitor the time of day regulation of root responses to simulated microgravity. For these experiments, we used the RPM, which consists of two independently rotating axes.

Evaluating multiple time points/lighting schedules simultaneously is challenging while growing samples on a microgravity simulator moving in three dimensions. The chief concern being that light from the "day" conditions must not be detectable by samples in "night" conditions. We have developed a customizable lighting solution and harvesting scheme for evaluating circadian clock effects on plants grown in simulated microgravity. This lighting rig and RPM mount are available for use at NASA KSC's lab.

1. Two pairs of independently controlled LED lights can be set manually to any light photoperiod or timing desired by adjusting power sources outside of the growth chamber.
2. Plants can be entrained to the light settings on the RPM while stationary or vertical plate mounts daisy-chained to the same power source. These vertical mounts can be used for ground control samples.

3. Plants should be grown in a vertical orientation for 7 days under the same lighting schedule as planned for RPM experiments.
4. After 7 days, plates can be transferred and secured to the RPM. RPM should be programmed to 0 g settings and run for 48 h.
5. For phenotype data, pre- and post- RPM images can be taken using a Raspberry Pi IR camera.
6. To image plants currently in a dark photoperiod, create a workspace that keeps plants in the dark. Green light filters placed over a headlamp can be used to see, with minimal disruption to the plant's light signaling responses. Plants are imaged in the dark using a Pi NoIR camera and IR light only. Immediately return plants to the dark.
7. To image plants currently in a light period, non-IR lights can be used for imaging.
8. To capture images after an RPM run, plants in a dark photoperiod must remain in the dark if you plan on harvesting samples for transcriptional analysis. Dark photoperiod plants should be imaged in a dark room using an IR camera and light source (*see* **Note 8**). Plants currently in a light photoperiod can be imaged normally.
9. Remove plants from the RPM in a dark room. A headlamp emitting green light can be used to see samples.
10. Separate root and shoot tissue with a single cut using a razor blade.
11. Gather root tissue using fine forceps, place in a cryotube, and immediately flash freeze in liquid nitrogen.
12. **Steps 9–11** can be done to harvest samples currently in a light photoperiod. However, overhead room lights can be used instead of a green headlamp (*see* **Note 9**).

3.8 Visualization and Analysis of Time Course Data

1. Plotting the response by time will enable visualization of differences based on the phase. If the responses show variation across the time of day, the significance can be tested using an ANOVA test. To identify which time points may be unique, a Tukey post hoc test can differentiate individual time points.
2. The data is scaled to a range between 0 and 1 to compare phase and period differences between conditions, such as between mutants and wild-type plants or between diel and constant environments. This is done by subtracting the minimum value of all the time points for that sample from each time point, then dividing by the range (*see* **Note 10**).

$$z_i = \frac{x_i - \min(x)}{(\max(x) - \min(x))}$$

4 Notes

1. Critical note: Here, since the focus is specifically on examining the effects of time of day and the circadian clock, sucrose should be omitted from the media. Sucrose is commonly added to plant growth media to ensure uniform seed germination and increase plant growth rates. However, the presence of sucrose, glucose, and fructose shortens the circadian period, while reduced sugar levels lengthen the circadian period [54, 64]. The presence or absence of sugar in the media also affects responses in diel conditions. Therefore, experiments focused on the clock are generally performed without sucrose [54, 64–66]. For researchers accustomed to growing plants with sucrose, other factors of the experiment will need to be adjusted: longer stratifications are required (up to 5 days for freshly harvested seeds) for consistent germination; longer growth times since the plants grow slower; avoid bleach and use an ethanol only sterilization method; visually inspect seeds and discard those that are misshapen, immature, or shriveled, particularly when working with the clock mutants to enhance germination consistency.
2. The ribbon cables are very fragile and damage easily. Use caution when moving around and positioning cameras to image plants to avoid stretching or pinching the ribbon cables. The metal contacts can become worn out due to repeated use (plug/unplug). Static electricity and bending of the cable should be avoided.
3. The IR light's wavelength must be greater than 880 nm to minimize inducing light responses in the plant. Furthermore, because IR lights tend to produce heat, it is essential to monitor temperature changes within the chamber and avoid cycling temperatures, which affect the circadian phase.
4. A discussion on selecting the optimal sampling time for circadian analysis can be found here [58, 59].
5. While manipulating plants during dark periods, take care when handling to avoid a light stimulus. Using a headlamp with a green light can provide sufficient light to see while limiting the induction of light-responsive transcripts. The temperature should also be monitored to ensure heat shocks are avoided when opening the chamber for sampling or from heat from the lights.
6. Consistency between experiments performed in cycling conditions and those performed in constant conditions is crucial.

7. The first night of constant conditions should be avoided as the plant undergoes a stressful change that complicates the responses.
8. Transcriptional responses to light exposure can be detected within minutes, so maintaining darkness is critical.
9. It is easiest to image and harvest plants in a dark photoperiod first so that the researcher's eyes have sufficient time to adjust. However, it is critical to control for harvest order as a factor. Imaging and harvesting should be completed within 5–10 min for all samples.
10. It is important to note that scaling the data between 0 and 1 will mask amplitude differences between samples, so those must be evaluated separately.

Acknowledgments

Current work on the circadian clock and gravity is supported by NASA grant 80NSSC18K1466 and North Carolina Space Grant.

References

1. Dodd AN, Salathia N, Hall A et al (2005) Plant circadian clocks increase photosynthesis, growth, survival, and competitive advantage. Science 309:630–633
2. Spoelstra K, Wikelski M, Daan S et al (2016) Natural selection against a circadian clock gene mutation in mice. Proc Natl Acad Sci U S A 113:686–691
3. Ouyang Y, Andersson CR, Kondo T et al (1998) Resonating circadian clocks enhance fitness in cyanobacteria. Proc Natl Acad Sci U S A 95:8660–8664
4. Ma P, Woelfle MA, Johnson CH (2013) An evolutionary fitness enhancement conferred by the circadian system in cyanobacteria. Chaos Solitons Fractals 50:65–74
5. Welkie DG, Rubin BE, Chang Y-G et al (2018) Genome-wide fitness assessment during diurnal growth reveals an expanded role of the cyanobacterial circadian clock protein KaiA. Proc Natl Acad Sci U S A 115:E7174–E7183
6. Horn M, Mitesser O, Hovestadt T et al (2019) The circadian clock improves fitness in the fruit fly, *Drosophila melanogaster*. Front Physiol 10:1374
7. Yerushalmi S, Yakir E, Green RM (2011) Circadian clocks and adaptation in Arabidopsis. Mol Ecol 20:1155–1165
8. Covington MF, Maloof JN, Straume M et al (2008) Global transcriptome analysis reveals circadian regulation of key pathways in plant growth and development. Genome Biol 9: R130
9. Mizuno T, Yamashino T (2008) Comparative transcriptome of diurnally oscillating genes and hormone-responsive genes in *Arabidopsis thaliana*: insight into circadian clock-controlled daily responses to common ambient stresses in plants. Plant Cell Physiol 49:481–487
10. Rawat R, Schwartz J, Jones MA et al (2009) REVEILLE1, a Myb-like transcription factor, integrates the circadian clock and auxin pathways. Proc Natl Acad Sci U S A 106:16883–16888
11. Fujita Y, Fujita M, Shinozaki K, Yamaguchi-Shinozaki K (2011) ABA-mediated transcriptional regulation in response to osmotic stress in plants. J Plant Res 124:509–525
12. Goodspeed D, Chehab EW, Min-Venditti A et al (2012) Arabidopsis synchronizes jasmonate-mediated defense with insect circadian behavior. Proc Natl Acad Sci U S A 109:4674–4677
13. Huang W, Pérez-García P, Pokhilko A et al (2012) Mapping the core of the Arabidopsis circadian clock defines the network structure of the oscillator. Science 336:75–79

14. Spoel SH, van Ooijen G (2014) Circadian redox signaling in plant immunity and abiotic stress. Antioxid Redox Signal 20:3024–3039
15. Fan G, Dong Y, Deng M et al (2014) Plant-pathogen interaction, circadian rhythm, and hormone-related gene expression provide indicators of phytoplasma infection in Paulownia fortunei. Int J Mol Sci 15:23141–23162
16. Harmer SL, Hogenesch JB, Straume M et al (2000) Orchestrated transcription of key pathways in Arabidopsis by the circadian clock. Science 290:2110–2113
17. Bläsing OE, Gibon Y, Günther M et al (2005) Sugars and circadian regulation make major contributions to the global regulation of diurnal gene expression in Arabidopsis. Plant Cell 17:3257–3281
18. Lu Y, Gehan JP, Sharkey TD (2005) Daylength and circadian effects on starch degradation and maltose metabolism. Plant Physiol 138:2280–2291
19. Graf A, Schlereth A, Stitt M, Smith AM (2010) Circadian control of carbohydrate availability for growth in Arabidopsis plants at night. Proc Natl Acad Sci U S A 107:9458–9463
20. Yazdanbakhsh N, Sulpice R, Graf A et al (2011) Circadian control of root elongation and C partitioning in *Arabidopsis thaliana*. Plant Cell Environ 34:877–894
21. Webb AAR, Satake A (2015) Understanding circadian regulation of carbohydrate metabolism in Arabidopsis using mathematical models. Plant Cell Physiol 56:586–593
22. Gnocchi D, Pedrelli M, Hurt-Camejo E, Parini P (2015) Lipids around the clock: Focus on circadian rhythms and lipid metabolism. Biology (Basel) 4:104–132
23. Fowler SG, Cook D, Thomashow MF (2005) Low temperature induction of Arabidopsis CBF1, 2, and 3 is gated by the circadian clock. Plant Physiol 137:961–968
24. Dong MA, Farré EM, Thomashow MF (2011) Circadian clock-associated 1 and late elongated hypocotyl regulate expression of the C-repeat binding factor (CBF) pathway in Arabidopsis. Proc Natl Acad Sci U S A 108:7241–7246
25. Wang W, Barnaby JY, Tada Y et al (2011) Timing of plant immune responses by a central circadian regulator. Nature 470:110–114
26. Ingle RA, Stoker C, Stone W et al (2015) Jasmonate signalling drives time-of-day differences in susceptibility of Arabidopsis to the fungal pathogen Botrytis cinerea. Plant J 84:937–948
27. Sharma M, Bhatt D (2015) The circadian clock and defence signalling in plants. Mol Plant Pathol 16:210–218
28. Zhou M, Wang W, Karapetyan S et al (2015) Redox rhythm reinforces the circadian clock to gate immune response. Nature 523:472–476
29. Gehan MA, Greenham K, Mockler TC, McClung CR (2015) Transcriptional networks-crops, clocks, and abiotic stress. Curr Opin Plant Biol 24:39–46
30. Li Z, Bonaldi K, Uribe F, Pruneda-Paz JL (2018) A localized Pseudomonas syringae infection triggers systemic clock responses in Arabidopsis. Curr Biol 28:630–639. e4
31. Yu W, Hardin PE (2006) Circadian oscillators of Drosophila and mammals. J Cell Sci 119:4793–4795
32. Harmer SL (2009) The circadian system in higher plants. Annu Rev Plant Biol 60:357–377
33. McClung CR, Gutiérrez RA (2010) Network news: prime time for systems biology of the plant circadian clock. Curr Opin Genet Dev 20:588–598
34. Filichkin SA, Breton G, Priest HD et al (2011) Global profiling of rice and poplar transcriptomes highlights key conserved circadian-controlled pathways and cis-regulatory modules. PLoS One 6:e16907
35. Lowrey PL, Takahashi JS (2011) Genetics of circadian rhythms in Mammalian model organisms. Adv Genet 74:175–230
36. Nakamichi N, Kiba T, Kamioka M et al (2012) Transcriptional repressor PRR5 directly regulates clock-output pathways. Proc Natl Acad Sci U S A 109:17123–17128
37. Liu T, Carlsson J, Takeuchi T et al (2013) Direct regulation of abiotic responses by the Arabidopsis circadian clock component PRR7. Plant J 76:101–114
38. Nagel DH, Doherty CJ, Pruneda-Paz JL et al (2015) Genome-wide identification of CCA1 targets uncovers an expanded clock network in Arabidopsis. Proc Natl Acad Sci U S A 112: E4802–E4810
39. Kamioka M, Takao S, Suzuki T et al (2016) Direct repression of evening genes by CIRCADIAN CLOCK-ASSOCIATED1 in the Arabidopsis circadian clock. Plant Cell 28:696–711
40. Ezer D, Jung J-H, Lan H et al (2017) The evening complex coordinates environmental and endogenous signals in Arabidopsis. Nat Plants 3:17087
41. Edwards KD, Anderson PE, Hall A et al (2006) FLOWERING LOCUS C mediates natural variation in the high-temperature response of the Arabidopsis circadian clock. Plant Cell 18:639–650
42. Michael TP, Mockler TC, Breton G et al (2008) Network discovery pipeline elucidates

conserved time-of-day specific cis-regulatory modules. PLoS Genet 4:e14

43. Más P (2008) Circadian clock function in *Arabidopsis thaliana*: time beyond transcription. Trends Cell Biol 18:273–281
44. Carbonell-Bejerano P, Rodríguez V, Royo C et al (2014) Circadian oscillatory transcriptional programs in grapevine ripening fruits. BMC Plant Biol 14:78
45. Covington MF, Harmer SL (2007) The circadian clock regulates auxin signaling and responses in Arabidopsis. PLoS Biol 5:e222
46. Fehér B, Kozma-Bognár L, Kevei E et al (2011) Functional interaction of the circadian clock and UV RESISTANCE LOCUS 8-controlled UV-B signaling pathways in *Arabidopsis thaliana*. Plant J 67:37–48
47. Grinevich DO, Desai JS, Stroup KP et al (2019) Novel transcriptional responses to heat revealed by turning up the heat at night. Plant Mol Biol 101:1–19
48. Weng X, Lovell JT, Schwartz SL et al (2019) Complex interactions between day length and diurnal patterns of gene expression drive photoperiodic responses in a perennial C4 grass. Plant Cell Environ 42:2165–2182
49. Edwards KD, Akman OE, Knox K et al (2010) Quantitative analysis of regulatory flexibility under changing environmental conditions. Mol Syst Biol 6:424
50. Flis A, Sulpice R, Seaton DD et al (2016) Photoperiod-dependent changes in the phase of core clock transcripts and global transcriptional outputs at dawn and dusk in Arabidopsis. Plant Cell Environ 39:1955–1981
51. Webb AAR, Seki M, Satake A, Caldana C (2019) Continuous dynamic adjustment of the plant circadian oscillator. Nat Commun 10:550
52. Seaton DD, Graf A, Baerenfaller K et al (2018) Photoperiodic control of the Arabidopsis proteome reveals a translational coincidence mechanism. Mol Syst Biol 14:e7962
53. Somers DE, Devlin PF, Kay SA (1998) Phytochromes and cryptochromes in the entrainment of the Arabidopsis circadian clock. Science 282:1488–1490
54. Haydon MJ, Mielczarek O, Robertson FC et al (2013) Photosynthetic entrainment of the *Arabidopsis thaliana* circadian clock. Nature 502:689–692
55. Panter PE, Muranaka T, Cuitun-Coronado D et al (2019) Circadian regulation of the plant transcriptome under natural conditions. Front Genet 10:1239
56. Kusakina J, Gould PD, Hall A (2014) A fast circadian clock at high temperatures is a conserved feature acrossArabidopsisaccessions and likely to be important for vegetative yield. Plant Cell Environ 37:327–340
57. Greenham K, Robertson McClung C (2013) Temperature and the circadian clock. In: Temperature and plant development. Wiley, Oxford, pp 131–161
58. Hughes ME, Abruzzi KC, Allada R et al (2017) Guidelines for genome-scale analysis of biological rhythms. J Biol Rhythm 32:380–393
59. Li J, Grant GR, Hogenesch JB, Hughes ME (2015) Considerations for RNA-seq analysis of circadian rhythms. Methods Enzymol 551:349–367
60. Velez-Ramirez AI, van Ieperen W, Vreugdenhil D, Millenaar FF (2011) Plants under continuous light. Trends Plant Sci 16:310–318
61. Wang ZY, Tobin EM (1998) Constitutive expression of the CIRCADIAN CLOCK ASSOCIATED 1 (CCA1) gene disrupts circadian rhythms and suppresses its own expression. Cell 93:1207–1217
62. Michael TP, Salomé PA, Yu HJ et al (2003) Enhanced fitness conferred by naturally occurring variation in the circadian clock. Science 302:1049–1053
63. Hicks KA, Albertson TM, Wagner DR (2001) EARLY FLOWERING3 encodes a novel protein that regulates circadian clock function and flowering in Arabidopsis. Plant Cell 13:1281–1292
64. Haydon MJ, Mielczarek O, Frank A et al (2017) Sucrose and ethylene signaling interact to modulate the circadian clock. Plant Physiol 175:947–958
65. Knight H, Thomson AJW, McWatters HG (2008) Sensitive to freezing6 integrates cellular and environmental inputs to the plant circadian clock. Plant Physiol 148:293–303
66. Feugier FG, Satake A (2012) Dynamical feedback between circadian clock and sucrose availability explains adaptive response of starch metabolism to various photoperiods. Front Plant Sci 3:305

Index

Elison B. Blancaflor (ed.), *Plant Gravitropism: Methods and Protocols*, Methods in Molecular Biology, vol. 2368, https://doi.org/10.1007/978-1-0716-1677-2, © Springer Science+Business Media, LLC, part of Springer Nature 2021

I

K

L

M

N

P

Q

R

S

T

V

CPSIA information can be obtained
at www.ICGtesting.com
Printed in the USA
LVHW060909201022
731087LV00008B/285

9 781071 616796